To Mick

Happy Easter

Mark [illegible]
4-14-17

TEST THE SHROUD

AT THE ATOMIC AND MOLECULAR LEVELS

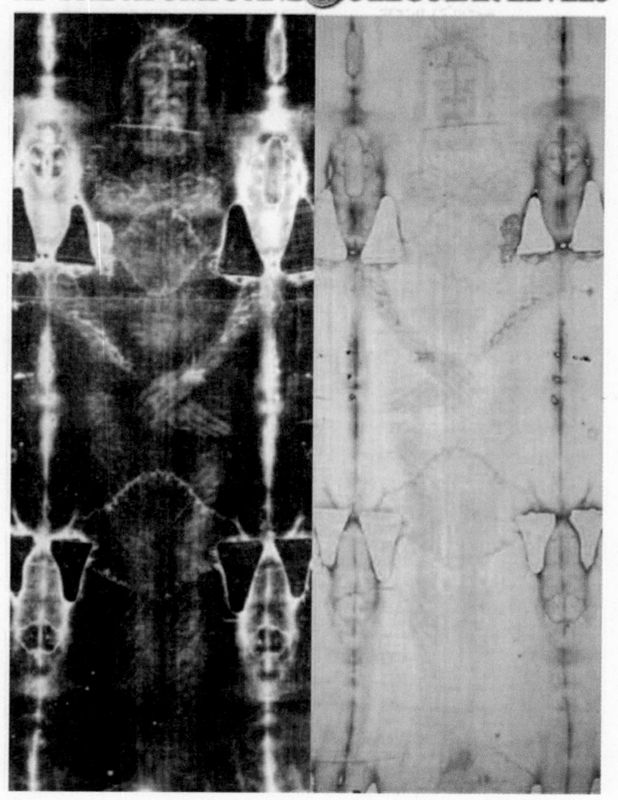

MARK ANTONACCI

Test the Shroud

Published by
LE Press, LLC

ISBN# 978-0-9964300-1-2

Book and Cover Design by Ellie Jones

Printed in the United States of America
Signature Book Printing, www.sbpbooks.com

TABLE OF CONTENTS

Eternity

The last great objective
desired by most individuals on earth
could become the next
great frontier and discovery for
all of humanity.

To Mary Rose,
my endearing and enduring wife

To Art Lind and Bob Rucker,
the smartest and nicest scientists I ever met

To Joe Marino,
who always helps — and can help in
more areas than anyone in this field

To Gary Habermas,
the first navigator in this expansive journey

To Paul Ernst, Dick Nieman and Pat Byrne,
whose financial help and friendship were absolutely critical

To Ellie Jones,
my talented and dedicated publisher

To Mike and Steve, our parents, children and relatives,
and to the dear spouses of all the people above

FROM THE AUTHOR

As an agnostic lawyer in 1981, I stumbled onto an overview of the findings from the first and only comprehensive scientific investigation of the Shroud of Turin. The evidence that was discovered on this burial cloth by scientists, physicians and other experts was not only astounding, it was new and original. It could only be described as unprecedented, as it contained many features that science had never seen before. It was not only new, it was very extensive, very unique and most of it was *unfakable*. As an attorney who relies on evidence, I was intrigued by the extent and quality of evidence that was found on the Shroud and wanted to learn more about it.

Compelling and consistent scientific evidence actually indicates that a miraculous radiating event occurred to the dead, crucified man who was wrapped within this burial cloth after he incurred a series of wounds, all of which are identical to those of Jesus Christ and under all the same circumstances. This miraculous event appears to have left distinct, full-length images of the corpse on the front and back of the cloth, even capturing all the wounds, bruises, swelling and 130 blood marks that the man incurred from the top of his head to the bottom of his feet on both sides of his body. These full-length images and blood marks have never been duplicated.

This book of course, will discuss all of this objective and independent evidence that was rather quietly accumulated in the last 117 years. My goal is to not only make all of this evidence known and understood, but to let the public know of all the evidence that could be acquired from further investigation of this burial cloth. Over the past thirty-three years, I have studied the Shroud extensively, yet I know that we have not begun to acquire most of the evidence that is contained on this cloth. It has become my goal – my life's work – to not only prove if it is indeed the burial garment of Jesus Christ, but to acquire as much evidence as modern technology can reveal regarding the events that occurred to this man before and after his burial within the cloth. Although thousands of tests and experiments have been performed on the entire cloth and its samples, the only way to prove the Shroud's authenticity and resolve all of its outstanding issues is to perform additional, non-invasive tests on the Shroud at the atomic and molecular levels.

Because both the evidence and the events are unique, I can't leave any of them out. By definition, you will be reading many scientific concepts that you have never read about before, but neither has most people in the world. Yet, understand that this evidence is more intriguing than it is complex. I struggled in college in the late 60s with Earth Science 101, 102 and 103. (Fortunately, that fulfilled the scientific requirements for a bachelor's degree in liberal arts back then.) Yet, even I can understand this scientific evidence because of thousands of hours of conversations with not only the scientists who performed the only comprehensive scientific examination of the Shroud, but with scientists from countries all over the world, and in particular, with physicist Arthur Lind and nuclear engineer Robert Rucker. These numerous tests and their incomparable findings can be appreciated by the world at their most fundamental levels. This understanding and appreciation not only applies to the scientific and medical investigations that have occurred to date, but also to the more important scientific testing that must be undertaken on the Shroud in the future. The basic principles of carbon dating and the creation of brand new atoms are given in an elementary manner so that all people can understand what information would be measured and acquired from the Shroud, how it got there, when it was embedded and any other significance it could have.

The sophisticated techniques and instruments available with this type of testing could not only prove the Shroud's authenticity, it could also *prove* that a miraculous event occurred to this dead human being that was the most important event in human history. This and the preceding series of events that occurred to the victim wrapped in this burial cloth would be relevant to every individual who will ever live. Although a great deal of scientific and medical evidence presently exists, a comprehensive amount could also be acquired that would directly address the most fundamental issues of life regarding the existence and nature of God, whether there is life after death and how we could acquire it.

So delight in the surprising, fascinating discoveries that have been found on the Shroud of Turin and enjoy the extent of its mysteries. There is not only extensive scientific and medical evidence on this cloth, but also a good deal of archaeological and historical evidence as well. I have not omitted any of it in *Test the Shroud* because I want the public to learn and observe how all of this comprehensive evidence, which spans centuries and millenniums, surprisingly and independently corroborates that the Shroud of Turin is indeed the authentic burial garment of Jesus Christ.

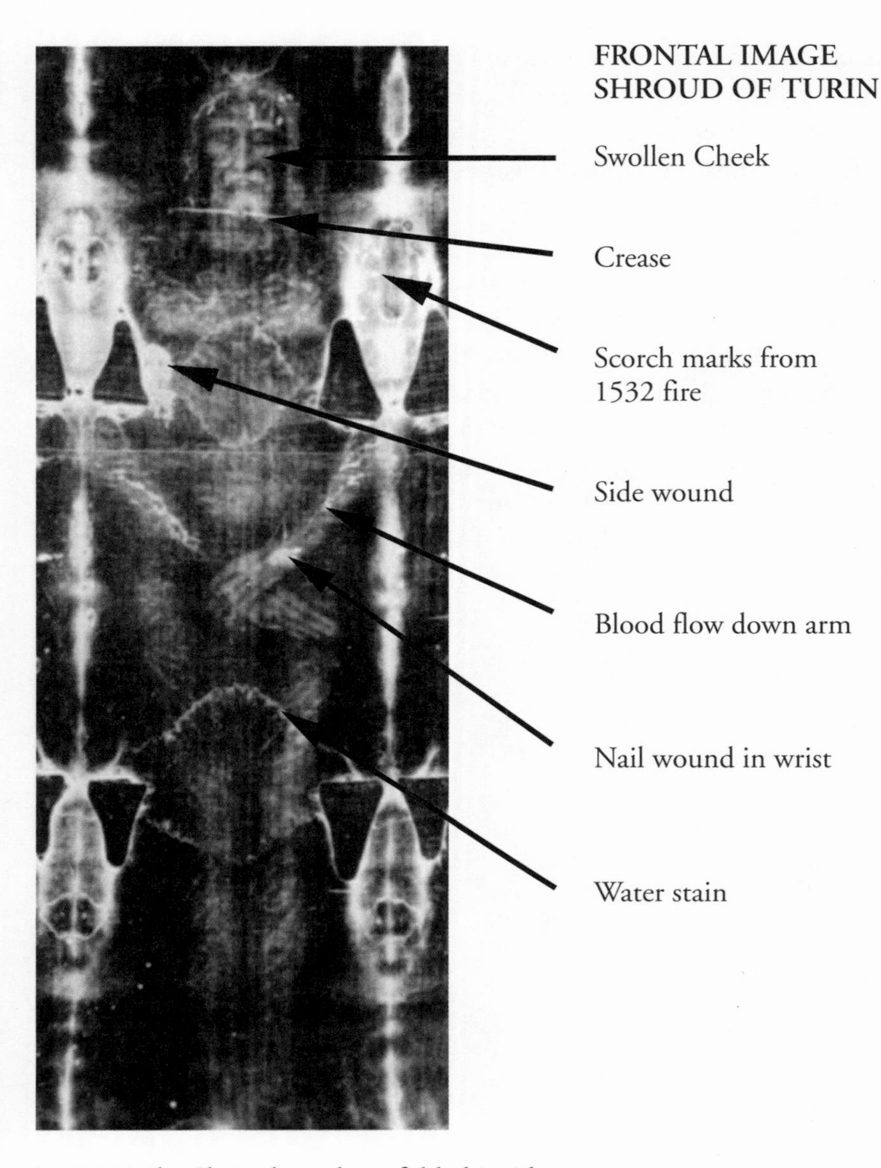

In 1532, the Shroud was kept folded inside of a reliquary, or container, which had a silver lining. This reliquary was kept in the niche of a wall within a sanctuary in Chamberry, France. When the sanctuary caught on fire, the corners, or parts of the large,

DORSAL IMAGE
SHROUD OF TURIN

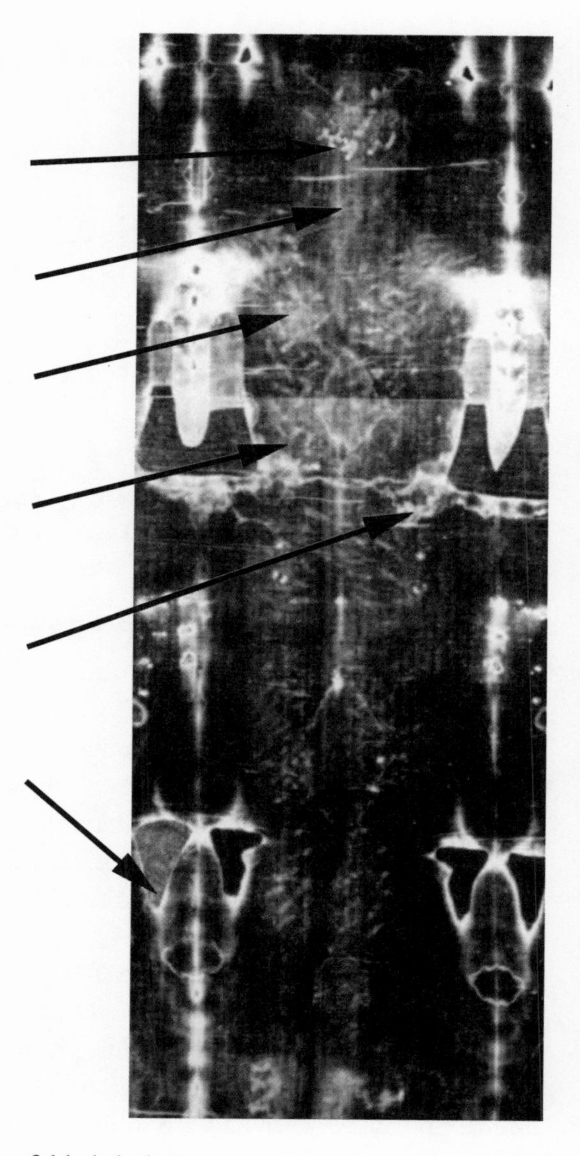

folded cloth that were most in contact with the silver lining (or that melted silver dropped onto) were burned completely through, while other parts in the folded pattern were scorched. Fortunately, this did little damage to the man's image or to the rest of the cloth.

PREFACE

In the 20th century, Winston Churchill concluded that despite its splendid virtues, science "does not meet any of the real needs of the human race." What good is scientific development, he asked, if it cannot answer the "simple questions which man has asked since the earliest dawn of reason – *'Why are we here? What is the purpose of life? Where are we going?'*"[1]

From the beginning of human existence, mankind has struggled with a number of inherent questions that affect every individual and every society throughout the world. We all clearly understand the one *absolute* in life — every single one of us is going to die. But what happens next? Is there life after death? If there is, how can we attain it? Does God exist? If so, why hasn't He made Himself known to us? These are among the most fundamental or universal questions of human existence.

Humanity, of course, has never been able to objectively or universally answer these questions. Throughout history, well-intended people have devised a variety of religious philosophies that attempted to answer these and related questions. However, neither these nor other philosophies have ever had any objective or independent evidence to support their central premises or answers. While these religions and philosophies have benefitted people in many ways; their number, variety and even their perceived lack of veracity have also made our world much more complicated and difficult.

People have also wondered: Why has God allowed so many religions with so many different premises to develop? Why has He allowed things to get so out of hand that wars and conflicts — in which religion has been a major cause or an underlying element — have continued for centuries throughout the world? And why has He allowed these conflicts to increase in number and in the alarming amount of destruction they inflict on humanity? It is difficult to fathom that God would want wars to continue for centuries on end, or to escalate further throughout the world, especially those caused or influenced by religious differences.

This book contends that God could have provided extensive and unprecedented evidence, thousands of years ago, for mankind to answer these questions. It contends that this evidence was left for all humanity by the very essence of God Himself. It contends that unfakable, far-reaching evidence was actually left by the body and blood of Jesus Christ — at the moment of his resurrection. This evidence also shows that every element of the passion, crucifixion, death and burial of Jesus Christ also occurred exactly as they were recorded in the most attested and reliable sources of ancient history. While this evidence is already extensive, much more irrefutable evidence could be acquired and demonstrated.

The proof for these events primarily consists of sophisticated scientific and medical evidence that was acquired only after the development of modern scientific technology. This scientific and medical evidence both confirms and is confirmed by existing archaeological and historical evidence. The present evidence is extremely consistent, though it is not unanimous. Further application of advanced scientific technology could remove the one item of lingering doubt, as well as make the case for the occurrence of these events irrefutable and virtually unanimous. Most people are unaware that hundreds and even thousands of tangible items of scientific and medical evidence now exist for the occurrence of these events; yet, this could just be the tip of the iceberg. Hundreds of thousands, even millions of items of *unfakable* evidence could also be available for all of humanity in the relatively near future.

This new evidence could cause some anxiety, as it initially did to me. Most people throughout the world are currently unaware that there is now objective and independent evidence that would allow them to analyze life's most perplexing questions. They are even less aware that a great deal more evidence could be acquired from further scientific investigation. Yet, when it is realized that all of humanity could answer its most difficult questions based on evidence and logic, as opposed to what others have alleged, the anxiety disappears. This evidence relates to every individual in the most universal and permanent of ways — and allows all of humanity to now recognize that their similarities far outweigh their differences.

1

MODERN TECHNOLOGY AND THE SHROUD OF TURIN

The objective and independent evidence referred to earlier is actually located throughout a burial garment long reputed to be the burial cloth of the historical Jesus Christ. This linen cloth, known as the Shroud of Turin, is more than 14 feet long and over 3-1/2 feet wide (4.34 m x 1.10 m).

Shrouds are long, cloth garments that were used to bury people in ancient times and are still used throughout the Middle East today. Jesus

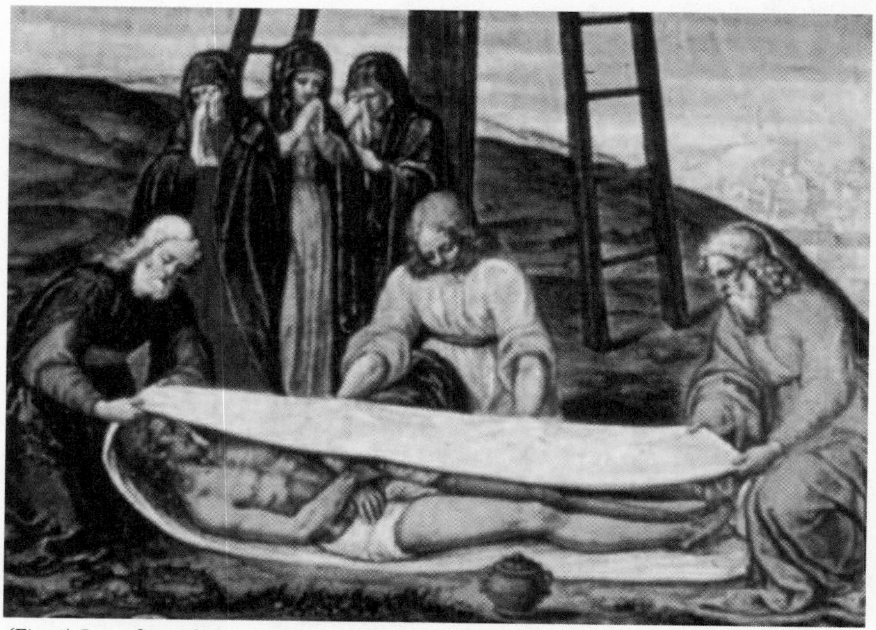

(Fig. 1) Part of a 16th century painting by della Rovere, shows the method of placing a shroud over the deceased body of Christ.

was buried in such a shroud. When a person is buried in a shroud, they are laid lengthwise on the lower part of the cloth, while the rest of the cloth is then folded over to cover and wrap the individual as seen above. The deceased are initially wrapped in shrouds before permanent burial in a casket or in some other manner. In Jesus' time, deceased Jews were frequently wrapped in shrouds and buried within tombs hewn out of soft limestone rock. After the passage of about a year, the tomb would be reopened and the deceased's bones would be collected and placed in a family ossuary or container and the shroud would be discarded.

Shrouds continue to be used for burial throughout the Middle East today and are frequently wrapped around many of the tragic victims who are killed during the widespread and ongoing violence in this part of the world. Unidentified victims of a brutal chemical attack outside of Damascus, Syria in 2013 are seen buried in shrouds in Fig. 2. Unlike the Syrian victims who were transported to a mass grave, Jesus' reputed burial shroud does not appear to have been tied with a rope around his body, which is consistent with a temporary, incomplete burial.

(Fig. 2)

The Shroud of Turin gets its present name from the city where it has resided for its last 437 years, Turin, Italy (Torino, Italia), located in the Piedmont region of northwestern Italy. (This is the same city where Pope Francis' father lived before immigrating to Argentina in 1929 to escape Mussolini's Fascist regime.) Previous names and references to this famous cloth throughout its history will be discussed in subsequent chapters. Archaeological and historical evidence also discussed in subsequent chapters show that both the size and weave of this cloth were consistent with first century weaving techniques and burial practices found in the Middle East. Scientific and medical evidence acquired throughout the 20th and 21st centuries has yielded a wide array of unparalleled evidence that this burial cloth clearly wrapped a dead human male, who left the most unique images and bloodstains ever known to history.

This man appears to have incurred the same series of tortures that happened to Jesus before he was also crucified and killed in the same manner. He also appears to have been temporarily buried according to detailed Jewish burial customs in the same rock shelf in which Jesus was

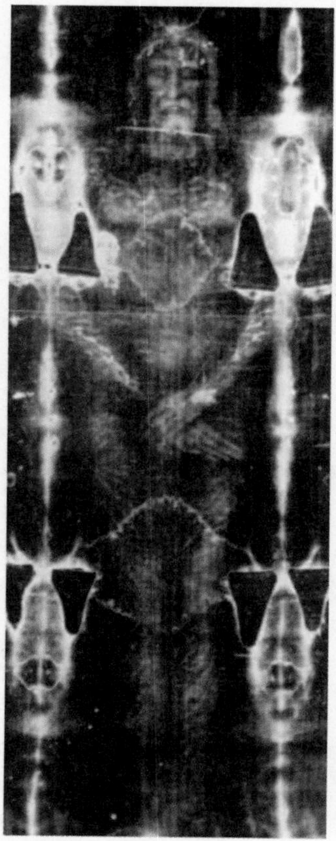

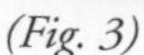

(Fig. 3)

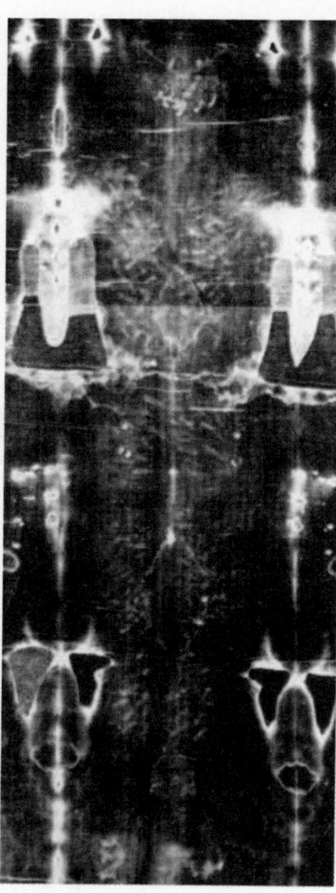

(Fig. 4)

reputed to have been buried. All of these events appear to have occurred in Jerusalem in the spring of the first century. Within two to three days of having been wrapped in the Shroud, however, the body mysteriously left this burial cloth.

An unprecedented event occurred to this body prior to or during its disappearance that caused the man's full-length frontal and dorsal body images and 130 corresponding blood marks to be encoded throughout this burial shroud. These full-length body images and blood marks are so unique they have never been duplicated in any era by any artist, scientist, physician or anyone utilizing any type of artistic, naturalistic or other method. Thousands of items of detailed evidence can be found within both of these full-length images and their corresponding blood marks that appear to be unfakable and to defy the laws of chemistry and physics. This evidence even appears to have resulted from a miraculous event that happened *to* the body of the historical Jesus Christ.

Initial Discussion of Body Image Features

Body Image is Three-Dimensional

Near the turn of the 20th century, French biologist Paul Vignon observed that the most intense areas on the frontal body image corresponded with those body parts that would have been in closest contact with a cloth lying over a reclined body: the nose, forehead, cheeks, hands, etc. Areas of the body that would have been farther away from the cloth left a lighter impression.[1] While scientifically testing this observation in 1976, physicist John Jackson and Sandia Laboratory image specialist Bill Mottern uncovered a much more profound correlation. They discovered that the man's body image actually contains *three-dimensional* information that was encoded onto this two-dimensional burial cloth. This had never been seen before in history.

The shading found throughout the man's frontal image correlates precisely with the subtle distances each part of his body was from the cloth. This feature was surprisingly demonstrated with computer imaging technology that displays relief or depth only when the lightness or darkness within a picture is directly correlated to its distance from the source of encoding.[2] For example, if you were to put a normal photograph into such a computer imaging devise, it would yield a distorted result such as seen in Fig. 5.

That is because the various degrees of intensity or the lightness/darkness of features such as the hair, skin, eyes, lips, shirt collar, etc. lack any correlation to their distances from the camera. However, when you apply this computer imaging technology to the image on the Shroud of Turin, you get an undistorted three dimensional image as seen in Fig. 6. The man's features appear undistorted because his nose, forehead, cheeks, etc. were all encoded in direct proportion to their various distances from the overlying cloth.

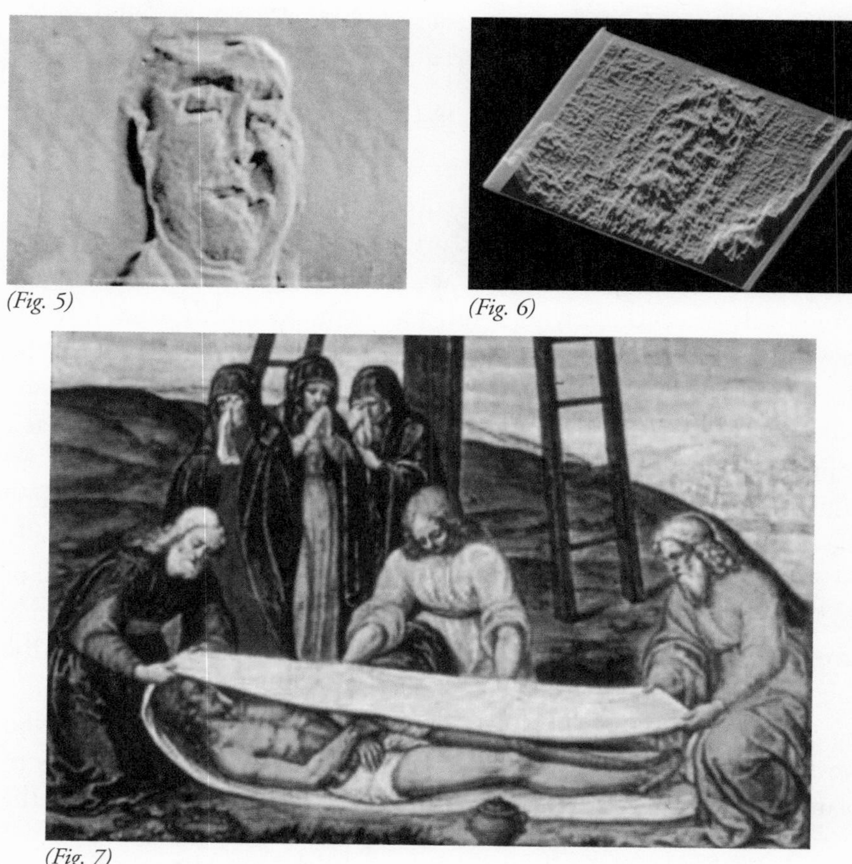

(Fig. 5)

(Fig. 6)

(Fig. 7)

When the two men at each end of the body let go of the cloth, Fig. 7, the top of it will roughly conform to the contours of the entire body.

Modern computer and other imaging technology not only confirmed Vignon's original observation, but illustrated that every shade of body image throughout the length and width of the draped cloth was directly correlated with their original distances from the underlying body.[3] True three-dimensional distance information had never been encoded before on a two-dimensional image. How could this have occurred centuries ago?

This feature indicates that radiation possibly could have been involved in the formation of the Shroud's full-length body image. Italian scientist Luigi Gonella, who studied the Shroud for decades and was centrally involved in negotiating and coordinating its scientific investigation, recognized that "an agent acting at a distance with decreasing intensity is, almost by definition, radiation."[4]

Body Image is Directionless

Another intriguing feature discovered by scientists Don Lynn and Jean Lorre working at the Jet Propulsion Laboratory in 1976 was that the body image on the Shroud of Turin lacks any normal directionality.[5] For example, if an artist had painted the Shroud, modern technology could easily detect underlying brush strokes on the body images whether they went side to side, up and down, or at any kind of angle on the cloth's surface. Since the 14th century, when the Shroud first became widely known in Europe, it was erroneously and angrily denounced as a painting by a neighboring Bishop Pierre d' Arcis, who claimed an unnamed artist had painted it and confessed to his forgery. This became the commonly accepted view of the Shroud until it was physically investigated for the first time ever by modern science toward the end of the 20th century. Because of this commonly accepted, yet completely erroneous opinion, the Shroud remained unknown for centuries by the vast majority of the people throughout the world.

Don Lynn and Jean Lorre were image processing specialists who worked in the mid to late 1970s on various NASA space missions to planets as near as Venus and as distant as Saturn and beyond.* At the

***Although Don Lynn lost his eyesight in one eye from a boyhood accident, it didn't keep him from becoming the supervisor of the Space Processing Group. He was the first STURP scientist that I talked to in person three years into my research. He also directed me to other STURP scientists and pathologists with whom I've discussed the Shroud for three decades.**

request of physicist John Jackson and a small but growing number of scientists interested in the findings that were emerging from the Shroud's image, Lynn and Lorre agreed to analyze the body image in their spare time. They scanned the best available photographs of the Shroud's body image with a microdensitometer, which digitally measures the density of photographic details. However, when this data was mathematically processed through a computer and displayed at high resolution on a television screen, the only directional features found were in the weave on the cloth itself as seen in Fig. 8.

(Fig. 8)

The white cross observed on the screen is simply the warp and weft (or the length and width) intersection of the weave of the cloth. In the body image areas the color was randomly oriented with a complete lack of two-dimensional directionality.[6]

The initial discoveries that a highly-resolved, three-dimensional body image was encoded randomly (without any two dimensional directionality) on a two dimensional burial cloth provided enormous impetus for the first comprehensive scientific examination of the Shroud of Turin. This examination was conducted by approximately 25 members of the newly-formed Shroud of Turin Research Project (STURP), consisting of the above and other scientists from some of the most prestigious institutions in the world. In 1978, these scientists were given unprecedented access to this burial cloth for 120 hours at the close of a rare public exhibition that was held in honor of the Shroud's four hundredth year in Turin. These scientists independently devised and applied a comprehensive range of non-destructive testing and examination of the entire cloth, including its body images, blood marks, and non-image areas, along with its scorch marks, burn holes, water stains and patches. After STURP shipped and assembled 72 crates of scientific equipment weighing approximately eight tons, the Shroud was magnified, illuminated, photographed, thermographed, vacuumed and arrayed with visible, infrared and ultraviolet light as well as radiation and X-rays. Fiber composition, discoloration and chemical content were also analyzed. Imaging specialists took between 5,000 – 7,000 photographs at various wavelengths of the light spectrum (including gamma rays) and photographed the cloth extensively through microscopes.

In 1978 STURP scientists also removed and brought back fibers from various areas throughout the Shroud. STURP scientists would receive additional threads from the cloth in 1984, along with other threads that were removed from the Shroud in 1988. Although the last two sets of threads were more plentiful than the fibers from 1978, they were also taken from the same vicinity as samples that were removed from the cloth and carbon dated in 1988. All the data collected from the cloth and from testing the various fibers and threads received or removed in 1978 and the 1980s, has been, and continues to be, examined at laboratories throughout the world for more detailed analysis. At the expense of sounding simplistic, hundreds of thousands of hours of scientific research that included countless mathematical calculations, measurements, and reconstructions were performed in support of these findings and results. These results have been published in scores of articles in peer-reviewed scientific journals and in academic and professional publications throughout the world. This analysis, and the interpretation of the results, continues today by scores of professionals within the general fields of science, medicine, archaeology, botany, textile analysis, art, history, religion and many other fields. The Shroud of Turin is easily the most studied relic in history.

Body Image Formed Along Vertical Paths Through Space

In the 1980s STURP scientists subsequently learned that the Shroud's body image was actually encoded in a *vertical* straight-line direction from the body to the cloth.[7] As previously noted, when the two men at each end of the body in photos 1 and 7 let go of the cloth, the top of it conforms roughly to the contours of the underlying body. Yet, regardless whether the contoured cloth was sloping downward, upward, or was relatively flat, all parts of the frontal body image were encoded

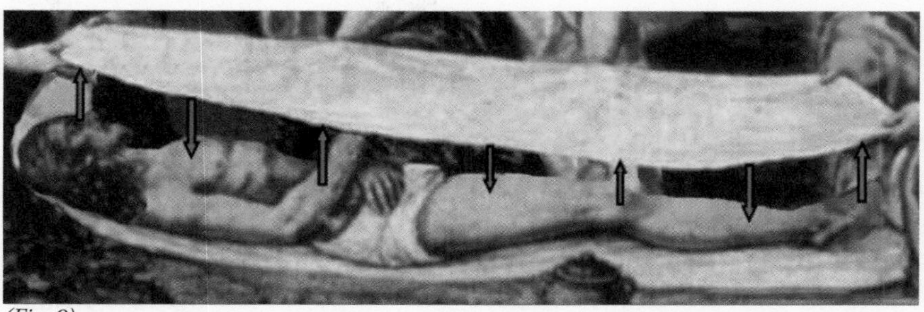

(Fig. 9)

in a vertical straight-line direction to their underlying points on the body.

Since both the left and right edges or sides of the body and the top or crown of the man's head were already vertically aligned on the body and were not *facing* the cloth, they did not become encoded on either the frontal or dorsal body images. Only the surface areas of the man's reclined (or supine) body that were facing upward with no obstructions to the cloth draped over them were transferred to the frontal image of the Shroud.

STURP scientists measured various directions that the image-encoding process could have taken to the cloth, and performed various computer modeling and scientific calculations with actual-size, three-dimensional models of the man in the Shroud. Scientists found that only the vertical, straight-line direction from the body up to the cloth (or the cloth down to the body) would correspond consistently between the observable points on the linen cloth and the underlying body. Any other paths resulted in blurred or distorted images, unlike the high resolution on the Shroud. Only a vertical encoding direction from the body to the cloth also resulted in an anatomically reasonable, three-dimensional body image such as that on the Shroud. This consistent vertical directionality was found throughout the length of the frontal image.[8]

Although this and many other image features will be explained in greater detail in subsequent chapters, STURP chemist Dr. John Heller correctly described the Shroud's unique vertical directionality when he observed: "It is as if every pore and every hair of the body contained a microminiature laser."[9]

The vertical and three-dimensional features on the Shroud's body image exist even where the cloth was not touching the body. The Shroud's body image, with all its unique features that we've just begun to discuss, was encoded through the various spaces between the draped cloth and the underlying body.[10] This, too, had never been seen before. It is difficult to imagine a process that could encode these features in a vertical straight-line direction through the space between the cloth and the body that did not involve light or radiation.

A clear initial indication of the *source* from which the full-length, frontal body image was encoded can be acquired just by considering its extraordinary vertical and three-dimensional information. This information is encoded throughout the inner part of the burial cloth that draped or lay over the entire underlying body. This information not only correlates *vertically* with those parts of the body lying immediately below the draped cloth, but also with the precise *distances* the various parts of the underlying body were from the draped cloth. Since the information for *both* of these *direct correlations* between the underlying body and the draped cloth are clearly encoded *on* the cloth — the source of this information could *only* be the underlying body.

Superficial Body Image

Another remarkable yet mutually inconsistent feature of the Shroud's body image is its superficiality. As seen in Fig. 10, its image resides only on the topmost fibers of the cloth's threads.

(Fig. 10)

By analogy, if a person's arm was a thread, the body image would lie only on the hairs on his arm. The thread fibers composing the body image are straw-yellow in color, yet the color does not penetrate into or between the threads. These fibers all consist of the same uniform color throughout the complete full-length frontal and dorsal images.[11] Where one fiber crosses another, it's white on the underlying fiber. When a fiber is cut into pieces, it's also white on the inside. Only the outer layers of the individually encoded fibers are colored.[12] Remarkably, the coloring is found 360° around each encoded or colored fiber. If one part of the Shroud's body image is darker than another, it's not because its fibers are encoded more intensely, rather it's because a greater *number* of colored fibers exist in that area. Neither this distribution nor type of superficial encoding had ever been seen before.

Body Image Does Not Consist of Any Material and Develops Over Time

These colored body image fibers do not consist of any pigments, powders, dyes or *materials* of any kind.[13] They actually consist of oxidized, dehydrated cellulose.[14] Cellulose is a natural material found in all plants and fibers and is the raw material from which linen is made. Linen naturally yellows (or darkens chemically) as it *ages*, or as it's exposed to air and light, by oxidizing and dehydrating. It will yellow or darken more over time if exposed to sunlight or radiation. A common example of this feature can be seen with a folded newspaper lying on a porch or sidewalk. Newspaper is one of the more inexpensive forms of manufactured cellulose. If left outside for a few days, exposed to air and sunlight, its top outer (or most exposed) pages will become yellow or darker than its inner white pages.

Thirty years ago, my scientific friends and I conducted elementary experiments that illustrated this yellowing feature in an analogous manner with linen, a more durable form of manufactured cellulose. We irradiated a square piece of cloth with a smaller circle of ultraviolet light for anywhere from 30 seconds to a couple of minutes. After we turned off the light and looked at the cloth, we couldn't see any difference between the irradiated area of the cloth and the rest of it. However,

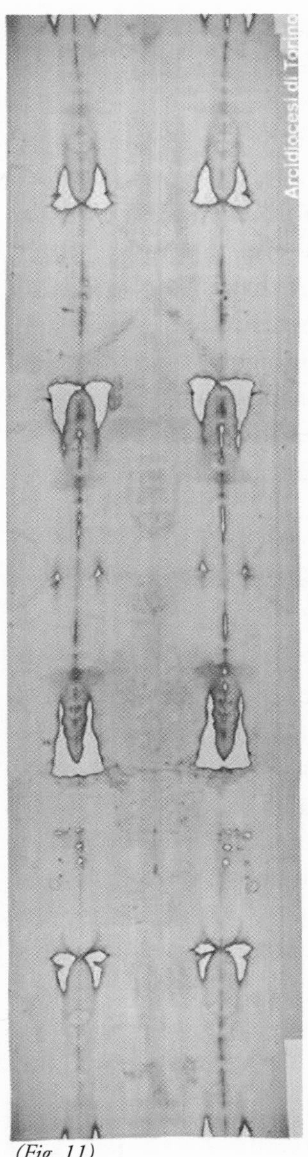

(Fig. 11)

after we artificially aged the irradiated cloth by baking it in an oven at a low temperature (which oxidizes and dehydrates the linen), an interesting result was visible. While the entire square cloth had darkened, the circle where it had been irradiated stood out because it was even darker or browner than the rest of the cloth. This is similar to what happened to the entire Shroud linen cloth. While its background has naturally yellowed as the cloth aged, something caused its two full-length body images to yellow even further as the cloth naturally aged over time.[15]

Because the straw-yellow color on the body image blends in with the background linen, the eye sees very little contrast, especially when standing near the cloth. In order to see the full-length body images with the naked eye one has to stand about 10-15 feet away from it.

Linen naturally consists of carbon, oxygen and hydrogen atoms that are single-bonded together. Something happened to the encoded body image fibers on the Shroud of Turin that caused many of its single-bonded atoms to break apart, thus allowing its carbon and oxygen atoms to then double-bond with each other. It is these double-bonded carbon and oxygen atoms that cause the straw-yellow coloration to appear on the Shroud's full-length, frontal and dorsal body images.[16] Like its connections with other body image effects discussed above — radiation can cause these single-bonded atoms to break apart faster than they would naturally over time — allowing double-bonded atoms to form and to reflect straw-yellow coloration throughout both full-length body images.

Body Image is a Negative

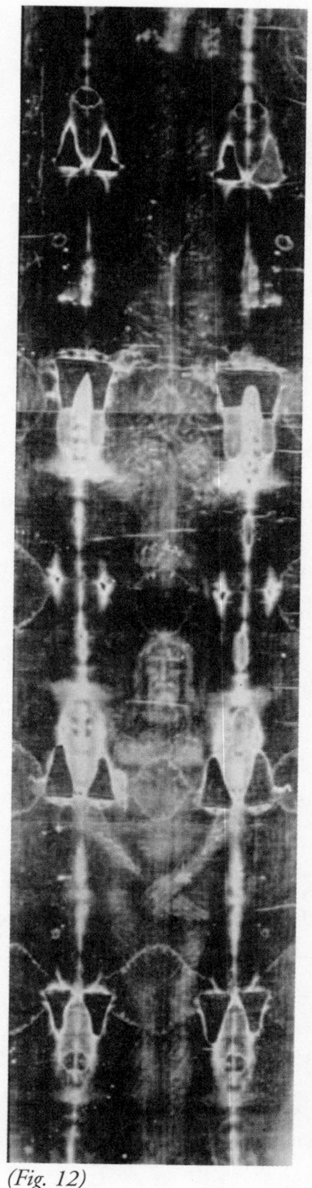

(Fig. 12)

Radiation can also cause a negative image, which is precisely what is observed on the full-length images on the cloth itself. Fig. 11 shows how the Shroud's body images appear with the naked eye. They have a sepia or straw yellow color that is vague and lacks clarity. However, when one takes a photograph of the Shroud, highly resolved, *positive* images of a human being become apparent on the photographic negative as seen to the left in Fig. 12. Light-dark reversal and left-right reversal are also found on the positive images.

Not until the Shroud was first photographed near the dawn of the 20th century was it accidentally discovered that both full-length body images on the cloth were themselves negatives. (A negative of a negative always yields a positive.) Photographs of the entire Shroud cloth within its large frame above the altar at St. John the Baptist Cathedral in Turin were first taken by Secondo Pia, a humble award-winning amateur photographer and attorney from the Piedmont region on the night of May 28, 1898. This occurred after, perhaps, hundreds of thousands of people had visited the Shroud during a rare public exhibition of the cloth. Later that night, alone in his dark room, in the initial stages of developing his photographs, Pia expected to see the normal shadowy, blurry outlines on his photographic negative of such things as the top of the altar, the Shroud linen cloth and the frame surrounding it. Each of these items did routinely appear on Pia's photographic negative. But, when the sharply-defined, full-length images of a tortured, crucified human, clearly recognizable to most as the historical Jesus Christ, unexpectedly appeared on the large photographic negative plate, Pia's hands began to tremble. He was so shocked he nearly dropped it.[17]

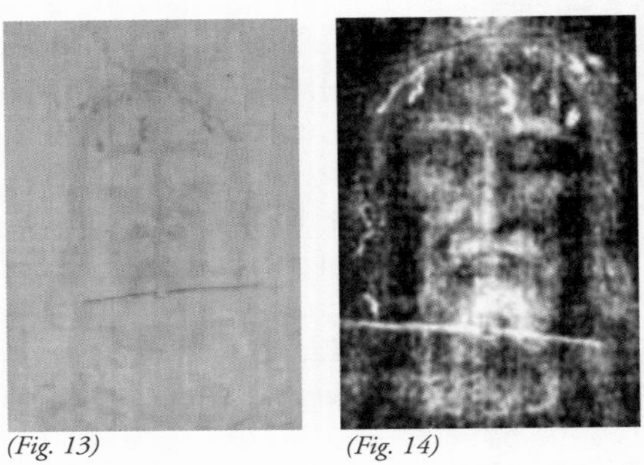

(Fig. 13) *(Fig. 14)*

The stark contrast in numerous distinct details and overall appearance can be seen above in the negative and positive facial images of the man in the Shroud.

This stunning revelation provided the earliest clue that a great deal of information invisible to the human eye was somehow encoded within the full-length body images. It would be another eighty years before scientists began to realize just how much more extraordinary evidence this burial cloth actually holds. Pia's startling discovery initiated the first limited scientific inquiry in 1900-02 based on his large photographic plates — just as the discovery that the Shroud's body image contained three-dimensional information served as the impetus for the only comprehensive scientific examination of the cloth itself in 1978.

It is difficult to imagine a process that could encode vague, negative full-length body images onto cloth — which reveals focused resolution throughout their positive images only after they are photographed — without involving some kind of light (or radiation). While in some respects it can honestly be said that the full-length body images on the Shroud are like photographs, this is an enormous understatement. Although the process that encoded the Shroud's body images clearly appears to have involved light or radiation, this unique process encoded many more features and is far more sophisticated than mere photography.

How could a medieval painter or forger, using any kind of artistic or other method, possibly have encoded all of these unparalleled features on a cloth? How could they possibly have occurred naturally? Yet we have only reviewed some of the insurmountable problems that a forger would have to overcome that were revealed from the initial applications of modern scientific technology to the Shroud. All of these findings clearly refute the 14th century publicly accepted assertion that the Shroud was painted by an unknown artist who confessed to his forgery. While I have listed the basic problems from our Chapter One discussion at the end of the chapter endnotes,[18] I can easily state that most of these features were not even *visible* to a 14th century forger. Nor could he have anticipated or utilized technology that would make these features visible for this technology would not be invented for another 500-600 years. Yet here's the clincher. Even 21st century scientists and artists who can now see the Shroud's many features with the benefit of modern technology cannot duplicate them.

A medieval artist simply could not have encoded all of the Shroud's above features regardless of the artistic or naturalistic techniques he attempted. (Chapters Four and Five of my 2000 book, *The Resurrection of the Shroud,* show that all artistic and naturalistic image forming methods proposed through the 20th century fail for many reasons to duplicate the Shroud's body images and blood marks. The numerous shortcomings of the methods proposed since then are discussed in Appendices C & D of this book.) We will see many more features throughout the Shroud's full-length, frontal and dorsal body images that neither medieval artists nor 21st century scientists can collectively encode. We can, however, begin to ascertain how the images were encoded by analyzing the evidence revealed by scientific and medical examinations.

We have seen that the presence and/or interaction of radiation within the process that encoded the full-length body images on the Shroud of Turin could account for their negativity; straight-line vertical directionality; being encoded through various spaces between the cloth and body; having double-bonded straw-yellow colored fibers that used to be single-bonded; that yellowed over time; with uniform coloring; very refined, superficially encoded fibers; without two-dimensional directionality; yet with precise distance or three-dimensional information.

Many more interesting features on the Shroud's full-length, frontal and dorsal body images will be seen in later chapters. From the microscopic to the macroscopic properties of these full-length images, we will continue to see that only radiation could have encoded these various features onto the body images, and even onto the background (or non-image) regions of this burial shroud. Interestingly, the radiation in this process does not appear to have originated from an outside source, such as a camera. Instead, the radiation or light appears to have originated from the body itself. The next chapter will take a more detailed look at the condition of the body at the time its unique features and images were encoded.

2

THE CONDITION OF THE VICTIM

Although accusations were made that Secondo Pia somehow forged his photographic negatives, photographs taken of the Shroud of Turin by the renowned photographer, Giuseppe Enrie, in 1931 not only confirmed this burial cloth contained negative images, but provided a new generation of even more detailed and resolved images. Subsequent photographs have done the same.

Numerous pathologists, anatomists and doctors have also studied the full-length images on the Shroud of Turin, beginning with Drs. Yves Delage and Paul Vignon in 1900-1902. The study of the Shroud was advanced further during the 1931 and 1933 Shroud expositions, which allowed modern physicians to see for themselves the realistic wounds and blood flows displayed on the cloth. Much greater scrutiny occurred in 1978 when more than two dozen American and Italian STURP scientists examined the entire cloth during its only comprehensive scientific examination, while taking thousands of photographs of the body images, wounds and blood marks in all wavelengths. The Shroud was displayed again in 1998, 2000, 2010 and 2015 where it was also observed and photographed by experts, as well as millions of visitors. In 2002, the burial cloth was examined by textile experts who removed its backing cloth and patches from 1534, along with the debris behind them. This also allowed for the first photographs to be taken of the Shroud's outer side (Fig. 15) before putting a new backing cloth on this side.

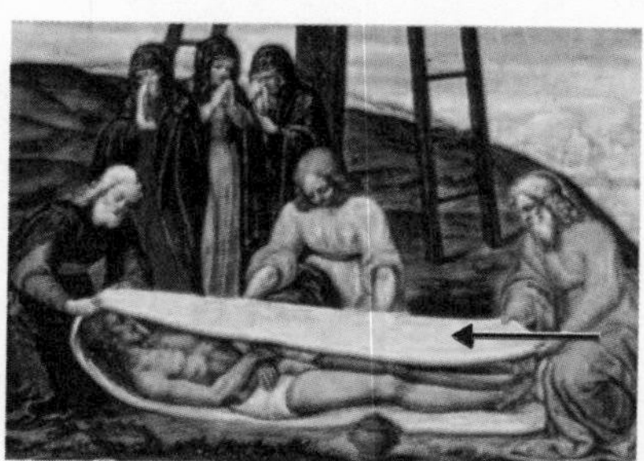

(Fig. 15) Arrow shows outer side of Shroud behind its backing cloth.

Based on a wide variety of findings, first hand examinations and countless photographs, experts are agreed that the Shroud depicts the unique full-length frontal and dorsal body images, wounds, blood marks and bodily reactions of an adult human male who was well proportioned and of average height and weight. This man incurred a series of wounds, died during his crucifixion and was wrapped in a linen burial shroud when his unparalleled body images and blood marks were encoded onto the inner sides of this cloth.

The man's blood flows and blood marks consist of real blood. More than 15 different tests confirm this.[1] Only two colors, black and white, appear on the Shroud's photographic negatives. Because the blood marks are denser than the body image and are reddish in color, they appear white on the light-dark reversed photographic negative. The Shroud's blood marks not only have different features than the body images, but were encoded in a different manner. For example, the blood marks penetrate into the cloth causing the fibers to stick and mat together, as seen in the photomicrographs below. (Figs. 16 and 17)

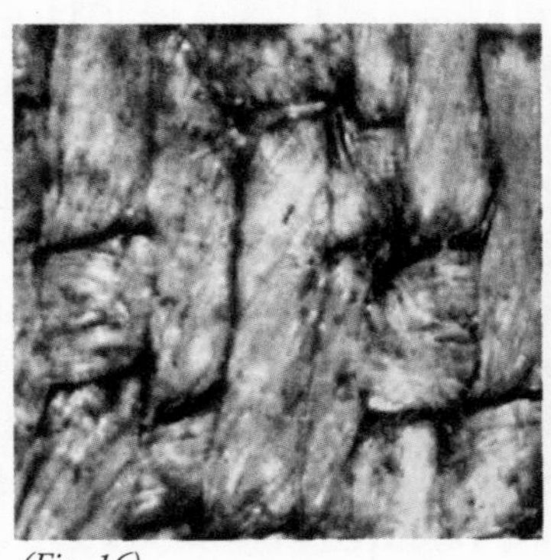
(Fig. 16)

(Fig. 17)

When seen with the naked eye, the blood marks on the Shroud still retain a reddish coloration. Yet, even if the Shroud is medieval, its blood should have long ago turned dark brown or black, like all other centuries-old blood. Blood will actually start turning dark within days, and certainly weeks of leaving the body and being exposed to air. The still-reddish color of the blood is not only apparent to anyone who has ever seen the Shroud over the centuries, but when the cloth is exposed to sunlight, it interestingly appears to be even redder.[2]

The blood marks vary in depth, size, shape and intensity. The man's wounds were inflicted in different ways, with different instruments and at different times. A variety of wounds, blood flows and blood marks are found on the man's head and face; arms and wrists; legs and feet; back and shoulders; and his chest. One set of wounds discussed immediately below is found throughout most of his body. Let us take a more detailed look at all the wounds, blood flows and the condition of the victim at the time his unique images were encoded on the Shroud.

Scourge Marks

The man's front and back, from his shoulders to his lower legs, are covered with an estimated one hundred or more scourge marks.[3] These dumbbell-shaped patterns, which are most noticeable on the dorsal image, generally run parallel and diagonal across the body in groups of two or three (see Fig. 18). Although all are approximately the same size, these scourge marks vary in intensity from light contusions to deep punctures, and close examination reveals the presence of blood in many of them.[4]

Because the scourge marks have two lobes (see Fig.19), these wounds must have been inflicted with a bifid instrument. The form and distribution of these marks led medical examiners to believe they were caused by a whip or cordlike device containing metal or some other sharp object at the end capable of tearing flesh. In particular, these wounds match, in size and shape, the Roman *flagrum* (shown in Fig. 20).[5] The flagrum, a whip used for flagellation, had pellets of lead (or sometimes bone) at the end of a pair of leather thongs. These unusual marks and their similarity to the flagrum led medical experts to conclude that the man in the Shroud must have been whipped or scourged.

Since the scourge marks are more numerous and visible on the dorsal view, physicians further believe the man was whipped from behind. Since the man's arms, head and feet seem to be the only areas that escaped scourging, we can assume that either his arms were elevated above his body during the scourging, or that his hands were tied to a post or pillar in front of him while he was whipped from behind. There are so many scourge marks on the back of the man that they are easily the dominant feature of the dorsal image.

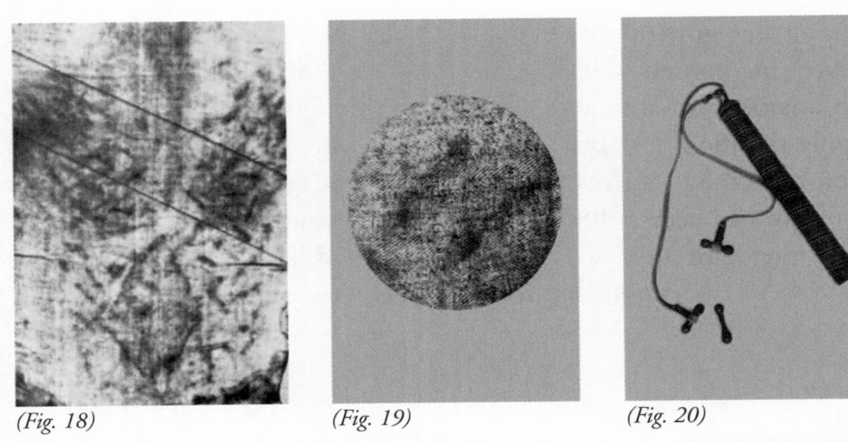

(Fig. 18) *(Fig. 19)* *(Fig. 20)*

Image on the left shows scourge marks and abrasions over the man's back and shoulders. Center image shows close-up of scourge marks and on the right is a Roman flagrum with lead pellets.

Modern technology and equipment such as photography, photographic enlargers, microscopes and ultraviolet lighting reveal many of the scourge marks that were invisible to the naked eye, and strikingly confirm that they occurred on a human body. When examined by such forms of modern technology, each scourge mark reveals slightly indented centers and upraised edges, just like a dumbbell-shaped or bifid object would leave on human skin.[6] Moreover, a "halo" of lighter color surrounds the scourge marks that chemical testing and photography under ultraviolet lighting confirms is blood serum.[7] According to STURP scientist Alan Adler, the upraised edges, indented centers and serum surrounding the scourge marks illustrate precisely the process called syneresis, which happens when a blood clot forms and then retracts.[8] When skin is abraded, the emerging blood remains whole for only a few minutes before it quickly begins to coagulate. During coagulation, red blood cells and serum separate; the red blood cells bond to form a blood clot, which then retracts in the wound. As the clot shrinks to a slightly smaller size, the serum is squeezed out and settles around the edge of the wound.

None of these intimate scourge mark features are visible on the Shroud with the naked eye. The indented centers and raised edges are perceptible only when these areas of the body image are photographed and enlarged, then examined under a microscope. Photography under ultraviolet light (fluorescence testing) is needed to show the serum-fluorescing borders and to observe that scratches and cuts invisible to the unaided eye accompany the scourge marks. These findings are highly significant because they prove that the Shroud could not have been created by an artist in the Middle Ages. A medieval artist would not have had access to photographic equipment, a microscope or an ultraviolet light source because none of the tools would be invented for several more centuries. There are more than one hundred scourge marks on the man in the Shroud. The inaccurate representation of just *one* of them would reveal an unnatural physiological reaction and expose the work as an artistic creation. We can only conclude that the scourge marks are not the product of an artist but are, instead, evidence of the natural processes of an actual body.

Facial and Head Wounds

The man on the Shroud has a mustache, beard and hair that falls to his shoulders from a central part. His cheeks appear swollen, and the area below the right cheek contains a triangular-shaped wound.[9]

On close examination, his nose, which is bruised and swollen, shows a slight deviation that indicates the cartilage may be separated from the bone.[10] Microscopic study also reveals that scratches and dirt are on the nose.[11] The areas above and below each eye, especially the right eye, look swollen and the face appears to have been beaten with a hard object (such as a fist or stick) and/or injured in a fall. The man's eyes are closed.

A number of wounds are visible on the top, middle and sides of the man's forehead. Altogether, more than a dozen blood flows have been counted on the front of the head alone (Fig. 21). The blood marks associated with the frontal head wounds seem to run in different directions from their points of origin, which suggests the head was in different positions as the blood was flowing.

Circling the top and middle of the back of the head is another series of blood marks (Figs. 4 and 22). Since these scalp wounds are covered by hair, the exact number of blood rivulets is difficult to determine, but Dr. Sebastiano Rodante has identified as many as twenty separate blood flows on the back of the head (Fig. 23), which brings the total number

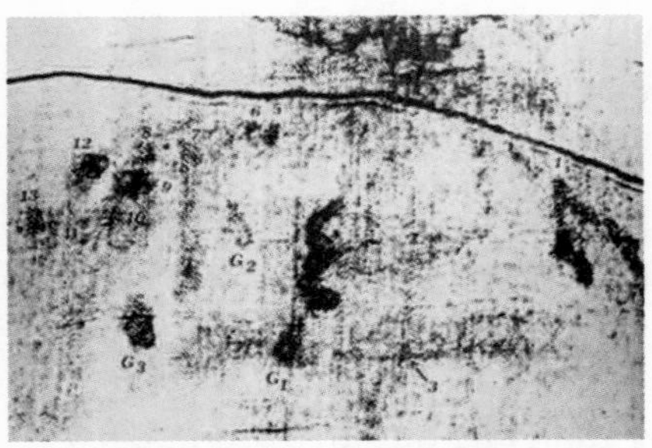

(Fig. 21) Number of forehead clots, reduced size.

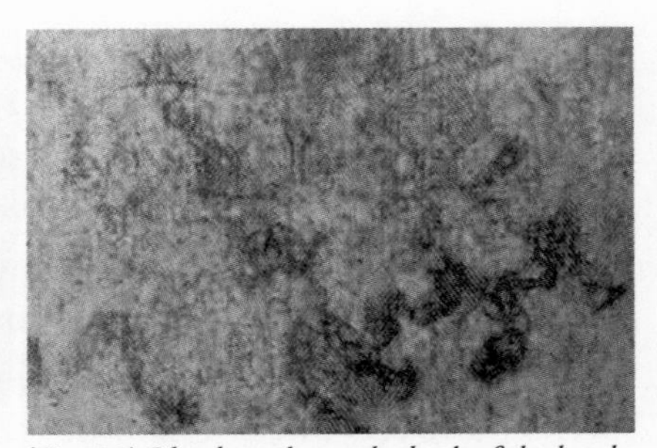

(Fig. 22) Blood marks on the back of the head.

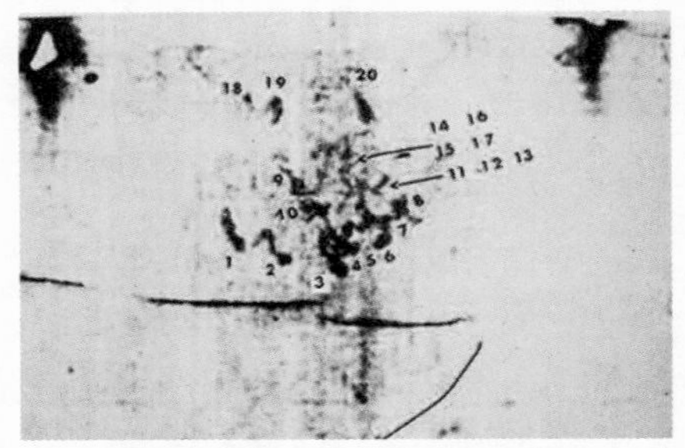

(Fig. 23) Number of blood marks on the top and back of the head in reduced size.

of head wounds to more than thirty.[12] Like the frontal image head wounds, the dorsal blood flows also run in different directions, until they seem to stop along a concave line just below the middle of the head. Above this line, wounds can be seen in the middle and near the top of the back of the head. When considered with the wounds evident on the top, middle and sides of the forehead, the head wounds give the impression that the man was wearing something like a cap made of sharp, pointed objects. Several physicians have noted that a cap made of thorns would produce head wounds identical to those on the man in the Shroud.[13]

Venous blood flows can even be distinguished from arterial blood flows in some of the bloodstains on the man's forehead. In general, venous blood appears denser and darker red, and it flows more slowly than arterial blood. In large wounds or wounds that puncture a vessel and produce a large blood flow, venous blood slowly thickens as it descends because it takes a few minutes for the coagulation process to begin and a clot to form. The large epsilon-shaped clot in the middle of the man's forehead is a good example of a large venous blood flow.

Smaller examples of venous blood marks are found in wounds numbered 9, 12 and G3 in Fig. 24. In contrast to blood from a vein, arterial blood spurts from a wound, driven by the pumping action of the heart. Wound number 1 on the right side of Fig. 24 is a good example of an arterial blood flow.[14]

Dr. Rodante, who has made one of the most extensive studies of the forehead wounds to date, has identified the origins of many of the head wounds based on the size or coagulation pattern of blood flows on the skin. (The arterial or venous origins of blood flows matted in the hair, and not free-flowing on skin, are impossible to determine.) As examples, the epsilon-shaped forehead clot lies over the frontal vein, while the tent-shaped arterial wound (number 1 in Fig. 24), which spurted blood as would an artery, corresponds with the frontal branch of the superficial temple artery.[15] According to Rodante, "The perfect correspondency of the forehead clots imprinted on the [Shroud], overlaying as they do the vein and the artery in mirror image, gives us the certainty that the linen covered the corpse of a man, who, while living, suffered the lesion of these blood vessels."[16] Numerous examples of distinctly venous and arterial wounds on the head of the man in the Shroud confirm that the various injuries evident on the man's images could have occurred only on an actual human body.[17]

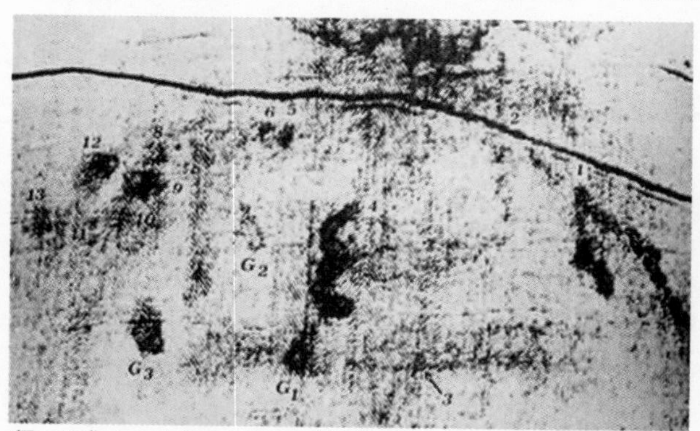

(Fig. 24) Arterial and venous bleeding has been identified on the forehead of the man in the Shroud.

Regardless of technique, no artist, especially one working in the Middle Ages, has ever represented the distinction between venous and arterial blood so accurately. In comparison to the Shroud's realism, Fig. 25, a medical illustration of wounds drawn in the 1400s, shows how poorly blood flows were understood at that time. In fact, the difference between arterial and venous blood was not even discovered until 1593, more than 230 years after some allege that the Shroud image was painted.

The epsilon-shaped clot on the man's forehead contains another realistic detail. As the blood flow descended, it broadened and changed course twice. Physicians believe this was because forehead muscles spontaneously contract when they are injured. The forehead, temple and scalp contain a web of nerves that is highly sensitive to pain.[18] Thus, contracting forehead muscles would be a naural reaction to the intense pain caused by having more than thirty head wounds.

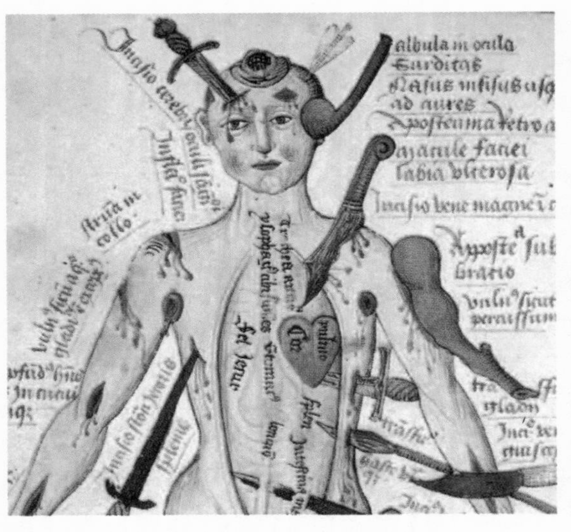

(Fig. 25) This medical illustration of wounds from the 15th century shows how poorly they were understood even a century after the Shroud was allegedly painted and first known in Europe.

Hand and Arm Wounds

The man's left wrist has been wounded or pierced, and blood flows from the wrist area toward the left elbow. Although the right wrist is covered by the left hand, similar blood flows are also visible extending along the right forearm toward the elbow. As shown in Fig. 26, both forearm blood flows run in two nearly parallel streams, with one stream measuring approximately 65° from the horizontal axis of the arm and the other stream measuring about 55° from the horizontal axis.[19] These unusual blood marks flowing from the wrists toward the elbows proved to be an important piece of evidence, helping physicians determine that the man on the Shroud had been crucified. During crucifixion, a victim's hands would be higher than his head. Since the blood flows from the pierced wrists toward the elbows of the man in the Shroud, we know that his arms were elevated, not hanging at his sides, while his wrists were bleeding.

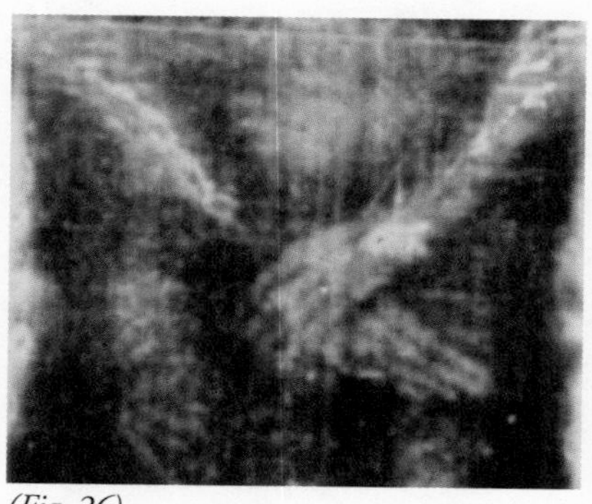

(Fig. 26)

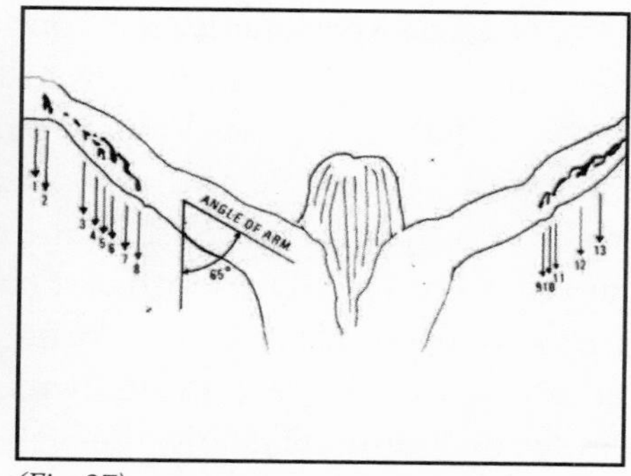

(Fig. 27)

The image on the left shows the blood flows on the arms and wrists. On the graphic on the right, the main angle of the forearms appears to be 65°, but there is evidence that at some stages, they were at 55°; indicating that the man in the Shroud sought to raise himself, probably continually, during crucifixion.

The two parallel streams running at slightly different angles from the horizontal are also significant. When a crucified victim hung suspended on the cross he was unable to breathe. Although he could take air in, he could not exhale unless he pushed himself up with his feet to raise his shoulders and expand the rib cage. This movement alters the horizontal axis of the arms by approximately 10° (see Fig. 27). Pushing upward in this fashion would temporarily lessen some of the constant pain in the victim's wrists and arms, but it would increase the pressure and pain in his feet. While this up-and-down motion was arduous, it did allow a crucifixion victim to breathe and forestall inevitable death — at least until he was too exhausted or in too much pain to push himself up anymore. Often, the executioners would break the legs of the crucified to stop this movement and hasten death.

Unlike the nearly unanimous portrayals throughout the centuries of Jesus being nailed in his palms during his crucifixion, this man's nail wounds are in his wrist. Many who have doubted the Shroud's authenticity pointed to the absence of hand wounds as proof that the man in the Shroud could not be Jesus. Artists who depicted the crucifixion were inspired not by an understanding of physiology but by the words in the Bible: the Gospels state that Jesus' "hands" were pierced.[20] In actuality, the original Greek word used in the Gospels is *cheir*, which also means wrist and forearm.[21]

Medical experiments first conducted in the 1930s and repeated since then have proved that it is anatomically impossible for a person to be nailed to a cross through the center of the palm. The upper extremities cannot be anchored because the body's weight causes the nail to tear through the flesh of the palm. In 1968, when excavating a site at Giv'at ha-Mivtar northeast of Jerusalem in a Jewish cemetery of the Second Temple period (the time of Jesus), archaeologists unexpectedly discovered the only known remains of a crucifixion victim. Professor Vassilios Tzaferis, the archaeologist with Israel's Department of Antiquities who excavated the site, determined that the man, called Yehohanan, had been nailed to the crossbeam through the flesh near the wrists, as evidenced by a scratch found on the wrist end of the man's right forearm radius bone. Tzaferis concluded: "The scratch was produced by the compression, friction, and gliding of an object on the fresh bone. This scratch is the osteological evidence of the penetration of the nail between the two bones of the forearm, the radius and ulna."[22] The friction and gliding to which Dr. Tzarferis refers resulted from the up-and-down motion discussed earlier: In order to breathe, crucifixion victims pushed themselves up and down until the executioners broke their legs. The remains found in the Jerusalem cemetery showed that both of Yehohanan's legs had been broken by a single strong blow.[23]

The nail wounds in the wrist of the man in the Shroud are not only consistent with the original wording of the Gospels, they are anatomically correct and consistent with known first-century crucifixion practices. The nail wounds in the wrist also indicate anatomical accuracy and uncopied originality in another important sense.

Christian critics of the Shroud contended that Jesus could not have been nailed at this location since the wrist contains numerous small bones and the Old Testament prophesized that not a bone would be broken on him.[24] Dr. Pierre Barbet, a pioneering autopsy surgeon and anatomist, who studied the body images and blood marks on the Shroud for three decades, experimented by driving nails into cadavers at the same location depicted on the Shroud. As he drove the nails he observed several interesting physiological reactions. The nail first diverts into what is called the Space of Destot pushing aside four small bones that surround this space and widening it, allowing the nail to pass freely through the flesh without breaking any wrist bones (Fig. 28). This provides a well-anchored location in which to nail a crucifixion victim to the cross beam of the cross.

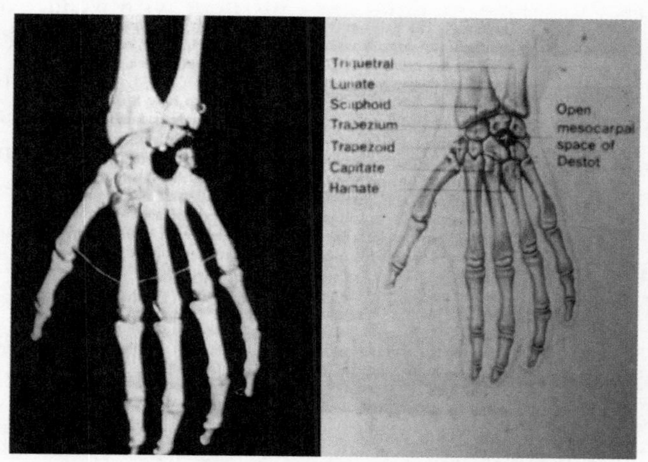

(Fig. 28) Location of the nail wounds in the Space of Destot.

When Dr. Barbet drove the nail further, he surprisingly discovered that the thumb contracts spontaneously inward toward the palm. He found a simple explanation for this previously unknown physiological phenomenon: When a nail is driven into the Space of Destot, the median nerve controlling the thumb is injured and stimulated, automatically causing the thumb to contract inward and lie across the palm. This is quite consistent with the man in the Shroud for his thumbs are absent. Injury to the median nerve, then, would also account for this anatomical reaction visible on the Shroud.[25]

The man's wrist wounds and the absence of his thumbs could also be unique points for the Shroud's authenticity as Jesus' burial garment — that no forger would ever have depicted. One can search through thousands of crucifixion paintings, carvings and statues that are still in existence, as this was easily the most popular subject in artistic history. And, while you may find a few where the wounds are depicted in the heel or wrist region of Jesus, *none* of them have ever depicted the wounds in the wrist *with* the thumbs absent or contracted into the palm. There is no reference, let alone a depiction, in all of history like this of Jesus or anyone else.[26]

Shoulder Injuries

Two broad excoriated areas are present across the victim's shoulder blades (Fig. 29). These scrapes are consistent with surface abrasions caused by contact between skin and a heavy rough object.[27] Because some of the scourge marks within this area are slightly different when compared to the clearly defined marks elsewhere on the body,[28] the scourging must have preceded the shoulder abrasions. We know that many crucifixion victims were forced to carry their crossbars to the execution site. Carrying such a large chunk of wood could easily have caused some of the man's shoulder wounds, especially if he fell under the weight of the beam and was struck by the wood falling on top of him. Cross beams were quite heavy (80 – 100 lbs.) and their weight could easily cause the victim to fall, especially if he was already whipped and beaten.

The Shroud contains evidence consistent with such a fall or falls. Scratches, lesions, and abrasions on the front of the man's knees have also been revealed by white light photos (Fig. 31)[29] and by ultraviolet fluorescent lighting.[30] Microscopic examination of the Shroud image also discloses particles of dirt on the front of the knees, nose and bottom of the feet.[31] It appears this man was unable to break his fall with his hands. Being struck by the crossbeam during a fall may also explain some of the wounds on the back of the man's head.

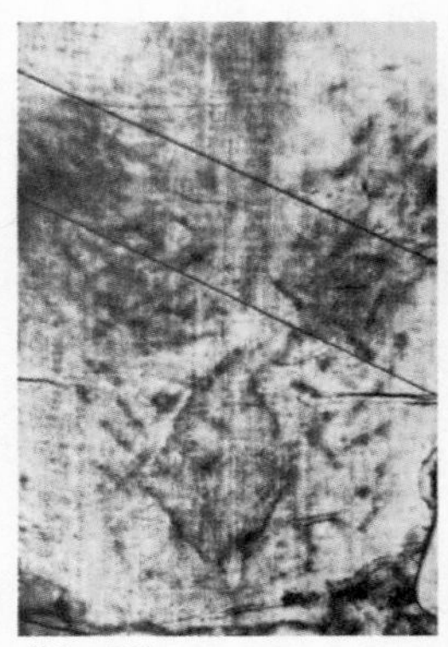

(Fig. 29)

(Fig. 30)

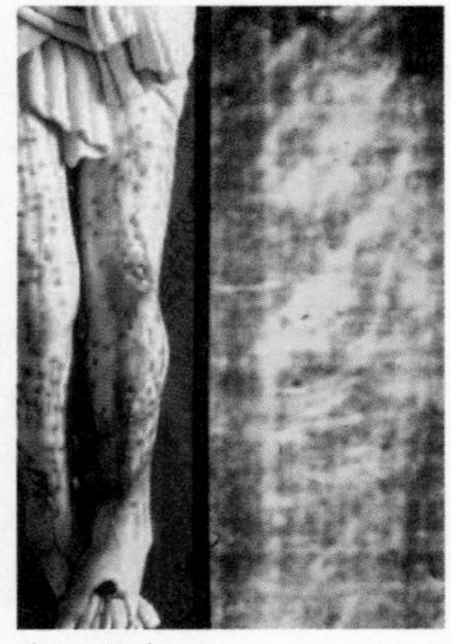

(Fig. 31)

Fig. 29 shows the back of the man on the Shroud covered with scourge marks and shoulder abrasions. Fig. 30 shows how the crossbeam could cause the abrasions on his back. Fig. 31 shows scratches and cuts on his left knee.

Some pathologists have identified another injury on the man that may or may not be postmortem. As he appears on the Shroud, the man's arms have been forcibly bent so his hands cover the groin. To accomplish this, the shoulder girdle would have had to be broken or dislocated, a practice common to morticians when positioning a body for burial.[32] In this procedure, the muscles between the neck and shoulder are massaged to release rigor mortis so the arms can be moved. Some medical investigators have noted that the man's right shoulder is about five degrees lower than the left, a feature most apparent on the dorsal view.[33] Dr. Barbet believes this indicates a dislocated shoulder, which may have occurred either during the hand-positioning at burial, when the victim fell, or when he was raising and lowering himself on the cross.[34] If the man's shoulder had been dislocated while he was still alive, that injury would have been another source of intense pain.

Leg and Foot Wounds

Detailed study of the lower extremities reveals two large blood marks on the front of the feet, the larger of which has a surrounding border that fluoresces under ultraviolet light.[35]

On the dorsal image, two bloodstained imprints of the feet are evident, with the right foot impression being more complete and showing the outline from heel to toes (Fig. 32).

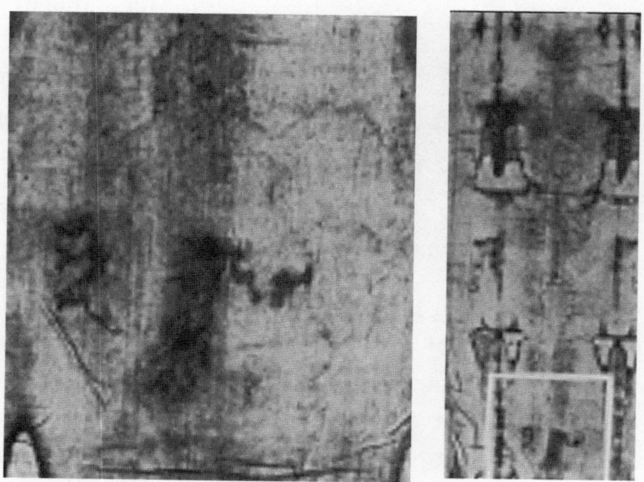

(Fig. 32) From close-up view and full-body image view

Some blood has flowed off the right heel area and onto the cloth. Medical experts agree that this large amount of blood resulted from a piercing wound to the foot[36]and STURP pathologist Robert Bucklin has identified the source of the blood flow: "a square image surrounded by a pale hole" in the metatarsal zone.[37] From this wound, some blood runs vertically toward the toes, but most flows toward the heels and horizontally onto the cloth.[38] This tells us the man was bleeding while in different positions — vertically while on the cross and horizontally when being carried after he was dead. The blood flow; deeper in color and running toward the heels and onto the cloth, has been identified as postmortem.[39] The most likely explanation for this postmortem bloodstain is that most of the blood that accumulated in the front and lower part of the foot while in the vertical position flowed from the wound after the piercing instrument was removed and the body was laid flat. The Shroud's medical examiners have concluded that this piercing instrument must have been the nail or spike typically used for crucifixion. Since this foot wound (shown in Fig. 33) is surrounded by the metatarsal bones, a large nail would provide the support necessary to prop up the victim's weight.

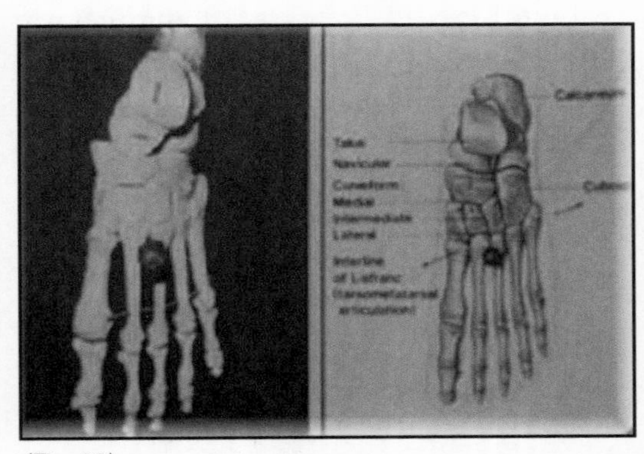

(Fig. 33)

When viewing the back of the man's legs and feet, we see that the left foot and leg images are less defined than the right ones. In addition, the left heel is elevated above the right. These facts indicate that the left knee was flexed to some degree. While this is most apparent on the dorsal view, the left leg visible on the frontal image also appears slightly raised. In light of these findings, most pathologists contend that the right foot was placed directly against a flat surface, while the left leg was bent at the knee and the left foot rotated to rest on top of the right foot. With a body in this position, a single nail driven between the metatarsal bones could affix both feet in a stationary position.

Unlike Yehohanan and the victims who were crucified with Jesus, neither leg of the man in the Shroud was broken. This was not necessary for he was already dead when he received a postmortem wound in the side of his chest.

Chest Wound

On the right side of the man's chest, a large side wound is apparent. This wound accounts for the most massive concentration of blood on the Shroud (Fig. 34), and all medical authorities cited agree that this large blood flow resulted from a postmortem wound. The blood from this wound appears to have oozed out and flowed due to the force of gravity, not driven by a pumping heart. Also, no swelling surrounds the side wound. Although this wound's bloodstain is partly hidden by one of the patches sewn on the cloth to repair damage from the fire of 1532, so much fluid poured from this wound that it collected in a puddle along the small of the man's back when he was placed in a horizontal position after death (Fig. 35). The side wound blood is darker and more copious than that from the other wounds, yet is interrupted by patches of a clear watery fluid.[40] While serum separated from blood may account for a portion of this stain, there is far too much watery fluid to be explained by the process of serum release from a blood clot. Forensic examination reveals that the blood and watery fluid from the side wound flows from an elliptical-shaped lesion approximately 4.4 cm long and 1.1 cm wide. The size and shape of this wound match excavated examples of the Roman leaf-shaped *lancea*, an instrument used by foot soldiers of the Roman militia.[41]

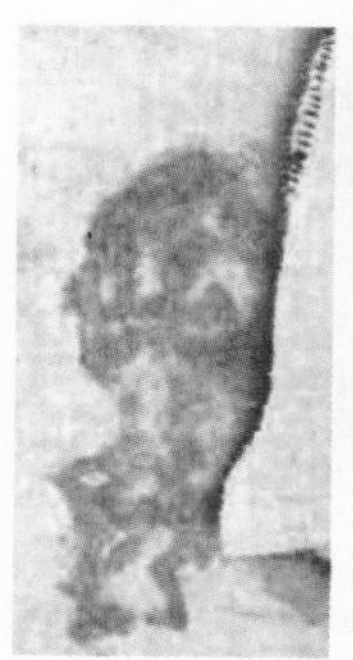

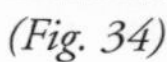

(Fig. 34)

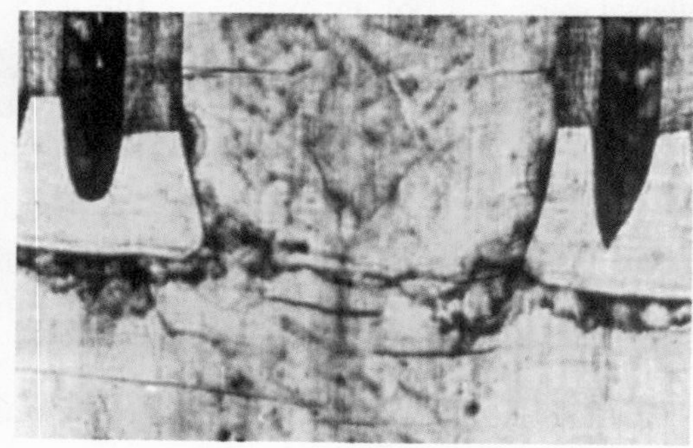

(Fig. 35)

Although medical experts differ somewhat in their explanations of this blood and water, most experts believe the blood came from the heart and the watery fluid from the pleural cavity in the chest.[42] STURP pathologist, Robert Bucklin, who summarized the slightly divergent opinions of pathologists, concluded that the most realistic interpretation is that the watery fluid came from the pleural space and, perhaps to some extent, from the pericardial sac surrounding the heart, while the blood came from a piercing wound to the right side of the heart.[43] The side wound is located between the fifth and sixth ribs. An instrument such as a lancea thrust upward into this rib cage region would have pierced the right auricle of the heart, which is only about three inches from the chest surface and fills with blood on death (Fig. 36).

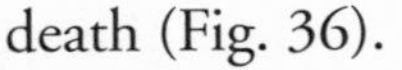

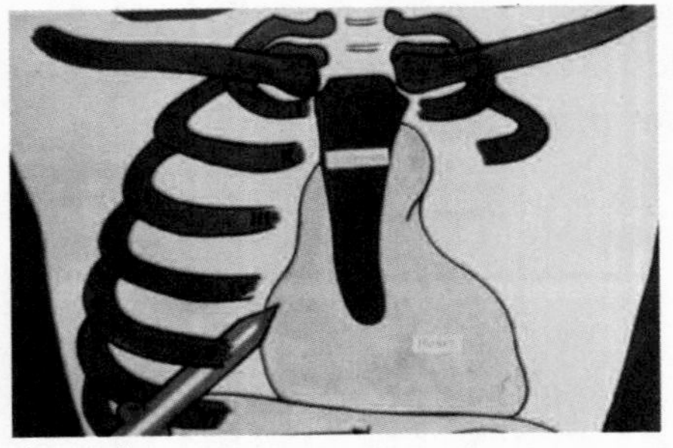

(Fig. 36)

The various wounds throughout both sides of the man's body total 130 or more conservatively. They not only reflect the various instruments, but also the locations, natural reactions and positions of a human body. From the tiny scourge marks to the large lesions, the blood marks and wounds evident on the Shroud all demonstrate the characteristic syneresis process that occurs when whole blood coagulates (or congeals) as a person bleeds. Many years before STURP went to Turin to study the Shroud with modern scientific equipment, physicians had strongly suspected that the wounds and blood flows were comprised of actual coagulated blood.[44] For example, in the 1930s, Dr. Barbet emphasized: *"The thing which immediately strikes a surgeon . . . is the definite appearance of blood congealed on the skin,* borne by all the blood-marks" (italics are Barbet's).[45] Barbet and many others hoped that one day rigorous testing could be performed on the Shroud to confirm his suspicions; specifically, he called for chemical testing and studies with spectroscopy, photography in all the zones of the spectrum, radiography "and everything else we could imagine."[46]

In 1978, STURP scientists conducted these tests, plus many others Barbet could not have imagined, which did much more than confirm

Barbet's observations. In addition to the halos discussed earlier around the scores of scourge marks covering the body, lightly colored and fluorescing borders surround many of the major bloodstains, specifically those wounds on or near the front and back of the head, the lower lip, the lower left wrist and arm, the side of the rib cage, the small of the back, and both sides of the feet.[47] These halos around the larger wounds have also been identified by scientists as blood serum. The presence of serum around the wounds means the Shroud's bloodstains are composed not only of real blood, but of whole blood. Other studies of bloodstained fibers confirmed the presence of bile pigments, serum-type proteins (such as albumin), and nonheme proteins adjacent to blood stains — all of which are indicators of actual *bleeding* wounds.[48] Subsequent tests by Fourier Transform Infrared (FTIR) micro spectrophotometry and ultraviolet-visible spectrophotometry confirm these finds by identifying the presence of bilirubin in blood samples and yellow serum-coated fibers from the Shroud.[49] Dr. Adler states that the ultraviolet photographs of the man in the Shroud reveal that "every single blood wound shows a distinct serum clot retraction ring."[50]

Spectrophotometric studies further revealed that the bloodstained areas possess the spectral characteristics of human hemoglobin.[51] Drs. Adler and Heller obtained positive results when they tested bloodied fibers for human albumin and human whole blood serum.[52] Working in Italy, Dr. Baima Bollone and colleagues used fluorescent antibodies to demonstrate the presence of human blood on threads removed from the Shroud. They even concluded that the blood is type AB. Using antiserums to test trace materials left in the fibers' test tubes, Bollone also learned that the Shroud contains human immunoglobulins.[53] It was announced in 1997, after examining two blood samples taken from the Shroud at the back of the head, that Dr. Victor Tryon of the University of Texas found human DNA with both X and Y chromosomes present in the samples. This also confirmed that the samples were those of a human male. In addition, he found that the DNA was very degraded, which is consistent with ancient DNA.[54] Summarizing the findings of STURP and other scientists and physicians, Dr. Heller stated the obvious when he noted: "It was evident from the physical, mathematical, medical and chemical evidence that there must have been a crucified man in the shroud."[55]

Death by Crucification

As stated earlier in this chapter, pathologists and other physicians are convinced the man in the Shroud was crucified. The wounds in the wrists and feet, along with the parallel blood flows on the arms produced as the man pulled himself up to breathe, confirm this. The identification of serum around the vertical blood flows at the wrist wound and arm also indicate the blood bled and coagulated while the man was in the vertical position. Other pieces of evidence contained on the Shroud also corroborate a crucifixion: the man's abnormally expanded rib cage and enlarged pectoral muscles appear drawn in toward the collarbone and arms;[56] the upraised left leg was bent so that the left foot was apparently placed on top of the right foot, after the man was beaten and whipped.

Several physicians have discussed the exact cause of death resulting from crucifixion. Most believe it involved asphyxia — that is, difficulty in exhaling — or some other respiratory problem caused by a lack of oxygen, accompanied by or related to cardiac failure or to complications from shock and pain. Muscle spasms, progressive rigidity and the inability to exhale would also contribute to the cause of death. When a man is hung by his arms, the pectoral muscles contract around the lungs; to exhale, the victim must strenuously and continuously work to raise himself up and expand those muscles. The pain from this pushing would be intense and constant, especially for the wrists, arms and feet. Tremendous pain can exhaust a person and is often accompanied by excessive sweating, which causes the body to lose vital fluids and minerals that regulate heartbeat. Eventually the victim goes into a state of shock. Hence, a series of different physiological events contributes to death by crucifixion. Everyone who has written about this form of execution describes it as a gruesome and horrible type of torture. If the victim was lucky, death would come within a few hours, but some victims lasted as long as two or three days before dying.[57]

Virtually every medical authority who has studied the Shroud agrees the man died while on the cross. This is first indicated by the right side wound, which we know was inflicted after death, as evidenced by the lack of swelling around it. While the side wound blood is darker and overflowing, it is also interrupted by patches of a clear watery fluid, both of which oozed out by gravity from an elliptical-shaped lesion between the fifth and sixth ribs. If an instrument, such as the Roman lancea mentioned earlier, was thrust upward at this location, it would pierce the right auricle of the heart, which fills with blood after death, allowing its postmortem blood to escape along with the watery fluid from the pleural cavity.

Further evidence of the man's death on the cross is found in the numerous identifications of rigor mortis apparent on the Shroud image.[58] Rigor mortis develops because of complex chemical processes that cause all body muscles to stiffen. The actual stiffening typically begins four to six hours after death and continues for another twelve hours. Once complete, rigor mortis gradually declines over the next twelve to twenty-four hours and the muscles relax again. The onset of rigor mortis can be accelerated by muscular exertion before death, an elevated body temperature or warm weather.[59] In cases where physical activity has been strenuous and intense as would be the case in a crucifixion, rigor mortis can set in immediately after death, especially in a hot climate.[60] If the corpse were then placed in a cool environment, such as a tomb, rigor mortis would tend to remain longer.

When looking at the back of the man's legs and feet, we see that his left leg is raised slightly and that both feet, especially the right one, are flat and pointed down. For the lower extremities to have remained in such an awkward position indicates that rigor mortis set in while the man remained crucified.[61] Moving up the back of the man, we notice that the thighs, buttocks and torso are not flat, but instead are stiff and rigid. If rigor mortis had declined and the muscles had relaxed, these parts of the body would appear flatter and wider.[62] On the frontal image we see the chin drawn in close to the chest and the face turned slightly to the right. For the head to remain in this position inside the burial cloth without rotating further to the side requires the presence of rigor mortis.[63] The man's expanded rib cage is a sign of asphyxia, and the enlarged pectoral muscles, drawn in toward the collarbone and arms, provide evidence that the man had been pulling himself up to breathe.[64]

For these parts of the body to remain in such positions further shows that the onset of rigor mortis occurred while the man hung suspended.[65] Rigor would also maintain the thumbs in the positions held during crucifixion.[66]

Another indication that the man died on the cross, or at least within 12 to 24 hours after he received his first open wounds, is that scabs are not observed with his coagulated blood marks. Scabs begin to appear on live human skin about 12 to 24 hours after open wounds are first inflicted, but this protective process stops after a person dies.

When we combine all the information about the wounds on the man in the Shroud with the knowledge that they were inflicted over a period of several hours, we can reconstruct what happened to him with some accuracy. Most likely, he was first beaten about the head, which caused swelling, bruises, and lacerations on his head and face. The scores of scourge marks all over his body attest to a whipping. Something made of sharp, thorn-like objects placed over his head caused numerous piercing wounds on the front, top and back of the head. Some of these wounds could have occurred from being struck on the head after the thorn-like objects were placed over it. Other scalp wounds may have resulted from falling and being struck in the head by the crossbeam, often carried by victims to the execution site, or from scraping his head against the cross when he pushed himself up and down to breathe. The shoulder abrasions could also have been imposed as he carried the crossbeam or later scraped his back during the up-and-down breathing motion. If some of these injuries were suffered while the man carried the crossbar, they may have occurred at the same time he apparently fell, as evidenced by the dirt in the nose and knee areas, as well as the scratches and cuts detected on his nose, cheek, knee and leg. Such dirt and scratches suggest the man was unable to break his fall with his hands. With a crucifixion crossbeam weighing an estimated one hundred pounds,[67] it is reasonable to assume that victims fell frequently, especially if one was in a weakened condition already.

The man's foot and wrist wounds were next inflicted by large nails driven through his flesh between the metatarsal and wrist or forearm bones to anchor him on the cross. The crucifixion alone would have taken several hours. From the two parallel streams of blood flows, the expanded rib cage, the enlarged pectoral muscles drawn in toward the collarbone and arms, and the taut legs and buttocks, we can observe this victim was pushing and pulling himself up to breathe. In addition, he likely incurred two more especially painful experiences. The median nerve, which controls the thumb, is not only a motor nerve, but is also a great sensory nerve.[68] It would be excruciating for this nerve to be lesioned by a nail, and the pain would only be aggravated further when the man pulled himself up to breathe.[69] Also, the forehead, temple and scalp contain a rich supply of nerves whose sensitivity is among the most painful in the body.[70] The infliction of more than thirty different wounds about the head would have contributed even more pain and agony than a typical crucifixion.

3

INTRICATE RELATIONSHIP BETWEEN THE CLOTH AND THE BODY WRAPPED WITHIN IT

Since the crucifixion alone would have lasted a few hours, physicians conservatively estimate that the bleeding from the various wounds went on for a period of four to twelve hours. This means the blood that transferred to the Shroud varied in age from relatively fresh postmortem coagulated blood to blood that was up to twelve hours old. Yet, amazingly, all of the bloodstains transferred from the body to the cloth in a way that has never been recorded anywhere else in history. The blood marks correspond perfectly with the natural physiological reactions of a bleeding human body that has been in various positions. Ranging from the large side wound to the tiny scourge marks, the realistic wounds represented on this burial cloth are anatomically flawless.

After examining the Shroud, Dr. Barbet was among the first to notice the unique transfer of bloodstains to the cloth. As a battlefield surgeon in World War I, Barbet had seen countless dressings removed from soldiers' wounds; these dressings covered injuries in various stages, ranging from fresh to several days old. In all of his experience, never had he removed a dressing that showed the exact form of the underlying wounds or blood marks with such clean outlines. Anyone who has ever removed a tissue or bandage from a cut can attest to this same phenomenon: the blood on the bandage is not a mirror image of the coagulated blood that formed on the body. Commenting on the unusual correspondence between the blood on the Shroud and the man's injuries, Robert Wilcox summarized efforts to reproduce the kind of blood marks seen on the Shroud:

> Paul Vignon and Pierre Barbet found, after many attempts, that it was impossible to transfer blood to a linen cloth with anything like the precision shown on the shroud. If the blood were too wet when it came into contact with the cloth, it would spangle or run in all directions along the threads. If it were not wet enough, it would leave only a smudge. The perfect-bordered, picture like clots on the shroud, it seemed, could not be reproduced by staining.[1]

In describing the distinctive blood encoded throughout the Shroud, Dr. Barbet noted it was comprised of "stains with clearly marked edges, which with such outstanding truthfulness reproduce the shape of the clots *as they were formed naturally on the skin.*" (emphasis added)[2] The full significance of the unique transfers of the victim's various bloodstains will be discussed in Chapter Twelve. For now, it is sufficient to state that a bloodied, crucified corpse had to have been wrapped within this cloth when both of its unique, full-length body images and their perfectly corresponding blood marks were encoded.

Further illustration that this burial cloth had to have wrapped a bloodied, crucified corpse when its body images and blood marks were encoded is seen with the man's side wound. A large postmortem flow of blood and watery fluid oozed downward from an elliptical-shaped lesion between the fifth and sixth ribs. The lesion is approximately 4.4 cm long, 1.1 cm wide and matches the ancient Roman leaf-shaped lancea described earlier. Computer imaging technology even shows this line at the top of the blood stain.

This lesion, however, was encoded on the man's *left* side as seen with the naked eye. This was the only image of the man that anyone ever saw until the dawn of the 20th century when the Shroud was first photographed. How could, and why in the world would a medieval forger encode a lesion with a blood flow and watery fluid so perfectly on the man's left side so that when photography and computer imaging technology were developed 500-600 years later, the wound would line up perfectly on the *right* side of the body (where postmortem blood would drain from the right auricle of the heart which fills with blood upon death)? Of course, such intricate encoding and alignment could only have resulted from this burial cloth originally laying over the right side of a man who incurred such a postmortem wound. Not only was computer imaging technology and photography unknown in the Middle Ages, but this kind of knowledge about blood and cardiology was also unknown in the 1350s.[3]

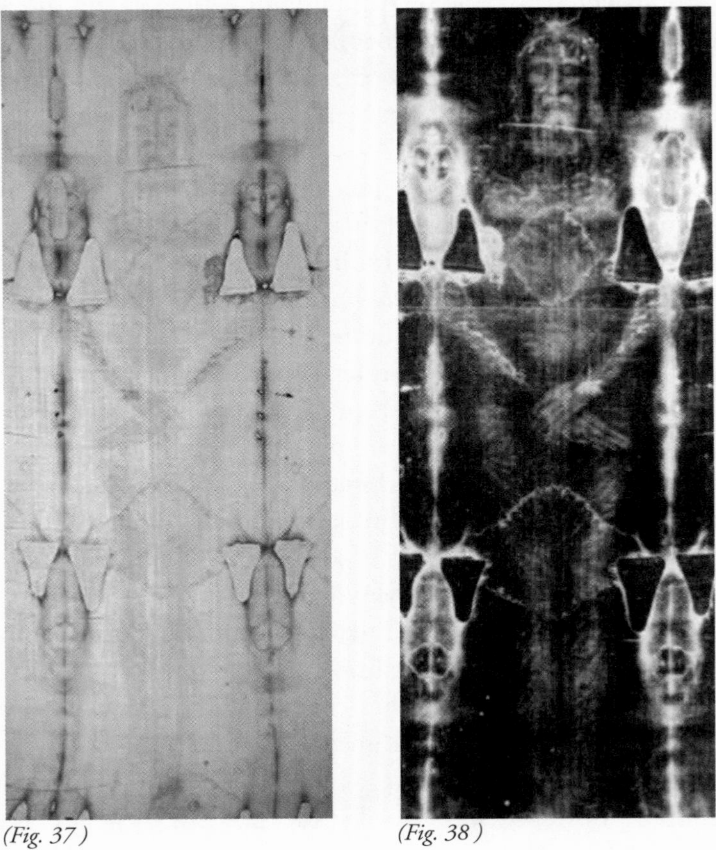

(Fig. 37) *(Fig. 38)*

The side wound on the negative image, Fig. 37, is just one of many that align with blood marks and blood flows on the man's positive image, Fig. 38.

Like the man's right side wound, the intricate relationship between this particular burial cloth and the body wrapped within it can easily be observed elsewhere in the precision of the body's anatomy, and the precision with which its wounds and its blood marks are encoded throughout the cloth. Beginning at the top of the head arterial and venous blood marks are distinctly encoded, corresponding in mirror image to the underlying artery and vein of the victim who incurred these wounds while alive. The nail wound at the Space of Destot aligns with an anatomical location that not only would have supported the weight of the victim, but would not have broken any of his wrist bones. This wound also lies over the median nerve that, when struck, would have contracted the victim's thumbs, which are absent from the image.

The upraised edges, indented centers and serum surrounding borders (that could not be seen until the 20th century) on approximately 100 scourge marks located throughout the body, also clearly reflect *intimate* contact with this cloth throughout both sides of the body. Serum surrounding borders can also be seen by modern technology around the variety of blood marks and blood flows encoded throughout both sides of this cloth. Furthermore, all of these blood marks have been encoded in this burial cloth in the same shape and form as when they flowed and formed on the *body* wrapped within it. An intricate relationship between this cloth and the man's anatomy, blood marks and wounds can be observed *throughout both* body images and both inner sides of the cloth.

All of the data gleaned from extensive studies of the pathology evident on the Shroud of Turin tells us this burial garment was wrapped around the human body of a man who incurred a series of wounds, was crucified and died while still nailed to a cross, all under circumstances identical to those of Jesus. We will see in later chapters that this man's burial in a linen shroud also occurred under circumstances like those of the historical Jesus. Neither modern scientists nor artists have been able to encode the numerous, yet intricate microscopic details found on thousands of colored fibers that collectively comprise the full-length body images. They have also been unable to encode the more than 130 perfectly corresponding blood marks that range from relatively fresh to several hours old, and vary from heavy flows (involving other fluids) to light scratches, all of which were intricately encoded with completely different characteristics than the body images.

Failure of Naturalistic Methods

Of the billions of people who have died throughout history, none have left any body images or blood marks on any kind of cloth that even begin to contain the many features encoded on the Shroud of Turin. Like artistic methods, naturalistic explanations for the Shroud's body images and its blood marks can be and have been tested by scientists and others throughout the 20th and 21st centuries. Needless to say, no forger, by any method, has ever come close to duplicating all the documented features on the body images and the variety of human blood marks contained throughout the frontal and dorsal sides of the Shroud of Turin. If a corpse could naturally leave body images or blood marks like those on the Shroud, it should have happened on countless occasions on other burial garments, blankets, sheets, shirts, trousers, soldiers' uniforms, bandages or other wrappings that came into contact with dead bodies throughout history.

Naturalistic methods have proposed that the Shroud's body images were caused by a variety of materials on the body or the cloth ranging from perspiration, myrrh, aloes, aloetin, sweaty blood, urea, saline and ammonia, as well as numerous solutions of these materials in various combinations. One of the most interesting attempts of a naturalistic direct contact method was made by Dr. Eugenia Nitowski, a Middle Eastern archaeologist, who tested this method the most fully by trying to duplicate, as much as possible, the actual conditions involved in the burial of a crucifixion victim in ancient Jerusalem.[4] Accordingly, she performed her experiments in the mid-1980s in a tomb complex at the École Biblique, which is part of the same rock shelf as that in the Holy Sepulcher and the Garden Tomb — the most likely candidates for the actual tomb of Jesus. The date of these tests also approximated the Passover/Easter season in which Jesus was executed.

Nitowski postulated that a crucifixion victim who had been scourged would experience a loss of body fluids (perspiration and blood) and undergo hematidrosis (bloody sweating), along with a lack of fluid intake and sufficient rest, all while enduring extreme physical exertion. The victim would become severely dehydrated, so his body temperature would rise. She further presumed that the severe trauma and emotional stress accompanying a crucifixion would produce an acidic condition in the victim's blood and perspiration.

In Nitowski's experiments, a sweat solution of normal saline and acetic acid was applied to a mannequin that had been filled with water warmed to a temperature between 110 and 115° F. A mist of the sweat solution was also sprayed on the mannequin just before it was covered with a linen shroud, which was then tied with rope around the neck, waist, and ankles so the cloth would remain next to the body. Variations of this experiment were also performed, including adding blood, myrrh and aloes to the sweat solution and using five different types of linen. While the mannequin was in the Jerusalem tomb, the air's humidity was constantly monitored. When humidity levels dropped below the percentages normally found in the tomb, the air was misted to raise the moisture level.

Despite the fact that Nitowski and other experimenters tried different combinations of liquid substances and atmospheric conditions in an effort to reproduce the Shroud's body images, none of their results even remotely approached the image characteristics found on the Shroud. Nitowski's published pictures of her work (below) illustrate the best results of her tests. Instead of verifying the direct-contact theories, the photographs highlight the numerous problems inherent in such naturalistic, image-formation methods.

(Fig. 39) Frontal image of Nitowski's naturalistic method.

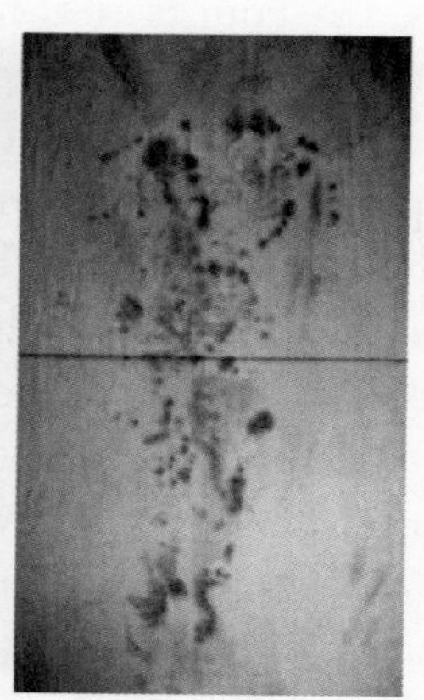

(Fig. 40) Dorsal image of Nitowski's naturalistic method.

Naturalistic and artistic methods have also been studied and tested by STURP scientists, who even proposed new variations and methods.[5] All of these methods, however, failed to duplicate the Shroud's many features. One of the fundamental problems for all naturalistic methods is that no traces of myrrh or aloes,[6] or the presence of perspiration, urea, bodily fluids*, organic liquid solutions or materials have been detected on the body images, or on the linen cloth.[7] (While it is theoretically possible that such molecules or material could have decomposed or oxidized over time, some residual traces of the responsible elements that formed the images should still be present.) No signs of decomposition, which would have started two or three days after death, are present on the body images or the entire linen cloth. It is also very doubtful that a uniform application of any of these materials could even have been achieved, let alone remained throughout the body until burial. Equally dubious is the likelihood that an organic stain mechanism would operate equally (if at all) over skin and hair. And, if these materials did get on the cloth, they would not just remain on the topmost fibers of the threads.

Another fundamental problem with naturalistic direct contact methods is that even when scientists overcame some of the noted problems by uniformly coating a plaster mold face with ink and draping a cloth over it, an image was imprinted only where the cloth and face made contact. No impression was left where the cloth and face were separated by space.[8] Four critical features will be absent from such an image. Without continuous shading throughout the face, this image will lack the high resolution found on the Shroud image. Additionally, the image will not possess the three-dimensional distance information that is also found on the Shroud image. Without continuous shading, this image could not have been encoded though the spaces between the cloth and the body, and could not have been encoded in a vertical straight-line direction from the body to all parts of the cloth draped over it.

These same problems existed with naturalistic diffusion methods. Even where scientists factored out the many chemical and physical problems inherent with materials or liquids associated with diffusion methods by using plaster body shapes and model cloth, they still did not duplicate the Shroud's features. Images with continuous shading, high resolution and three-dimensional information encoded along straight line vertical paths through the spaces between the plaster body

---

*Except those associated with blood marks throughout the frontal and dorsal body images (along with a watery fluid that emerged from a wound on the right side of the chest).

shape and the model cloth could not be produced under the diffusion mechanism. Scientists have repeatedly tested and rejected this method for these and other reasons.[9] Vignon spent decades attempting to produce images by the diffusion (vapograph) method, but never achieved publishable results. Even where he claimed to produce a decent hand image with this method, the image went well beyond the most superficial fibers of the cloth.[10] (There are many other shortcomings to these and other naturalistic and artistic methods that are discussed in Chapters Four and Five of my 2000 book, *The Resurrection of the Shroud* and in subsequent chapters and appendices in this book.)

In another attempt to remedy one of the fundamental problems of encoding discoloration only at the points of contact, scientists pressed the cloth onto a uniformly coated plaster face mold where contact had not been made. Yet, the resulting image was grossly distorted (see Fig. 41). The three dimensional relief of this direct-contact image yielded even more pronounced distortions (see Fig. 42).[11]

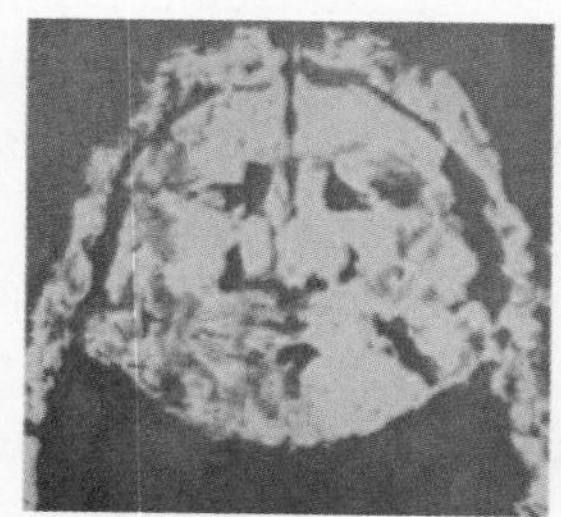

(Fig. 41) Pressed direct contact image.

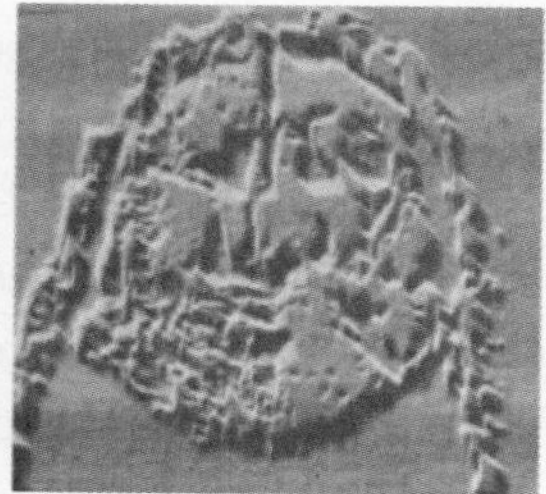

(Fig. 42) VP-8 relief of direct contact image.

While all naturalistic and artistic image forming hypotheses have severe shortcomings, naturalistic hypotheses have one key advantage over artistic hypotheses. By definition, a human body or corpse is assumed or postulated in any naturalistic hypotheses. At least this assumption contains one of the mandatory requirements. The body images and blood marks on the Shroud of Turin are those of an actual human body who incurred a series of wounds and was crucified and killed before he was wrapped in his burial cloth. The two full-length, frontal and dorsal body images, their various wounds and the 130 blood marks scattered throughout them, show they all involved some form of extensive and intimate contact with a dead human body after it was wrapped in this burial cloth.

The Shroud's body images were also encoded independent of contact pressure or body weight. Recall that Nitowski's very poor images or stains were more encoded on the dorsal side of the cloth. Even though most of the man's weight would have pressed down on the bottom half of the cloth, the Shroud's frontal and dorsal body images have nearly equal intensities. Just as the thousands of body image fibers throughout the Shroud were encoded with the same uniform intensity, nearly equal intensities are also found with both frontal and dorsal body images.[12] Such uniform, widespread intensity throughout both frontal and dorsal images could seemingly only be accounted for by radiation. (We will see that the Shroud's blood marks were also encoded independent of pressure or weight.)

Naturalistic methods involving pressure or contact with a body could not produce intricate and extensive uniformity on a cloth for many reasons. Even when STURP scientist Sam Pellicori put oil, lemon juice and perspiration on his fingers and gently pressed them against a cloth (in contrast to the full weight of a body) capillary action was also observed, in which the transferred liquid spread within the core of the fibers.[13] Capillary action was also observed at cloth contact points when molecular diffusion techniques were tested.[14] Some naturalistic diffusion or direct contact image forming methods include soaking the cloth in aloetin or in solutions of aloes and myrrh.[15] Capillary action would clearly be present with these methods. These results have significance because none of the fibers throughout the full-length frontal and dorsal body images show any signs of capillary action.

The last feature and, perhaps, some subsequent one(s) may appear a little mundane, especially when compared to some of the unique features on the Shroud that were discussed earlier. Yet, these features, too, serve a dual purpose. They are necessarily part of the image features that a forger would have to encode if he was going to duplicate the Shroud's body images. They also help to lay a foundation for understanding the leading explanation for the Shroud's various features and its unique body images — which could be tested during another scientific examination of the Shroud of Turin.

Significantly, many of the Shroud's blood marks would not even have been in contact with the cloth when it originally draped over the body, and others would only have been in partial contact. For example, the scourge marks on the back of the upraised left leg would not have been in contact with the cloth below it. Since the body is in rigor mortis with the right leg stretched out, pointing down and flat toward the ground, and since the left leg is upraised and also pointed similarly as the right foot, both the small of the back and the lower part of the back

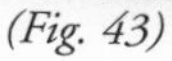

(Fig. 43)

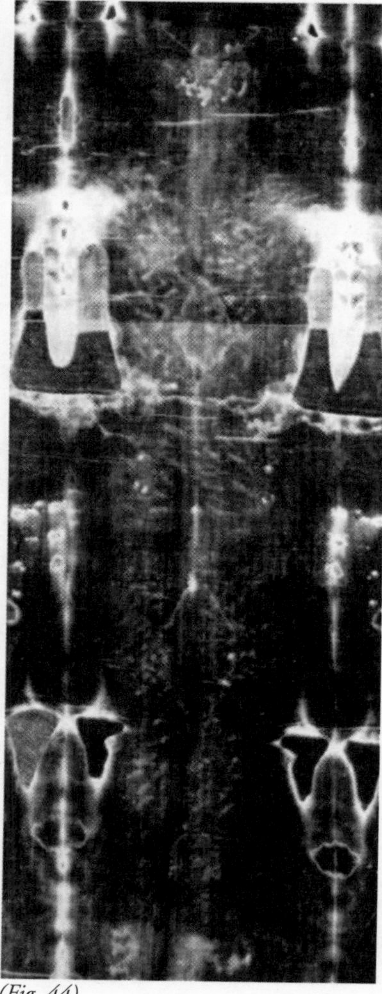

(Fig. 44)

will arch farther, preventing them and their scourge marks from contacting the underlying cloth. This is especially true at the spinal column, yet these parts of the back and their scourge marks are clearly encoded. There are also scourge marks located above the shoulder blades and abrasions at the very top of the shoulders near the neck. Because the man's arms were placed over his groin, the top part of the shoulders and their accompanying scourges would not have been in contact with the underlying flat cloth. (A simple experiment of lying in these positions on a hard floor verifies the points of contact.)

In addition, some of the man's wounds may have only been in partial contact with the cloth. One example is the side wound, where the upper arm could have prevented the cloth from draping around the entire bloodstain. Similarly, the bloodstains on the back of the head cover a very wide area. Since the head is round, it would not be in direct contact with the flat cloth along the entire width of these wounds. Yet, regardless of contact, all of the man's blood marks appear intricately encoded in this cloth in the same shape and form as when they flowed on the body.

Even if the cloth had been pressed or tied against the body to make contact at these various locations, neither the body images nor the blood marks at these pressed locations would have resembled those on the Shroud of Turin. These areas would have appeared like those on other burial shrouds and would not have left any recognizable images or realistic blood marks. The body images would have probably resembled Nitowski's stains. (Even after scientists pressed cloth onto a uniform plaster mold face that was uniformly colored in ink, its body images were flat and distorted, unlike any areas on the Shroud.) The blood marks would also have been altered by pressing against them.

To insure intimate contact with all the man's scourge marks, blood marks and blood flows would have necessitated pressing the cloth to the body at all points over and around each individual blood mark. This would be difficult to uniformly achieve by using hands or rope or something else, especially since the blood marks that lacked contact or only had partial contact were on both sides of the body. Even if this were achieved, pressing the cloth in this fashion would easily alter the jelly-like, coagulated blood marks. Furthermore, the perfectly bordered edges on the Shroud's blood marks, especially the intimately encoded edges on all the scourge marks, would have been altered.

Moreover, the coagulated blood marks on the Shroud of Turin are *embedded* in the cloth. They are so embedded, they can be seen on the opposite (or outer) sides of the burial linen that draped over and laid under the bloodied corpse. In 2002, the Shroud's patches and backing cloth from 1534 were removed. This revealed the full outer (or opposite) side of the cloth for the first time in five centuries. This allowed all 14 feet of the inner and outer sides of the Shroud to be photographed for the first time ever. (Afterward, a new protective backing cloth was sewn onto the Shroud's outer side, replacing the old one.) Most of the man's blood marks are clearly visible on the opposite or outer side of this linen cloth in approximately the same shape and form as those on the inner side. The numerous blood flows from the front of the man's head, his side wound, his wrist, both of his arms and the front of both of his feet are clearly visible on the outer side of the cloth that draped over his body. The blood marks and/or blood flows from the back of the man's head, the small of his back and the back of both of his feet

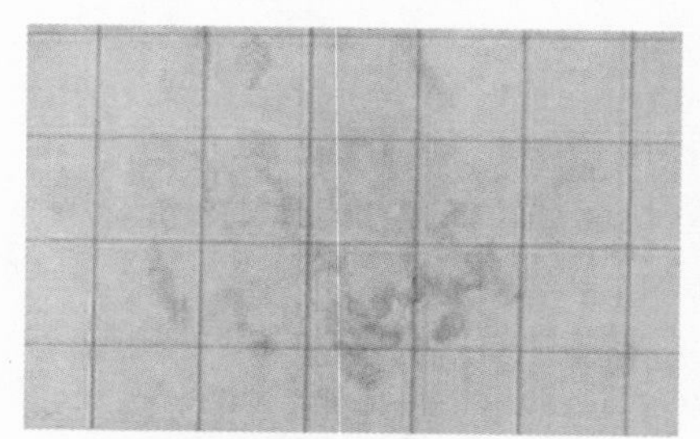

(Fig. 45) Blood marks from the inner side of the Shroud at the back of the head.

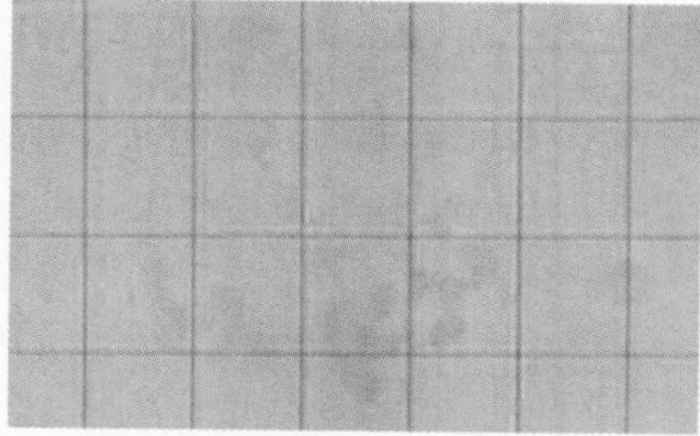

(Fig. 46) Blood marks from the outer side of the Shroud at the back of the head.

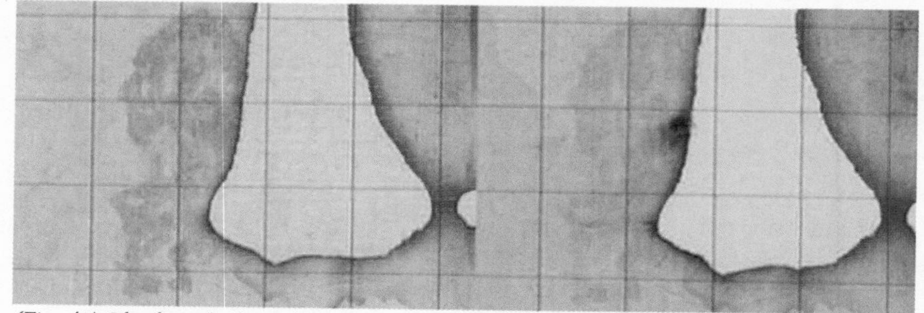

(Fig. 47) Blood marks from the inner side of the Shroud at the side wound.

(Fig. 48) Blood marks from the outer side of the Shroud at the side wound.

are also easily visible on the opposite or outer side of the burial cloth. Many of the scourge marks on the back of the man's calves and legs and on his back are also visible on the outer side of the cloth that laid under the man.

Many of these same blood marks would not even have been in contact with the inner side of the cloth, yet are visible on the outer side of this burial garment. Even where there was contact, these jelly-like, coagulated blood marks could not have been embedded *into* the cloth and onto its outer side by pressing against the blood marks or on either side of the linen. This would only break, smear and alter the blood marks and their edges.

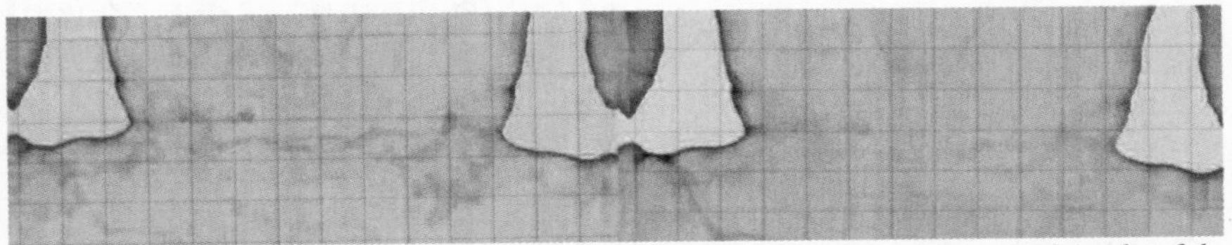

(Figs. 49 and 50) Blood flows are even visible on both the inner (left) and outer (right) sides of the Shroud at the small of the back.

Scientific and medical examination of the Shroud of Turin, various tests and experiments in laboratories and in tombs, as well as countless natural burials, demonstrate that naturalistic direct contact, diffusion or other methods do not begin to explain the Shroud's extraordinary features. Yet, the unique and intimate details revealed by modern scientific and medical investigation demonstrate that a bloodied, crucified corpse had to have been wrapped within this burial cloth when both of his unique, full-length body images and his perfectly corresponding blood marks were encoded within it. Some extraordinary event occurred to this dead body that caused its blood marks and its delicate, full-length negative images to appear on the cloth, yet would reveal its most remarkable details only as various forms of modern technology developed. This extraordinary event occurred within two to three days of being wrapped in this burial cloth.

Had the corpse been there longer, decomposition stains would be present on the cloth, but the Shroud contains no signs of bodily decomposition. Since the body was also in rigor mortis (postponing staining due to decomposition) when both images were encoded, the unprecedented encoding process had to have been completed before any signs of bodily decomposition.

The bloodstains also remain pristine and unsmeared having been encoded onto and *within* the *cloth* in the same way they appeared on the *body*. Jelly-like coagulated blood has never been known to have transferred experimentally or naturally from a body to cloth, let alone become embedded in the cloth. Yet, even if the blood marks somehow became uniquely encoded in this fashion — if the cloth was then removed by any human, mechanical or natural manner — some, most or all of the blood marks and their edges would have been broken, smeared or altered. A unique process(es) seems to have been present that encoded the full-length, frontal and dorsal body images and their corresponding blood marks onto this burial shroud. The same or another process could also have been involved in the body's removal from the cloth.

Additional Shroud Features

The application of modern technology and the only comprehensive scientific investigation of this particular burial cloth have given scientists and the public much greater insight into the depths of the Shroud's encoded body images and their corresponding blood marks. What we have seen thus far, however, is just the tip of the iceberg. There are more remarkable features encoded throughout this man's images that we haven't even discussed yet. Moreover, there is an unprecedented depth of evidence and information that could be acquired by adapting and applying sophisticated 21st century technology to the images, blood marks and background areas of this burial cloth.

As modern technology such as photography, microscopy, ultraviolet lighting and various forms of computer imaging are applied to the Shroud of Turin, the quality of the resulting images and the number of features found within them actually improves. For example, start with the vague, watery body image visible with the naked eye and apply a camera to it. The resulting image is highly detailed and resolved, having photographic quality. Next apply computer imaging technology that accurately interprets the brightness of every part of the image to its original distance from the camera and the result is an image with true three-dimensionality.

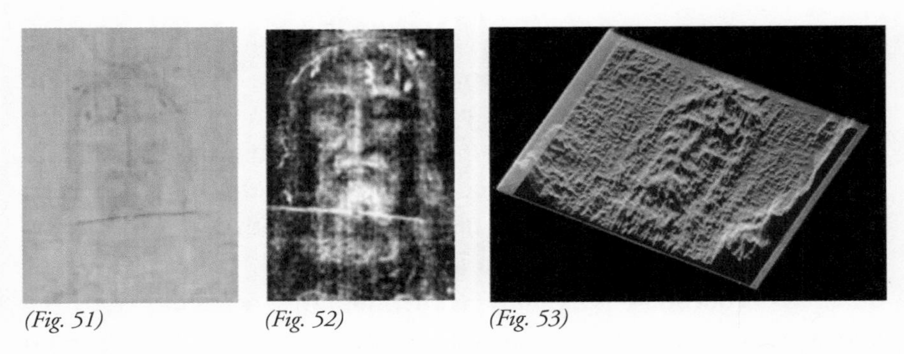

(Fig. 51) *(Fig. 52)* *(Fig. 53)*

If famous paintings or historical images are well cared for, the original representation can be preserved, but neither they nor their resulting images *improve* with time or technology. The Shroud images, however, not only improve, but make *quantum leaps* in development as time and technology progresses. No other relic or work of art has taken such strides. This remarkable feature is not due to the development of technology, but is due to the remarkable way in which these images were encoded centuries ago.

Unlike natural processes, the image encoding agent on the Shroud operated over different materials or surfaces such as the man's skin and the hair on his face and head. We will see in the next chapter that features from an ancient coin or *lepton* minted during the reign of Pontius Pilate in 29-32 A.D., along with flowers that collectively grow only in Jerusalem and the surrounding area, may also be encoded over the man's right eye and around his body. If so, the encoding process would have operated over even more diverse surfaces. Unlike natural processes, radiation can operate effectively over various surfaces. One particular form of radiation could account for these coin and flower images, if they are present on the Shroud. New proposed scientific testing could determine whether these coin and flower features are indeed faintly encoded on the cloth. If so, they would represent two more examples of the intricate relationship between this burial cloth and the body wrapped within it. If present, they could not only confirm the location where the man's images were encoded, but could indicate when this unique event occurred with far more specificity than any scientific test.

Blood flows from the perfectly aligned wrist wounds further illustrate the intimacy in which the blood marks and the body images were encoded. Real coagulated blood marks flow along the man's arms toward his elbows. His hand, wrist and arm images are comprised of oxidized, dehydrated cellulose. As the blood flows toward the elbow joint, however, each arm contains a blood clot that consists of blood on the left side of the blood mark, yet also appears to contain a blood image comprised of oxidized, dehydrated cellulose on the right side of the same blood mark. An entire blood mark located even closer to the man's right elbow joint appears to be comprised of oxidized, dehydrated cellulose or body image. As the man laid on his back, the parts of his arms containing these blood marks would have been lower than and at an angle from his hands and wrists, which were placed over his groin.

As we saw, the full lengths and widths of both body images on this burial cloth are comprised of oxidized, dehydrated cellulose. Since its first arrival in Turin, the Shroud was kept rolled on a spool for almost 425 years until the early 21st century when it was then placed flat in a long rectangular-shaped, atmospherically controlled container. Over the previous centuries the cloth was undoubtedly kept folded, as indicated by scientific and historical evidence. The countless times that this cloth was rolled and unrolled and folded and unfolded over many centuries have undoubtedly caused flakes of blood to fall off or translocate on the cloth. Where such flaking is observed, however, body image (oxidized, dehydrated cellulose) is not found on the fibers that were once covered by blood.

The observation that some blood flows on the arms were comprised of both coagulated blood marks *and* body image wasn't even made until 1987 when STURP physicist John Jackson examined photographs taken under ultraviolet fluorescence and by transmitted light (from the outer side of the cloth). Jackson also examined these blood marks in color on the photographic positives and in black and white on the photographic negatives.[16] A medieval forger could not have seen, let alone known how to encode blood marks that were comprised of both actual blood and oxidized dehydrated cellulose. Nor could they occur naturally.

These particular blood flows on the arms are analogous to some of the man's scourge marks. While the scourge marks are comprised of real coagulated blood located throughout the body, some of them fade into lines that are visible only under ultraviolet light. These particular scourge marks and the above blood marks on the arms are like small microcosms of various features on the entire Shroud. They could not have been encoded naturally nor could a forger have encoded them so precisely.

As we saw, only the topmost two or three fibers on the threads of the cloth were encoded as oxidized, dehydrated cellulose. These superficially encoded body image fibers are located throughout the lengths and widths of both full-length body images. If one part of the body image is darker than another, it's not because its fibers are encoded more intensely, it's because a greater *number* of colored fibers exist in that area. Something had to have reduced the intensity of the image encoding agent. Shroud scientists often use the word *attenuate* to describe this dilution.

Even where the cloth was clearly touching the body at the highest points on a reclined or supine body — with the image encoding process acting in a vertical straight-line direction — the body image discoloration does not penetrate more than a couple of fibers into the threads of the cloth. This is the same distance that the image encoding agent penetrated at the parts of the cloth farthest away from the body. When the Shroud's body images were encoded, an extremely delicate process transpired throughout the short spaces between the length and width of both sides of the body and burial cloth around the body.

Yet, the image encoding process abruptly broke off at all the edges of the body wrapped within this cloth. Neither the inside nor outside of either leg, the sides of either hip, the top of either side of the man's head nor either side of his forearms are visible on the Shroud's frontal or dorsal body images.

Dr. Giles Carter, Professor Emeritus, Eastern Michigan University, has investigated the Shroud's images and conducted years of experiments with X-rays. He first suggested in 1984 that the man's finger bones are visible on the photographic negative images.[17] In addition, he notes that the bones extending into the hand, over the region of the palm, could also be visible, helping to explain why the man's fingers appeared so long. Since then, other scientists and physicians have confirmed the identification of these finger and hand bones.[18]

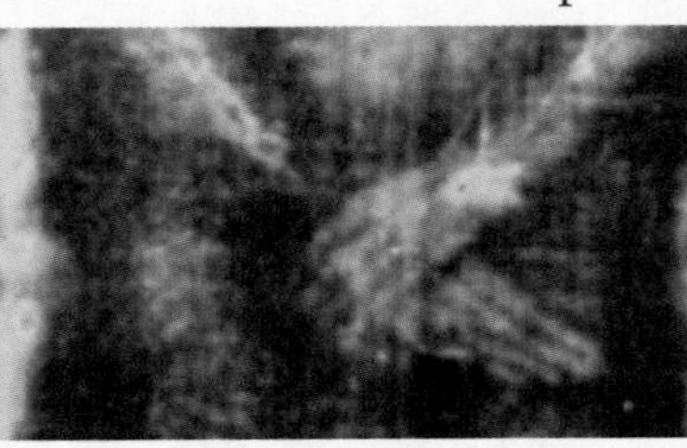

(Fig. 54) Skeletal features identified on the man's hands and fingers.

Scientists and physicians have identified other possible internal skeletal features on the man in the Shroud. Dr. Jackson has noted that part of the skull at the forehead may be visible on the man. Surgeon Alan Whanger, utilizing his modified Polarized Image Overlay Technique with the Shroud's negative and positive images, has also identified features from the skull, as have Dr. Carter and Dr. August Accetta.[19] Dr. Accetta, a physician, has also conducted experiments concerning radiation-imaging of skeletal and other bodily features. Dr. Jackson and Dr. Accetta have further identified faint images of the curved and inverted thumb under the man's left palm.[20] Carter, Whanger, and Accetta have stated that images of the man's teeth could be partially visible, especially on the right side of the man's mouth.[21]

Dr. Carter also first stated that, "Part of the backbone may be visible on the dorsal image . . ." of the man in the Shroud.[22] This identification has also been confirmed by Dr. Whanger.[23] Subsequently, I enlisted the services of Dr. Joseph Gerard and Dr. Cheri Ellis, who, in their profession as chiropractic physicians, make and view more X-ray images of the spinal column than almost any other profession. After studying quality photographic negatives of the dorsal area, they were able to specifically identify numerous vertebrae in the neck and backbone (and even a few pedicles of the vertebrae with disc spaces prevalent).

All of these skeletal and dental features lie near the surfaces of the frontal or dorsal sides of the man that was wrapped in this burial shroud. All were encoded correctly, yet none were visible for hundreds of years until the development of modern technology. These features, too, could not have been encoded by a medieval forger regardless of the method he utilized. These and many other features testify to the intricate relationship that existed between the cloth and the body at the time its body images and its blood marks were encoded.

The insides of the Shroud's body image fibers still retain their original white color. The body image fibers are only encoded on the outer layer of each fiber, 360° around the fibers. Their uniform straw-yellow coloring is visible because something broke apart the three naturally single-bonded atoms (carbon, oxygen and hydrogen) within the outer layer of these cellulose fibers. This allowed the carbon and oxygen atoms to then double-bind with each other, a combination which reflects straw-yellow in visible light. Thousands of these uniformly colored fibers comprise the very subtle, negative images on this burial cloth that were encoded independently of the weight of the body and in places that weren't touching the body. Yet, these images break off sharply along the entire lengths of all four sides of the body (and at the tops of the heads) on both frontal and dorsal images.

While the process may have acted over a variety of surfaces, such as hair, skin, bones and teeth, it only encoded features that were on or near the surfaces of the man's body, such as his beard, his mustache, the hair on the front and back of his head, his skeletal and dental features and the skin throughout both sides of his body. All of the man's features were encoded as oxidized dehydrated cellulose that formed over time within the negative images on this cloth. As modern technology developed and was applied to the Shroud, it revealed that astonishing details with anatomical precision had been uniquely encoded onto this cloth from the body enveloped within it.

Continuous degrees of shading are found throughout both of these vague negative images, which yield focused positive images of the man's appearance and his anatomy. The degrees of shading within the Shroud's body image were encoded in direct proportion to their various distances from the underlying body. The various degrees of shading were even encoded in vertical straight-line directions from the body to the various parts of the cloth draped over it (whether it was sloping upward, downward or relatively flat).

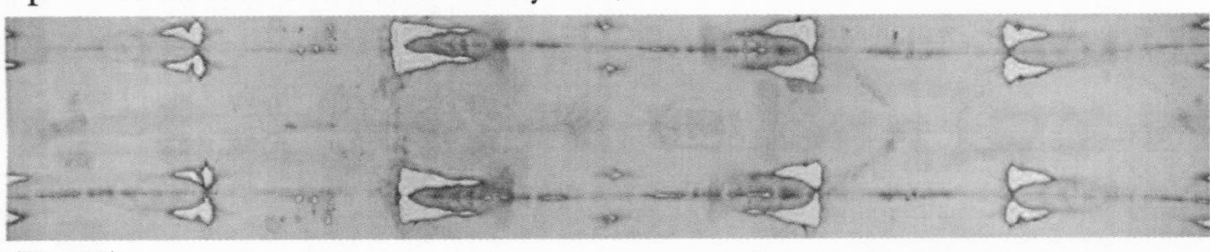

(Fig. 55)

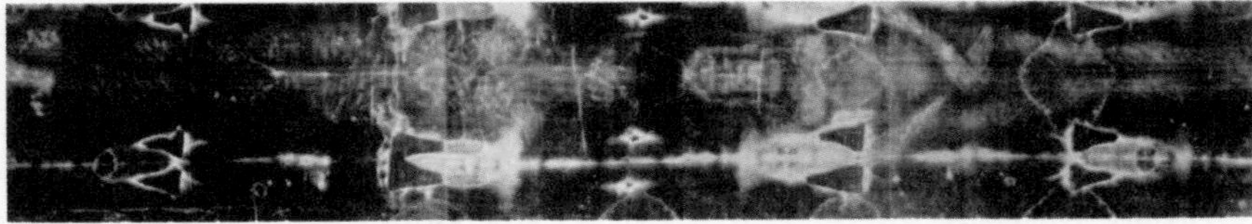

(Fig. 56)

These body image features reveal an extremely *intimate and complex* relationship between the cloth and the body wrapped within it. This intricate relationship occurred within the very short spaces or distances between the cloth and the body imaged within it.

The blood marks also reveal an intricate relationship between the body and the Shroud linen cloth. The blood marks display a very natural relationship to the body from which they flowed and then congealed. Both the detailed body images and their corresponding blood marks allow us to ascertain what happened to this victim. They indicate such things as what parts of his body were injured, what kind of injuries he received, the positions of his body, the types of wounds and instruments used, the types of bleeding that occurred, the amounts of blood or other fluids involved, whether the bleeding was premortem or postmortem and whether it was whole or coagulated blood. That is because these blood marks appear in this *cloth* in the same shape and form as when they flowed and coagulated on the *body*.

These wounds, blood marks and blood flows have been discussed extensively in this and previous chapters. Very revealing information can be acquired by studying their correspondence and appearance on both the negative images on the cloth itself and on the man's positive images. The alignment and correspondence of the blood marks become even more remarkable when they are left-right reversed and then seen on the man's highly-resolved positive images, which show how he appeared during the encoding process while he laid in this burial cloth.

It's as if the man's picture was taken as he laid in this burial cloth, except this intricate process left so much more complex information than photography ever could. It's as if the cloth not only served as the medium for all the high-resolution information, but also for the three-dimensional, vertically-directional, superficial and all of the other body image features discussed in the first three chapters so far. The very same parts of the cloth also contain about 130 actual human blood marks from this body that have never been encoded so diversely, intricately and accurately in all the ways that we have also talked about so far.

Everything about these body images and blood marks indicates they all had an intimate and complex relationship with the front and back of the multi-wounded, crucified corpse *as well as* with the burial cloth in which he was wrapped. All of the facets of this intricate and complex relationship all occurred within the very short spaces between the cloth and the body wrapped within it.

In Chapters Five, Eleven and Twelve, we will learn of even more body image features that were encoded on this burial cloth as a result of its intricate and unprecedented relationship with the bloodied, crucified corpse contained within it. These chapters will illustrate how the many body image features and the blood marks were encoded within the very short distances between the cloth and the dead body once wrapped within it. Before we do, let's acquire some context from other professional fields that have investigated the Shroud and the surrounding circumstances of the victim's executioners, when and where the events occurred and the man's identity.

4

CORROBORATING ARCHAEOLOGICAL EVIDENCE

—~—

Let's take a break from the scientific and medical evidence and acquire additional perspective from the archaeological evidence. As we will see, this evidence relates further to the numerous events that happened to the man in the Shroud, who performed them, where they occurred, when they happened, the unique features contained on this burial cloth and the identity of this man.

Textile Studies

The provenance of the Shroud has also been indicated by numerous textile studies. If kept in dry climates and rarely exposed to air, textiles can survive thousands of years. While the Shroud is easily the most extraordinary cloth ever known, it is not the oldest surviving cloth of its kind. At the Museum of Egyptology in Turin, I observed numerous cloths from Egyptian dynasties that predated Jesus and the Shroud by many centuries; and certainly other museums in the world hold similar mummy and other ancient cloths. STURP scientist Dr. John Heller observed a cloth at the archaeological site of Diuropus that was five thousand years old and in good condition.[1] Numerous other textile artifacts — dating back four thousand to five thousand years — have also been found in good condition in the Andean highland.[2] In 1981, a seven thousand year old textile was reported to have been found at the Paloma Village site in the deserts of the Peruvian coastal plain.[3] While there are cloths that are older than the Shroud, I'm not aware of any that are as large and in as excellent condition.

Whether the Shroud is going on seven hundred years old or two thousand years old, everyone who has examined it has noted its remarkable condition. As STURP scientists Roger and Marion Gilbert remarked quite some time ago in an observation section of a scientific paper, "The cloth [Shroud] is in excellent condition, extremely soft and pliable with no apparent degradation of strength."[4] Needless to say, ancient cloth is rarely, if ever, described as "in excellent condition." The only friable parts are its topmost two to three fibers, which contain the superficial body image. We will see in subsequent chapters how one unique form of radiation will strengthen the cloth and give it greater resistance to degradation, aging and mildew reactions.

The Shroud is much thinner and more flexible than most people realize. STURP scientists not only described it as being in excellent condition, but having a similar feel as a T-shirt. I have observed many ancient shrouds over the years, many of which were older than the Shroud of Turin, but I have yet to find one that is in such good condition. Many of these medieval burial cloths, found in a variety of locations, also showed obvious signs of decomposition and staining, yet none had the flexibility and lack of degradation as the Shroud. Throughout the Shroud's history, the cloth has been kept in the ideal environment for preservation: dark surroundings with little or no moisture, either folded or rolled inside a container or sealed inside a wall. With the exception of several public viewings each century, the linen has rarely been exposed to sunlight or open air. In subsequent chapters, we will see how an extraordinary event early in the cloth's history could have contributed to its excellent condition.

In 1973, Gilbert Raes of the Ghent Institute of Technology in Belgium was allowed to examine two small cloth portions and two small threads from the lower left corner of the Shroud. He found all four samples to be made of linen. Interestingly, the fabric of the Shroud would have been very expensive to produce at the time. Most burial shrouds of the day were woven in a plain "one over, one under" fashion. The Shroud of Turin was woven in a three-to-one (3:1) herringbone twill pattern and spun with a Z twist.[5] In the herringbone weave, the warp (vertical) thread passes over three weft (horizontal) threads, under one, over three, and so forth for each run of the warp thread across the loom. The next warp is offset by one, and then the next, and a twill is formed. After a few threads the offset is reversed, and this forms the herringbone. The resulting appearance is that of a herring fish bone.

It has been argued that the complex weave pattern on the Shroud could not have been produced in the first century A.D., however, similar weaves have been dated well before the time of Christ. An example of this was a late-Bronze-Age cloak, found at Gerumsberg, Germany.[6] Many other complex weaves have also been found well before or contemporaneous with the time of Christ.

Italian textile expert Franco Testore notes the Shroud's twill was utilized in Egypt as far back as 3400 B.C., especially in the use of mats.[7] Fabrics with a 4:1 twill were found in the burial wrappings of the mummy of King Thutmes II (c. 1450 B.C.).[8] The burial scarf of King Seti 1 (1300 B.C.) contained a border with a 1:3 weave[9] and a piece of fabric from the tomb of Queen Makeri (1100 B.C.) had a 1:3 twill bordered with a 1:10 twill.[10] A mummy cloth of the high priest Nessita-neb-Ashir from the same period contained weaves with 1:2 twill, 1:3 twill, and 1:6 twill.[11]One particularly striking fabric is a linen girdle of Ramses III (1200 BC.). This seventeen-foot-long cloth, is woven with threads of five colors with a 3:1 twill alternating with a 4:1 and 5:1 pattern weave.[12]Textile expert, Dr. Mechthild Flury-Lemberg discovered leggings that were woven in a 2:2 herringbone twill that dated 800-500 B.C.[13] Further, bands of twill weave linen from Egypt dated A.D. 136 and 200 have also been found,[14] as well as those found in the ancient city of Palmyra from the first to third centuries A.D. which featured a 3:1 pattern.[15]

Herringbone twill examples were also found in silk. These fabrics, thought to be Syrian and dating from A.D. 250 and A.D. 276, were found in Syria as well as in England.[16] According to Italian sindonologist Emanuela Marinelli, this weave originated in Mesopotamia or Syria and was known in the Middle East at the time of Jesus.[17] Without a doubt, the herringbone weave pattern that is found on the Shroud linen could have been produced in the Middle East in the first century A.D.

The two small cloth portions examined by Professor Raes came from both sides of a seam that runs the length of the Shroud approximately three inches from one of the long sides of the rectangular cloth. The extent of this seam along the left side of the length of the Shroud can be seen in the photo below taken at the time of the Shroud's 2002 textile restoration.

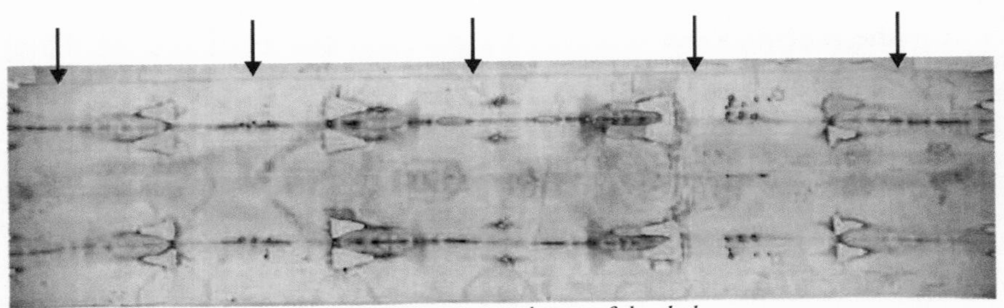

(Fig. 57) The side strip can be seen running across the top of the cloth.

Dr. Flury-Lemberg, who conducted the 2002 textile restoration, has now examined the Shroud more than any other textile professional. Her detailed inspection of the entire cloth, its selvages and the seam led her to conclude the three inch strip between the cloth's long edge and the seam consists of the same original fabric as the rest of the Shroud, all having been manufactured at the same time and originally part of a wider piece of cloth from the loom.[18] This three inch strip was cut from the original wider cloth (ancient Egyptians used looms of a width up to 350 cm) and then expertly sewn back onto the larger remaining piece at the seam when the Shroud was originally fabricated. At the Third International Dallas Conference held in September, 2005, Dr. Flury-Lemberg and several others who participated in the Shroud's 2002 restoration stated that when some of the threads from the seam were removed, the side strip fell away from the cloth, clearly indicating it was attached at the seam.

Dr. Flury-Lemberg notes the complexity of the seam connecting the three inch strip with the larger Shroud "clearly displays the intention to make the seam disappear on the face of the cloth as much as

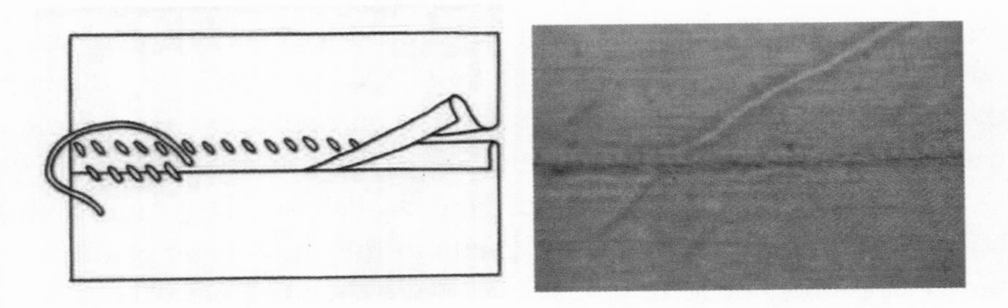

(Figs. 58 and 59) The Shroud's seam has been folded onto and sewn from the reverse side so it appears flat and hardly noticeable on the front side.

possible."[19] She notes the cloth was both folded onto the back and was sewn from the reverse side of the fabric as seen in Figs. 58 and 59.

Although this caused the reverse or outer side of the Shroud to have a raised or rolled effect, Flury-Lemberg further notes that the stitching in the Shroud's seam is distinctively similar to the hems of cloths found in the tombs of the Jewish fortress of Masada, destroyed in 73 A.D. [20] as illustrated below.

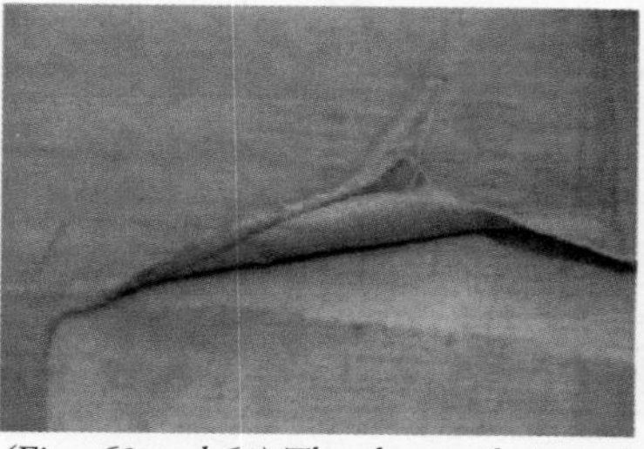

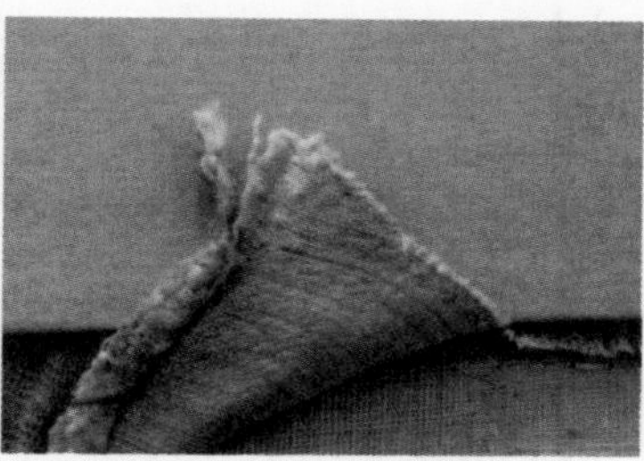

(Figs. 60 and 61) The above technique, however, causes a doubled or rolled effect on the reverse side of the Shroud on left. This effect can also be seen at right on a textile fragment recovered from the fall of Masada in 73 A.D.

Dr. Flury-Lemberg states that the Shroud's selvage uncommonly contains two double threads that also occur in the textile findings of Masada. From Flury-Lemberg's analysis of the entire cloth and all of its components, she concludes that "the linen cloth of the Shroud of Turin does not display any weaving or sewing techniques which would speak against its origin as a high quality product of the textile workers of the first century A.D." [21]

Using a microscope and polarized light on the Shroud samples, Raes identified unmistakable traces of cotton fibers in the portion taken from the main body of the Shroud. This indicates that the Shroud linen was woven on equipment that had been used at some point for weaving cotton.[22] This cotton, which was introduced into the Middle East by the seventh century B.C.,[23] was of the *Gossypium herbaceum* variety and was distinctive of the Middle East.

It is interesting to note that there were no microscopic traces of wool on the Shroud of Turin even though traces of cotton were found. As noted in Leviticus 19:19 ("Neither shall a garment mingled of linen and woolen come upon thee") and Deuteronomy 22:11 ("You shall not wear mingled stuff, wool and linen together"), Jewish law (Mishnah) did not allow linen and wool to be mixed together. Although wool and linen were forbidden, neither Deuteronomy, Leviticus, nor the Mishnah prohibited mixing cotton and linen together.[24]

In 1532, the Shroud was kept folded in a locked container or reliquary in the wall of a sanctuary in Chambéry, France, known then as the Shroud of Chambéry. When a fire started in the sanctuary after midnight, three heroic monks ran into the burning sanctuary and carried the reliquary containing the Shroud outside, where they then doused it with water. The reliquary contained a silver lining which became very hot and partially melted. The molten silver likely dripped onto the folded Shroud and/or the cloth came into contact with the hot interior lining, thereby causing a pattern of scorch marks, burn holes and water stains that are clearly visible on the Shroud. Fortunately, the scorch marks, burn holes and water stains did little damage to the Shroud's body images or blood marks. In 1534, the Poor Clare nuns from a nearby order sewed patches over the holes and attached a backing cloth to the exterior side of the Shroud for protection. These patches and the backing cloth would remain on the Shroud until 2002, when they and the charred material behind the patches were removed and replaced with another backing cloth.

As we shall see further in subsequent chapters, STURP scientists were able to learn a number of things by studying the various fire-related features and their locations on the Shroud. One indication that the Shroud had a much older date than the Middle Ages was found in studies of the patches and backing cloth sewn onto the Shroud after the fire of 1532. A series of transmission photographs were taken where the Shroud was illuminated from behind while photographed from the front. These photos showed that the patches were much lighter than the Shroud, and even though there were distinct differences in the thickness of the Shroud and the patches, STURP scientists stated that, "This will not adequately explain the difference in the transmission coloration."[25] Their findings revealed that "This is an indirect indication that the Shroud is a great deal older than the patches and implies that the Shroud probably has a history prior to the known A.D. 1350 date."[26]

Similarly, after studying X-radiographs of the Shroud, the late John Tyrer, who for more than 25 years was head of textile investigations at the Manchester Chamber of Commerce Testing House and Laboratories in England, stated:

> I am very interested in the comparison that can be made between the altar cloth used to patch the Shroud, the Shroud itself and the backing cloth....The altar cloth and the backing cloth are plain woven and are much better products than the Shroud. They seem to contain less weaving faults whilst the Shroud is a very poor product by comparison. It is full of warp and weft weaving defects, many mistakes in "drawing in." The impression I am left with is that the cloth is a much cruder and probably earlier fabric than the backing and patches. This I think lifts the Shroud out of the Middle Ages more than anything I have seen about the textile.[27]

At conferences in Paris in 2002 and in Dallas in 2005, Aldo Guerreschi and Michele Salcito presented new analysis and evidence that also indicated the Shroud may derive from a much earlier time.[28] Using transmission photographs and x-rays of the Shroud taken by STURP, while recreating the cloth's folding patterns, they demonstrated that the cloth contains two sets of water stains from two different occasions. One set of smaller stains is clearly associated with, and very near, the pattern of scorch marks running the length of the cloth in parallel lines on the outer edges of the body images. (A in Fig. 62) When the Shroud is folded so that its burn holes and scorch marks from the fire of 1532 line up on top of each other within the folds, these smaller water stains also line up with each other within this folding pattern. However, a much larger set of water stains found along three of the cloth's four edges and in its middle on the entire frontal and top half of the dorsal image does not align within this folding pattern (B in Fig. 62). In addition, the smaller

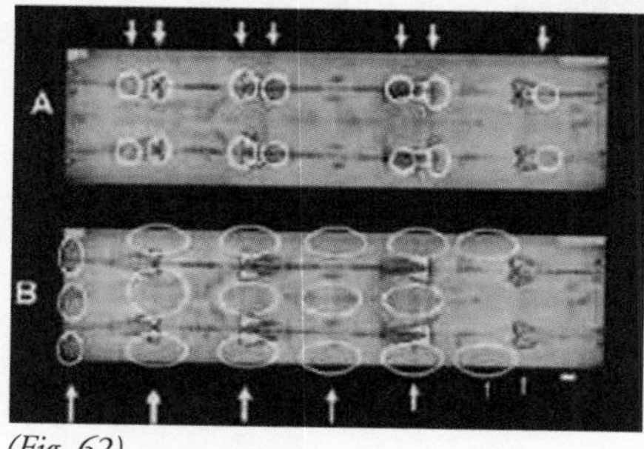

(Fig. 62)

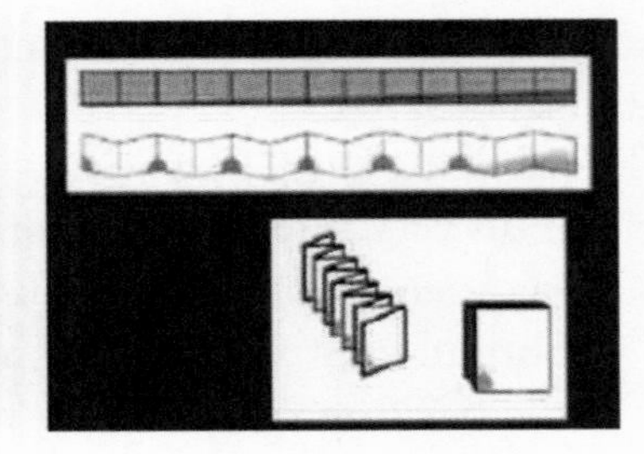

(Fig. 63 and 64)

stains have well-defined linear edges, perhaps from the transport of pyrolysis products, while the larger stains display less intense edges.

This can only mean the Shroud was in two different folding configurations at the time it received the two different sets of water stains. Noticeably, they found that if the Shroud is folded twice lengthwise and then in an accordion style as seen in Figs. 63 and 64, that this fold pattern reproduces all of the Shroud's large water stains if its container received or contained a small amount of water on the bottom. Interestingly, earthenware of the size in which a 52 segmented cloth would snugly fit was very commonly used in ancient times, such as the kind illustrated (Fig. 65) that were discovered at the archaeological site of Qumran.

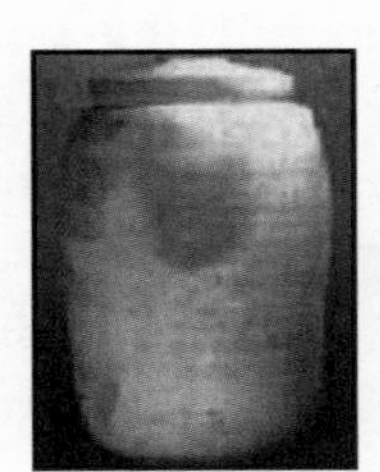

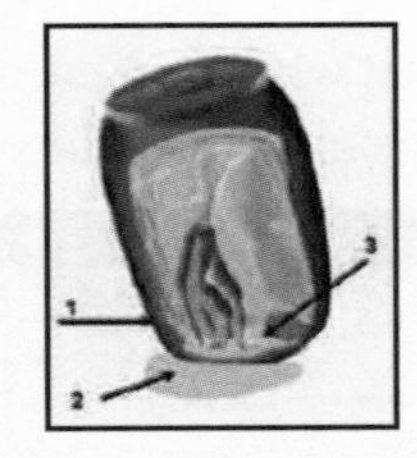

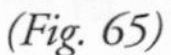

(Fig. 65)

Roman Executioners

A strong indication for the age and origin of the Turin Shroud can be found by examining other archaeological artifacts from the first century A.D. One example of these artifacts is the Roman flagrum. The body of the man in the Shroud is covered from head to feet on the front and back with dumbbell-shaped marks. These significant marks match the size and shape of marks which would have been made by the Roman *flagrum*, an instrument that was not typical of any other culture.[29] This instrument, which is used to whip the victim, consists of *plumbatae*, or balls of lead or bone located in groups of two or three and attached to the end of thongs. A Roman flagrum such as this was excavated in the eighteenth century. It had been found in Herculaneum, the sister city of Pompeii, which was destroyed in A.D. 79. This instrument, occasionally depicted on Roman coins as well, dug deep, contused wounds on the flesh and caused hemorrhaging and extreme weakening of the body.

Even though Jewish law limited to forty the number of strokes that could be administered during a scourging, the Pharisees conveniently reduced the number of strokes to thirty-nine. Roman executioners, however, had no such limitations and only had to assure that their victim was alive for his final crucifixion. Since it was such a severe form of punishment, scourging by the use of a flagrum was not allowed on Roman citizens,[30] which would indicate that the man in the Shroud was not a Roman citizen.

In Chapter Two, we discussed the *lancea*, a weapon commonly used by soldiers of the Roman military garrisons responsible for guarding Jerusalem in the first century A.D. This spear was of varying length with a long, leaf-like tip that thickened and rounded off toward the shaft. This spear was designed for repeated use. Other weapons of the day were designed to break off inside the victim's body so they could not be reused against the Romans. These included such weapons as the *hasta*, which is a long, heavy spear with points of various design. The *hasta velitaris*, another weapon of the day, was a short javelin designed with a very thin, long point. Similar to the *hasta velitaris* was the *pilum*, a spear also used by the Roman infantry, characterized by a long, thin point, but twice as long as the hasta[31] and much heavier. The elliptical-shaped lesion on the side of the man in the Shroud corresponds in detail to several excavated examples of the Roman *lancea*, and does not match any of the other weapons.

The wound found on the man in the Shroud also corresponds to the Roman military custom of stabbing the victim on the right side, just below the armpit. The lance that pierced the side of Jesus, as noted by Origen, a second and third-century theologian, was inflicted in this manner.[32] Roman adversaries would customarily wear shields on their left side, leaving the right side unprotected.

The executioners of the crucified were recruited from the ranks of the Roman Army and comprised the Roman military guard, which was commanded by a centurion. These executioners carried out all crucifixions ordered by the government and were responsible not only for standing guard at the cross until the condemned was dead, but also for carrying out a variety of other functions, including walking the victim to the crucifixion site with the crossbar across his shoulders. The act of having to carry one's own cross beam to his execution was probably intended to cause further humiliation as the victim suffered the taunts and jeers of onlookers along the way.

Usually, only the crossbar was carried by the victim for the vertical part of the cross, the *stipes crucis*, was permanently affixed in the ground. The *patibulum* or crossbar was placed horizontally across the shoulders, with the victim's arms likely tied to it near the middle (Fig. 66) and/or near the hands with the arms completely outstretched on the crossbar. (The victim's hands could also have been tied to the patibulum while he carried it perpendicularly over his shoulder, as one would carry a beam.) Under either scenario, if the victim fell, he would not have been able to break his fall. The man in the Shroud clearly fell numerous times as the cuts, scratches and dirt on his face and knees indicate. Anyone carrying a one hundred pound crossbeam, after having been repeatedly beaten and scourged, would have been in an extremely weakened condition making it very difficult to walk, much less while carrying a heavy crossbeam.

(Fig. 66) Crucifixion victim carrying a cross over his shoulders.

The Roman military guards who performed the executions were allowed, and sometimes encouraged, to mock and torment the victims. The swelling around the eyes and cheeks of the man in the Shroud, as well as the abrasions on his head and face, are indications that he was beaten and tortured by the guards. Of particular interest is that he had sharp, thorny objects placed upon his head in the configuration of a cap that covered his entire head. In the East, a mitre was traditionally used for a crown.[33] A mitre is a cap-like structure that encloses the entire skull and is not like the "wreathlet" that serves as a crown in the West. Since this victim was obviously condemned, its use would likely have been a form of mockery.

Since it was against Jewish law to leave crucifixion victims on the cross after sunset, the Roman executioners would defer to this law. In order to enhance the time it took for crucifixion victims to die, however, the Romans would break their legs. By doing this, the victims would asphyxiate because they were not able to push themselves up to breathe. The only known remains of a crucifixion victim, who was also crucified by the Romans in Jerusalem around A.D. 50-70, exhibited broken tibias and a broken fibula.[34] The legs of the man in the Shroud were not broken, however, and according to Gospel accounts of Jesus' crucifixion, his legs also remained intact.

While crucifixions continued for many centuries afterward, this particular type, involving instruments and techniques of the Roman military guard executioners, was banned around 315 A.D. by the great Emperor Constantine. This indicates that the man in the Shroud was crucified before this time.

Possible Pontius Pilate Coin Features

Perhaps one of the more dramatic instances of the Shroud's images revealing more information as technology is applied to them can be found from the possible features of a Pontius Pilate *lepton* minted between 29 and 32 AD. These features possibly appear over the right eye of the man in the Shroud. As we shall see, these features would not only indicate the geographic location, but would provide some of the most specific confirmation of all as to when the events occurred to the man and when his images were encoded onto the burial cloth. Only when photographic enlargers and microscopes were applied over the facial area on the highly resolved and detailed photographic negatives were these faint features discerned. The Greek letters UCAI (part of the inscription for Tiberius Caesar) may be indicated around the 9:30 to 11:30 position, a staff or *lituus* could appear in the middle, along with a clipped coin margin. All of these features were distinctive to this coin. Archaeological excavations have only recently revealed in the latter part of the 20th Century that the use of coins at Jewish burials was a common feature in Jerusalem, but only in the Second Temple Period, or from approximately the 1st Century BC through the 1st Century AD.[35] Some of these coins were found to have been placed directly over the deceased eyes as faintly indicated on the man in the Shroud.[36]

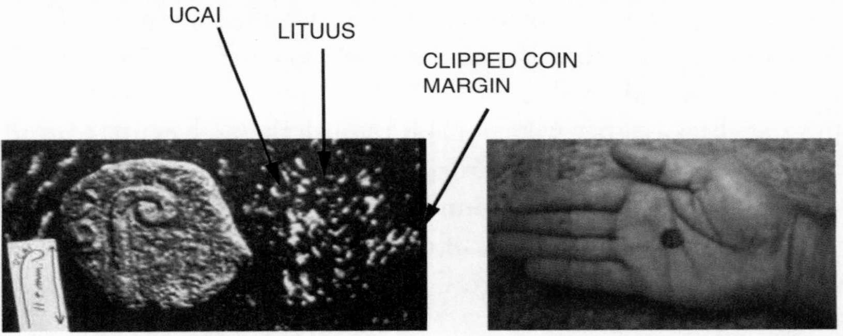

(Fig. 67) The letters, lituus and clipped coin margin possibly indicated on Shroud (right), compared to existing lepton whose letters have mostly worn off.

(Fig. 68) shows the small size of a Pontius Pilate coin.

Existing Pontius Pilate coins contain the same letters UCAI, at the same location, with the same corresponding height as those possibly found over the right eye of the man in the Shroud.[37] These letters are located around the curve of an astrologer's staff, or *lituus*. The lituus was used as a constant motif on coins minted by Pontius Pilate after 29 A.D. Following the rule of Pilate, the lituus was not used again by a ruler in Palestine, nor anywhere in the Roman world, as a central independent symbol.[38] While the lituus is not as clear as the inscription on the Shroud image, it is also consistent with those on Pilate leptons in that it is turned to the right, or clockwise, and it corresponds to the correct height.[39] While graceful curves are found on the lituus stem of most

Pilate coins, a cruder appearing lituus without graceful curves has been found on Pilate leptons, again possibly matching that seen over the right eye of the man in the Shroud.[40]

Further comparison of the enlarged area over the right eye of the man in the Shroud with the enlarged Pontius Pilate lepton reveals that the sizes and outlines of both are quite similar.[41] The right side of the rim of each appears to have been clipped at the 1:30 to 3:30 clock position, which is typical of this poorly minted coin (see Fig. 67). The late Francis L. Filas, S.J., of Loyola University in Chicago, along with several numismatists working with him, made these observations in the early 1980s following the Shroud's first comprehensive scientific examination a few years earlier. Filas summarized the many points of comparison: "To sum up, there exists a combination of size, position, angular rotation, relative mutual proportion, accuracy of duplication… and parity [i.e., turned in the proper direction]. This combination concerns at least six motifs: a *lituus* or astrologer's staff, four letters, 'UCAI,' and a clipped coin margin."[42]

Pictures of the enlarged areas over the eyes were processed in a Log E Interpretation System (Fig. 69), as was the photographic negative from which all of these features were found. This system, which is very similar to a VP-8 Image Analyzer, also revealed the letters UCAI, the *lituus*, and the clipped edge at the 1:30 to 3:30 clock position which also arguably appear on the imaging system photo seen below. According to Filas, for the first time the clarity of the boundary of a coin over the left eye also became visible. The possible appearance of these same features on the Log E Interpretation as found on the photographic negative, only points further toward a coin with the same inscription, motifs and designs.

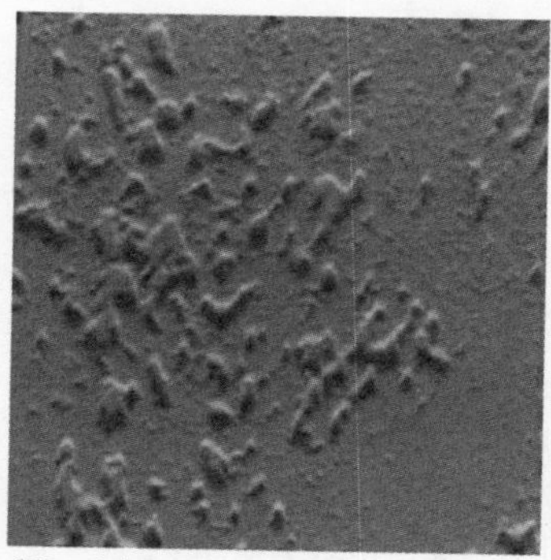
(Fig. 69) Log E Interpretation features

Filas argues that critics who feel that the presence of these features could be a coincidence, a mistake or a chance pattern in the weave of the cloth have not considered that these possibilities are very remote. Even though one aspect of the features could produce the appearance of one motif, such as a Greek letter, how does that account for the other features? All four letters are upright (not sideways, upside down, or backward); the letters appear to be side by side and match the spelling found on other coins minted by Pontius Pilate, and the lituus and coin margin are consistent with a genuine coin. It would be too much of a coincidence that these findings would appear on the 14'3" length and 3'7" width of the Shroud at the exact location over the victim's right eye. According to Filas, the odds are astronomical that these things occurred by coincidence, and not from an inscription.[43]

The letters UCAI form part of the inscription TIOUKAICAPOC,[44] an abbreviation of TIBEPIOUKAICAPOC ("Tiberiou Kaisaros," Greek for Tiberius Caesar). Both inscriptions have been found on Pilate coins. Prior to the possible identification of the above letters on the Shroud, an interesting point concerning these Pilate coins had never been known to numismatists. This may constitute an even more convincing point of authenticity that these features resulted from an inscription on a Pilate coin placed over the man's right eye and not from the work of a forger or a chance pattern. UKAI is actually the correct spelling of the word. UCAI is a misspelling that was more than likely because "Caesar" in Latin was pronounced the same as "Kaisaros" in Greek; both have a hard "K" sound (although the Latin S sounded like the Greek C). When it was discovered that the spelling on the coin over the man in the Shroud's right eye was misspelled, the spelling on Pilate coins was checked. At least four Pilate coins currently exist that exhibit this misspelling.[45]

Although Pontius Pilate coins could have been used as currency in any part of the Roman Empire, they were circulated and used most during the time and in the region that they were minted: first century Palestine. Coins used in connection with burials in ancient Jerusalem and the surrounding area have enormous relevance. Only in the last few decades have archaeological excavations in Jerusalem and the surrounding area revealed that the use of Roman and other contemporary coins was actually a common feature at Jewish burials in the Second Temple period. There are now many known instances in which the use of contemporary coins at burials has clearly been found, primarily in Jerusalem and the surrounding area around the time of Christ.

The use of contemporary coins at Jewish burials in Jerusalem during the time of Christ was dramatically illustrated in the early 1990s when a burial cave outside Jerusalem's old city boundaries was discovered containing six undisturbed ossuaries in their original positions, one of which "in all probability" contains the name and remains of "the high priest who presided at Jesus' trial — or at least a member of his family."[46] Two of these elaborate and ornate ossuaries contain three different inscriptions of *Caiaphas*, the well-known family of high priests of this period. Two almost identical inscriptions on one especially beautiful ossuary read, "Joseph, son of Caiaphas."[47] Dr. Reich, of the Israel Antiquities Authority who conducted the etymological study, further states that these inscriptions "may well be understood as Joseph of Caiaphas."[48] The New Testament refers to the high priest who presided at Jesus' trial by the single name Caiaphas; however, the first-century historian Josephus gives his proper name as Joseph Caiaphas and "Joseph who was called Caiaphas of the high priesthood."[49] As the high priest in Jerusalem from A.D. 18-36, Caiaphas charged Jesus with blasphemy, causing the council to attack him. This started a long series of bruises and wounds and Jesus' ultimate execution — all of which can be found on the man in the Shroud.

A bronze coin of Herod Agrippa I dating to 42 or 43 A.D. was found in the skull of an adult woman in one of the six undisturbed ossuaries at the Caiaphas family burial site. Jews placed skulls and other bones of the deceased in ossuaries after the corpses had lain in shrouds for approximately a year, and their bodies had decomposed. Ossuaries are large rectangular chests or containers usually made of limestone and often decorated or inscribed. Secondary burials in ossuaries were rare in Jewish tombs after the Roman destruction of Jerusalem in 70 A.D.; however, during the Second Temple Period, Jews were the only group to utilize this practice.[50]

Many other coins have been found at Jewish burials in Jerusalem and the surrounding area during the Second Temple Period. Thirty-six coins were found at the foot of the deceased at Jason's Tomb in Jerusalem with six more coins close by. Forty-one of these coins date to the period of 5/6 A.D. to 30/31 A.D., while the other belongs to the reign of Alexander Jannaeus from 103-76 B.C.[51] Two more coins from the latter period were also found in separate tombs located in Jerusalem and northwest of Jerusalem, as was an Agrippa I coin dating to 6 A.D. located in the debris of an ossuary in Talpiot.[52]

A coin "of the Second year of the Revolt" (the first Jewish Revolt lasted from A.D. 66 to 70) was found in an undisturbed tomb, along with "a coin of Tiberius" and "two unidentified coins."[53] Perhaps most interesting of all, because it is like the one possibly encoded on the Shroud image, was the discovery of "a Pontius Pilate coin in a Jewish tomb on Jabel Mukaber, south of Jerusalem."[54] These brief descriptions were taken from an overview of coins found at Jewish burial sites of the Second Temple Period in and around Jerusalem by archaeologists Rachel Hachlili and Ann Killebrew. At the conclusion of their overview the authors state, "Among tombs dating to later periods, coins are only occasionally found."[55]

In 1979, Rachel Hachlili described her findings from excavations of tombs that were hewn out of rock in the hills overlooking Jericho. These tombs, also dated from the Second Temple Period, were required by Jewish law to be located outside of the city limits of Jericho and were composed of both primary and secondary burials. One of the tombs contained a skull with two bronze coins of Agrippa I (A.D. 37-44). Another tomb contained a bronze coin of Herod Archelaus (4 B.C. to A.D. 6) that was discovered in a damaged coffin-interred skull. A second coin of Yehohanan Hyrcanus II from 63-40 B.C. was discovered in the rubble by the door of the tomb. Immediately after describing all found coins, Hachlili stated, "The coins originally must have been placed on the eyes of the deceased...." a practice often followed during that time.[56]

In 1970, it was first reported that a man was uncovered at the site of an excavated fortress in the Judean Desert at ᶜEn Boqeq, who had silver coins from c. 133 A.D. placed over both of his eye sockets.[57] Archaeologist William Meachem thinks in all probability, this man was a Jew. A Bar Kokhba coin (A.D. 132-135) was found nearby, and hidden Bar Kokhba documents were found less than twenty miles north of there near ᶜEngedi. This area was known as a traditional refuge where the last stand of Masada occurred and also where David hid from Saul. From A.D. 135 to 220, Jews were excluded from living in Jerusalem and a zone surrounding it, but ᶜEn Boqeq lies south of this excluded zone.[58] Meacham recognizes "...most importantly, the ᶜEn Boqeq burial establishes that the coin-on-eye ritual was found in second century Judea...."[59]

Amazingly, the skull that was discovered had not been disturbed and remained in excellent condition even after almost two thousand years. Fortunately such discoveries, although very uncommon, allow scientists to study how these ancient, tiny coins were used. With the man in the Shroud, we can look at his burial cloth and see his entire position within the cloth. (While a coin could also have been placed over his left eye, presently there is only evidence of possible coin features visible over the right eye.)

The scores of coins found at a wide variety of tombs in Jerusalem and the surrounding area strongly indicate they were used in some way at burials during the period from the 1st century B.C. to the 1st century A.D. Archaeologist, Zvi Greenhut, who excavated the Israel Antiquities Authority Caiaphas tomb site states, "I believe we must now regard coins discovered in the context of Jewish tombs from the Second Temple Period to be elements connected to the burial ceremony, despite the fact that they have not always been found in direct relation to the skulls or bodies of the deceased.[60]

The Roman coins, and some of the coins of Jewish rulers, have an important significance when they relate to the Shroud of Turin. It has only been within the last few decades that the suggestion of a coin over the right eye of the man in the Shroud has been raised. The advent of photography and three dimensional reliefs, which can be enlarged and studied, suggests that a coin could have been placed over his right eye. These features further support the theory that the Shroud is not a forgery – it would be impossible to encode features on this burial cloth that could not even be seen by the naked eye. It is well known that such coins were used for burials in the first century A.D., and that Pontius Pilate leptons were minted during the 29-32 A.D. period. When added to the other extensive evidence that has been discovered, use of the coin would further support that the man in the Shroud was not only Jewish, but buried in Jerusalem according to Jewish burial customs.

Limestone Flower and Pollen Analyses Confirm Jerusalem Origin

When a limestone sample was recovered from the foot region of the man in the Shroud and compared with limestone from the Ecole Biblique tomb in Jerusalem, it was yet another indication that the events encoded on the Shroud of Turin occurred in Jerusalem. The Ecole Biblique tomb provided researchers with access to the same rock shelf as the Holy Sepulcher and the Garden Tomb, which are considered the most likely tombs of Jesus Christ to be in existence. The limestone tombs in the Palestine/Transjordan region are known to be wet and malleable and will rub off at the mere touch of a finger.[61]

Limestone mainly consists of calcium carbonate. The limestone in the Ecole Biblique tomb in Jerusalem was determined to be in the form of travertine aragonite (with strontium and iron in lesser amounts) rather than the more common travertine calcite.[62] The conditions under which Aragonite develops is not as common as calcite.[63] Samples of calcium from the Jerusalem tomb were compared to samples from the foot of the Shroud fiber. The sample from the Shroud contained strontium and iron in small amounts and was formed with aragonite, not calcite.[64]

Dr. Ricardo Levi-Setti of the Enrico Fermi Institute at the University of Chicago confirmed this match.[65] Using a high-resolution scanning ion microprobe, Dr. Levi-Setti analyzed and compared the calcium from the Jerusalem tomb with that of the Shroud fiber. There were tiny pieces of flax on the Shroud sample (which had adhered to the calcium sample) that produced an organic variation, but other than that, the samples were an unusually close match.[66] Dr. Levi- Setti additionally analyzed limestone samples from other tombs from nine various test sites in Israel — but the only match was the one removed from the tomb in Jerusalem.

In the 1970s, Dr. Max Frei, an internationally known criminologist, botanist and expert in Mediterranean flora, was permitted to use adhesive tape to remove dust samples from in between the threads of the Shroud. Frei, with his long list of credentials, founded the renowned scientific department of the Zurich Criminal Police and served as president of the United Nations' committee that investigated the death of its

Secretary-General, Dag Hammarskjöld, in 1961. STURP scientists used a torque applicator to take the tape samples, which limited the pounds-per-square-inch pressure on the Shroud, taking only surface samples.

Dr. Frei, was able to identify fifty-eight pollen grains on the Shroud of Turin. Pollen grains can last millions of years, and even though they are very small, they all have distinct shapes and features that represent the plant species that they come from. See Fig. 70 to observe some of the pollen examples.

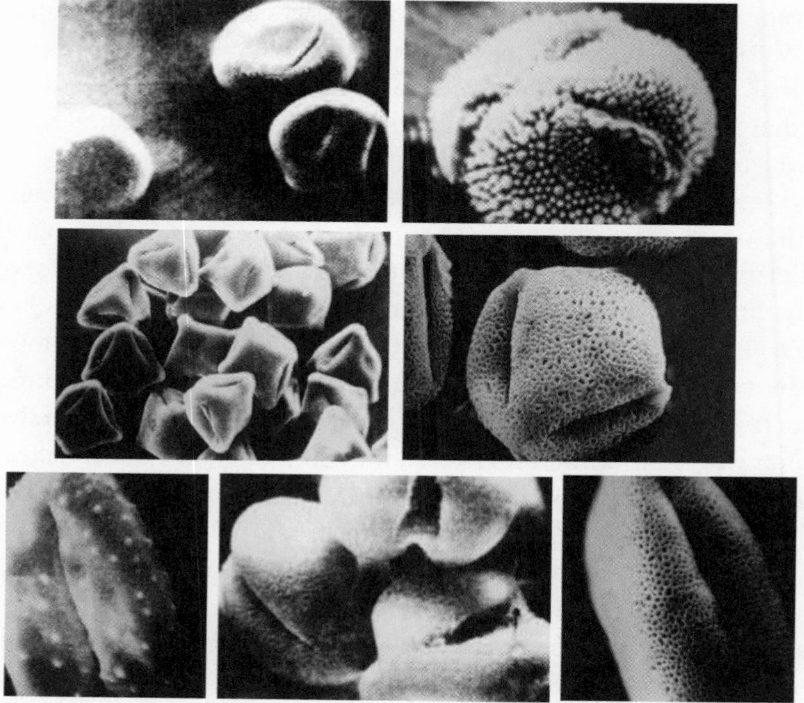

(Fig. 70) Microscopic photographs of pollen indicating the individual natures of each pollen.

The last nine years of Frei's life, which included seven trips to study different floral seasons, were spent identifying different pollens and the global locations of the plants from which they derived. Using an optical microscope, at magnifications ranging from 60x to 1200x, and under a scanning electron microscope (SEM), Frei compared ripe, Middle-Eastern pollens with Shroud pollen grains. Since these pollens had not been registered in various herbariums or other botanical manuals, these trips were necessary to personally collect and compare these samples.

When Frei studied the Shroud, he found sixteen plant pollens that are only found in salty soil or sandy deserts (similar to those found at the Dead Sea). These plants, however, are not found in Italy or France where the Shroud has been kept since the 1350s.[67] Also found on the Shroud were pollen grains from seven plants that are known to grow in rocky terrains, similar to Palestine and surrounding areas,[68] and six from plants that were grown in Anatolia, Turkey and areas between Iran and the eastern Mediterranean.[69] None of these plants exist in Italy or France, which indicates that the Shroud's history encompasses areas outside of these two countries. As we shall see in Chapter Ten, there are many reasons to conclude that the Shroud spent several centuries in both Edessa and Constantinople, both of which are located in Turkey.

Dr. Frei's extensive study of the pollen grains and plant species found on the Shroud revealed some interesting facts. Although the majority of the species were non-European (45 out of 58), there were only three of those specimens that did not grow in Jerusalem.[70] Of the three exceptions, two are found in Edessa, Turkey,[71] and one can only be found in Constantinople.[72] Since the majority of Shroud pollens were from Jerusalem, (three times greater than those grown in Italy or France where the Shroud was stored for 650 years), the only conclusion that can be made is that not only does the Shroud have a historical link to the Middle Eastern region, the cloth itself originated in Jerusalem.

The discovery of plant pollens on the Shroud of Turin is significant and easily explained by the very process of pollination itself.[73] When plants are pollinated, the pollen falls on land or water as it is being transported to other plants. The Shroud linen, made of flax, would have been soaked in a river or lake as it was being retted. When the linen was subsequently bleached, it would have been laid on the grass in the sun for extended periods of time, and intermittently sprinkled with water. The sticky pollens that had been deposited in the water and on the grass would have adhered to the linen. In ancient times, if a whiter linen was desired, the cloth would be exposed to the water and grass for a longer period of time. The Shroud is known to be of a very high quality, indicating that it was bleached for a long time. According to the Gospels, if the finished Shroud was indeed the burial garment of Jesus Christ, it would have been purchased by Joseph of Arimathea.

Dr. Alan Whanger of Duke University performed a variety of additional studies of the Shroud pollens, concentrating his efforts on the cloth's off-image areas. He used technology to underexpose photographs of the Shroud, allowing images that were encoded on the cloth to be seen with additional detail. His first discovery, in 1985, was what appeared to be a flower image near the victim's head. In 1983, Oswald Scheuerman had made a similar observation. Upon noticing other similarities which he thought might be relevant, Whanger began an intense study of Israel's botany. For four years, he compared drawings from a 6-volume set of botany books with the images on the Shroud photographs, using a Polarized Image Overlay Technique to superimpose one image over another. Four years later, 28 species of plants grown in Israel were tentatively identified by Whanger.[74] His results were confirmed by world-renowned flora authority, Dr. Avinoam Danin, Professor of botany at Hebrew University in Jerusalem, who went on to discover several more flower images that Whanger had not found.[75]

Twenty-seven of the 28 plants grow within the close vicinity of Jerusalem, where four geographical areas containing different specific climates and flora can be found. (The 28th plant grows at the south end of the Dead Sea.) All 28 of the plants studied by Whanger would have been available in Jerusalem, either at marketplaces, in fields or just growing along the roadside or in nearby fields. Although all were found in Jerusalem, "half are found only in the Middle East or other similar areas and *never* in Europe" (italics added).[76] One of them grows only in Israel, Jordan or the Sinai. Professor Danin therefore concluded that the only place in the world that all 28 of the Shroud's flower species could be found was Jerusalem[77] and interestingly, they all bloom in March and April — at Easter.[78]

(Fig. 71) The left side of the image shows a chrysanthemum found on the Shroud, and the right side shows a drawing of this flower.

In addition to the pollens we have just discussed, several more pollens from Israel were discovered. Dr. Uri Baruch, an expert on Israeli pollens, along with Danin and Whanger, also discovered pollens from three thorny plants, one of which had more than 90 pollen grains that were found only in Israel and the immediate surrounding area. Dr. Max Frei had earlier identified pollens from 25 of the 28 plants discovered by Whanger, although other scientists believe Frei should have identified the pollens by family or genus as opposed to species. Frei was, however, able to separate the pollens from the sticky tape and independently mount and rotate these samples under SEM at 1200x magnification. Although Baruch and Danin could not identify as many pollens as Frei, the images of the flowers and pollen grains that they studied clearly support Frei's evidence.

Capparis aegyptia is one of the plants that blooms in the spring and is found in Jerusalem. This plant independently supports the events of Jesus' crucifixion, as noted by Drs. Whanger, Danin and Baruch: "*Capparis aegyptia* is also significant as an indicator for the time of the day when its flowering stems were picked. Flowering buds of this species begin to open about midday, opening gradually until fully opened about half an hour before sunset. Flowers seen as images on the Shroud correspond to opening buds at about 3 to 4 o'clock in the afternoon. This was confirmed by a two day experiment with, first,[79] *Capparis aegyptia*, and later with *Capparis spinosa*."

Whanger went on to examine the flowers as they wilted after being picked. He concluded that the images left on the Shroud were most likely from wilted flowers that were 24 to 36 hours old.[80] This time frame is consistent with the formation of the body images, which occurred no later than two or three days after the body was placed within the Shroud. These findings would indicate that the formation of the flower images occurred at the same time.

If these flower images are indeed present, they are most likely reflective secondary images that were encoded from having been placed on or around the man's body. What is most interesting is that Whanger and Danin have always considered that some kind of radiative process, possibly from the body, best explains the presence of the flower images.[81] We shall see in later chapters that one particular form of radiation emitted from the body of the man in the Shroud not only explains the faint coin and flower images, but how his body images and blood marks were formed, his skeletal features, the cloth's excellent condition, its still-red, pristine blood marks and its aberrant, erroneous 1988 radiocarbon dating.

In some of Dr. Danin's subsequent presentations, he has emphasized that eight species indicate the geography and blooming seasons of the flowers and thorn plants that he identifies on the Shroud.[82] His plant identifications, along with the possible coin features, necessarily involve elements of subjectivity. Danin has identified hundreds of flowers and thorns on Shroud photos taken over a span of more than a century, as well as on the cloth's ultraviolet fluorescent photos. He has even been able to identify with binoculars two floral images on the Shroud itself.[83] Like myself, most sindonologists are not familiar with flora or plants from the Middle East, and do not observe nearly all the flora that an expert like Danin does, even if we may agree that we can observe the features of some. The plants, coins, and even the pollens observed or identified by these experts should be studied further by botanists, numismatists, palynologists, scientists, sindonologists and others. The implications of the plant identifications alone are enormous. They could confirm the Jerusalem location, the time period as the spring or Easter season, that different types of flowers and thorns were involved, that the flowers were picked around 3 to 4 o'clock in the afternoon and that the images were encoded before two days had elapsed.

Jewish Characteristics of the Victim and Jewish Burial Customs

The physiognomy of the man in the Shroud appears to be of Middle Eastern origin. Former Harvard professor and world renowned ethnologist, Carlton S. Coon, noted, after analyzing numerous photographs, that the man in the Shroud was definitely "of a physical type found in modern times among Sephardic Jews."[84] This is further supported when the man's hair and beard are studied.

Several attributes of the man in the Shroud support the theory that he was Jewish. His hair, worn long, parted in the middle and having what looks to be a long, unbound pigtail in back is typical for Jewish men of ancient times.[85] Clearly this style was not worn by

Greco-Roman men of the day. Professor Werner Bulst S. J., notes "Of the numerous portraits we have of Greek and Roman origin, there is not one of a man with hair parted in the middle and falling to the shoulders."[86] With few exceptions, Roman men were also clean shaven, where Jewish men tended to have beards.[87]

Another Jewish custom was to use a chin band to bind the jaw of the deceased to keep the mouth closed.[88] There is a suggestion of such a band on the man in the Shroud. On both the positive and negative images, it appears as though the man's beard has been pushed up. This is also evident on the three-dimensional image which shows the left side of the man's face has an upturned beard that drapes over something that is not visible in any of the images.[89]

The position in which the man in the Shroud was buried resembles numerous skeletal remains that were found at the c. 200 B.C.– 70 A.D. Jewish monastic Essene sect at Qumran, where the Dead Sea Scrolls were discovered. These bodies were positioned on their back, face pointing upward, elbows protruding at their sides and hands crossed over the pelvic region.[90]

Further evidence of ancient Jewish burial tradition is the single linen burial shroud. Fine linen cloth, however, was unusual on a shroud for a crucifixion victim. The type of linen that was used to make the Shroud of Turin would normally have been used to show respect for someone who was wealthy, had family ties or was in some sort of esteemed position, but not for a common criminal. According to the Jerusalem Talmud, Kilaim 9:32b, Judah the Patriarch, who was the compiler of the Mishnah and lived in the late second century A.D., "was buried in one linen shroud (without any other garments)."[91] This tells us that not only was a linen shroud used to bury his body, he was naked for burial.

One observation about the Shroud is the various measurements that had been recorded in different books over the years. Researcher Ian Dickinson from Canterbury, England thought that the routinely stated measurements of 14'3" x 3'7" were interesting. He began researching the unit of measurement at the time of Jesus, the cubit, and discovered that the Assyrian cubit of 21.4" was the standard. The Assyrian cubit has been recorded as 21.6 plus or minus 0.2. When measured by this standard the Shroud is 8 cubits (171.2") by 2 cubits (42.8") which suggests that the linen cloth could have been measured by this standard unit.[92]

Another group of writings, the Apocrypha, is worth mentioning here. Although it contains mostly legendary material that is usually not considered credible, they do accurately reflect the customs and traditions of their times. These writings reveal that shrouds were used almost exclusively as burial cloths at that time.[93] Jewish scholar Maimonides in the twelfth century wrote about corpses being wrapped in white linen cloth.[94] Clearly the man in the Shroud was not buried according to the Roman custom of cremating the dead or the Egyptian custom of disemboweling and pickling their dead before swathing their bodies in bandages.

Most crucified victims were left on the cross after they died or cut down and thrown in a pile on the ground where wild birds and scavenging animals would devour their bodies. The Romans, however, did allow crucified Jews to be buried in a mass grave and, in the first century, sometimes allowed families to take the body for individual burial if they desired. Professor Bulst notes, however, that this practice only occurred in a known interval of time.

> The practice of burial of a crucifixion victim was confined to a very short period: in Palestine, in the first century of our era. In 6 A.D. Augustus removed the Jewish King Archelaus, son of Herod I, and installed a Roman procurator for Judea and Samaria who had the authority of the death sentence....At the same time, however, the Jewish government was still in existence, which required burial before sunset according to the Jewish law. This exceptional double rule was finished with the Jewish War in 66 A.D.[95]

One such demonstration that the Romans adhered to the Jewish custom of burial was found in the remains of the crucifixion victim, Yehohanan, whose remains were discovered northeast of Jerusalem in a tomb at Giv'at ha-Mivtar, in a Second-Temple period Jewish cemetery. He appears to have been crucified between the beginning of the first century A.D. and A.D. 70.[96] He was average in height, 5'6" tall (167 cm) for a Mediterranean man,[97] and he was more than likely crucified by the Romans, who had conquered the area at the time. It was against the law for a man to be put to death by a Jew,[98] and after conquering the area, the Romans had become the sole executioners.

Yehohanan's remains were further evidence of the torture that occurred during a crucifixion at the time. A nail spike that was about 7 inches (17-18 cm.) in length was found, and thought to have attached his heel bones. Fragments of olive wood (thought to have been from the cross), were attached to the nail. It was clear that the man's right tibia and left fibula had been fractured (a process mentioned earlier to keep the victim from pushing himself up to breathe). Since Jewish custom was to bury their dead before nightfall, which likely was in deference to Deuteronomy 21:22-23, these bones were clearly broken to hasten Yehohanan's death. Remember, the man in the Shroud's bones were not broken.

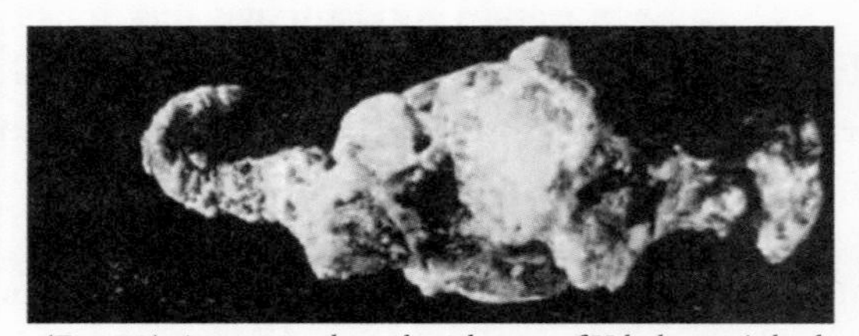

(Fig. 72) A seven-inch nail with part of Yehohanan's heel bone that was found in an ossuary in Jerusalem.

There is some controversy about the height of the man in the Shroud which is estimated at about 5'10", somewhat taller than "the mean height for Mediterranean people of his time."[99] If we take into consideration the cloth's drape during image formation with the fact that it was rolled for hundreds of years on a spool, causing it to stretch somewhat, the man's height could well have been 5'9" or 5' 9 1/2."[100] In a burial area recently excavated in a Jewish settlement in the Galilee region, where burials took place from the first century B.C. to the fourth century A.D., humerus bones of victims reveal that the average height of the men was 5'9" tall.[101] According to Talmud interpretation, the height of the man in the Shroud corresponds to the ideal height of 4 ells (176 cm or 5'9.29").[102]

One criticism that was mentioned over the years is that the man in the Shroud could not have been Jesus because his body had not been washed. Although Jewish burials usually included washing the body with water and anointing it with oil, the evidence clearly indicates that neither the body of the man in the Shroud nor Jesus' body was washed at burial. The Code of Jewish Law states that if the body has blood flowing before and after death, it should be buried without washing or undressing it.[103] The evidence on the Shroud clearly shows that the man wrapped within it was buried according to Jewish burial customs or law.

The Jewish burial customs that refer to the proper way to deal with a corpse's blood have been studied and documented extensively by Doctors Lavoie, Klutstein, and Regan. The positions of the Mishnah, the Talmud and the Code of Jewish Law, as well as those of the Old Testament, led them to conclude:

> The Mishnah and Talmud specifically quantitate the minimum amount of blood which is necessary to become unclean (a quarter-log of blood) [An amount equal to about one and one-half eggs] and the period at which it becomes unclean (at the time of death). This blood is described as an uncleanness of the first order and should be buried with the corpse.
>
> In the case of Jesus, who died a violent death, it can be stated that there was blood on his body that flowed during life and after death. Furthermore, as to the quantity of blood that flowed after death, an accumulation of at least a quarter-log of blood can be easily inferred by John's description of Christ's wounds, especially when considering the wound on the side. The blood on his body was, therefore, mingled blood which could not have been washed off the body because it had to be buried with the corpse in order to comply with Jewish custom.[104]

The three doctors studied laws and practices that had been handed down from century to century by Jewish law, and what they found was consistent throughout. "If a man dies a violent death and blood is shed, the blood is not washed from the body. He is simply buried in a white linen sheet with his clothes not removed for fear of losing blood that has flowed from the man at the time of death. This blood flowing at death is considered life-blood."[105]

Further studies by London University Faculty of Laws Member, Victor Tunkel, confirms that Jewish burial rites unequivocally prohibited washing the body of a man who died in the same circumstances as the man in Shroud. The Jewish scholar regards the absence of evidence that the body of the man in the Shroud was washed as evidence for the authentic "Jewishness" of this cloth.[106]

The Gospels also indicate that Jesus' body was not washed. In the first place, the burial that Jesus received was hurried and incomplete. Furthermore, the Sabbath was quickly approaching and all work, such as the burial, had to cease before sundown. To make matters worse for the buriers, this particular Sabbath was one of the holiest of Jewish occasions, the Passover.[107] Some scriptural scholars assume that Jesus' body was washed after his crucifixion because some translations of John 19:40 state that Jesus was buried in accordance with Jewish burial customs, which typically call for washing. Actually, this passage states that Jesus "was prepared" for burial (Appendix B of *The Resurrection of the Shroud*). So the implication of washing his body is not warranted.

The same arguments relating to premortem and postmortem blood being present on the man in the Shroud would obviously apply to the same blood flows on Jesus. Thus, washing his body in this instance would violate Jewish burial customs. Observe also how Luke simply states in Acts 9:37, "In those days she fell sick and died; and when they had washed her, they laid her in an upper room." If this had been the case with Jesus, any of the Gospel writers could have easily mentioned it. None of the four Gospels indicate that Jesus' body was washed.

When Joseph of Arimathea requested Jesus' body from Pilate (Matthew 27:57-58, Mark 15:42-43), evening was rapidly approaching. Between the time taken to make the journey (probably at least 10 minutes or longer), getting permission to speak with Pilate, and the fact that Pilate had to wait to make sure that Jesus was indeed dead (Mark 15:44-45), the burial would have been very rushed. Further delay would have occurred waiting for any legal documents that needed to be completed, especially since the Jews had already asked for the remaining Jewish crucifixion victims' bodies to be removed (John 19:31).[108] Upon Joseph of Arimathea's return to Golgotha, he would also have needed to stop and purchase a linen shroud, help remove Jesus' body from the cross and take it to the burial tomb. The women followed to see the tomb and how Jesus was laid in it, and then left to return and anoint him after the Sabbath.[109] Luke called the approaching dusk *epephosken* (23:54), or "lighting-up time" as the stars came out and lamps were being lit against the rapidly approaching darkness.[110] It is possible that by the time the body of Jesus was entombed, the Sabbath and Passover had already arrived.

The Gospel accounts of Matthew (27:59), Mark (15:46) and Luke (23:53) all clearly state that Jesus was buried in a linen shroud. John refers to linen cloths, which also includes a cloth that was placed around Jesus' head at his burial (19:40, 20:7 and Appendix C of *The Resurrection of the Shroud*). Several writers postulate that the large and abundant quantity of myrrh and aloes described in John 19:40 (75-100 lbs.) were in powdered, granulated or dry block form.[111] Such a large amount would have been used to temporarily postpone putrefaction until his burial could be completed after the Sabbath and Passover.[112] The myrrh and aloe could also have been in the form of leaves or plants. (It is very doubtful that liquids would be described in pounds or weight or that fragrant liquids would be sold in such huge amounts). If these materials were contained in bags or blocks, such a large amount could very well have been parceled or sold by weight. These open bags or dry blocks could have been placed around the Shroud, or anywhere in the tomb, to help overcome the expected odor of decomposition when the women returned to anoint the body.[113] If there was even time to utilize the myrrh and aloes, perhaps, as leaves, they were poured over the top of the enveloped Shroud after Jesus was wrapped within it. While STURP did not find any trace of myrrh, aloes or their residue on the Shroud, they only had access to the image or inner side of this cloth. The outer side of the cloth has never been scientifically examined with the backing cloth removed.

The Mishnah sets forth a series of steps that could be taken to preserve a corpse when a full burial is not allowed to be performed on the Sabbath. Although one could not close the deceased eyes, objects could be placed over the eyes to keep them closed.[114] Similarly, while the jaw could not be raised, the chin could be bound up so that it would not sink lower.[115] Both of these steps may have been taken at the burial of Jesus and the man in the Shroud as discussed earlier. If otherwise permitted, washing and anointing the body was also allowed as long as the buriers did not move any member of it. Yet, there would have been so many blood marks over Jesus and the man in the Shroud that this last act would not have been possible.

John 11:44 talks about a *soudarion* being around the face of Lazarus as he left the tomb. The *soudarion*, a Latin term for sweat cloth or handkerchief, was described as covering the head of Jesus in John 20.7.[116] As stated by New Testament scholar, Dr. and Rt. Rev. John A.T. Robinson, "The only position, I submit, which fits both these descriptions, assuming that they are referring to the same custom, is of something tied crossways over the head, round the face and under the chin. In other words it describes a jaw band…."[117] A jaw band could have possibly been placed between the man's hair and his temples or over the hair as depicted in Fig. 73.

(Fig. 73)

The use of a jaw band may be indicated on the man in the Shroud. Bands along the side of the face and the upturned beard are clearly visible on the positive image. The upturned beard is also visible on the three-dimensional image produced by the VP-8 Image Analyzer. On this 3-D image, the hair on the left side of the face seems to drape over an invisible object.[118] As in so many other respects, the Gospel accounts of the burial of Jesus and the image on the Shroud of Turin are completely consistent.

Clearly, Jewish laws were adhered to with the burial of Jesus regardless of whether Joseph of Arimathea and Nicodemus worked up until or through the beginning of the Sabbath. These men would have known and understood such burial customs since they were both members of the council that promulgated such rules. [119] This detailed knowledge of Jewish burial customs is yet one more confirmation that the death and burial of Jesus and that of the man in the Shroud are astonishingly congruent.

The Identity of the Man in the Shroud

While there is much more evidence to examine, we cannot avoid the obvious initial question of the identity of the man wrapped in the Shroud. Before we discuss his identity, however, we should make some observations about the archaeological, as well as much of the scientific, medical and historical evidence on this burial cloth.

Archaeology is the study of ancient periods of history based on the examination of their physical remains. Yet, never has a physical object revealed so much information as the Shroud of Turin. We saw in the last chapter how the images obtained from the Shroud actually improve as time and technology advances. As new technology is applied to this burial garment, its unparalleled images actually make quantum leaps in development. If historical artifacts or images are well cared for, the original representation can be preserved; but neither they nor their resulting images will improve or make quantum leaps in development as time and technology progresses. This remarkable Shroud feature, however, is not due to the development of technology, but is due to the unique manner in which these images were encoded centuries ago.

Modern technology reveals extensive information at both the macroscopic and microscopic levels of the Shroud. Computer imaging technology not only illustrates the three-dimensionality of the frontal body image, but also its lack of two-dimensional information and the fact that it was encoded in a straight-line vertical direction from the cloth to the body. Photography reveals detailed and focused information throughout the length and width of both sides of the man's body. His human appearance, muscle forms and body contours all become visible. His blood flows and blood marks are now visibly aligned throughout the human corpse, as are swelling, scratches, cuts, abrasions, lines and some scourge marks that weren't even previously visible.

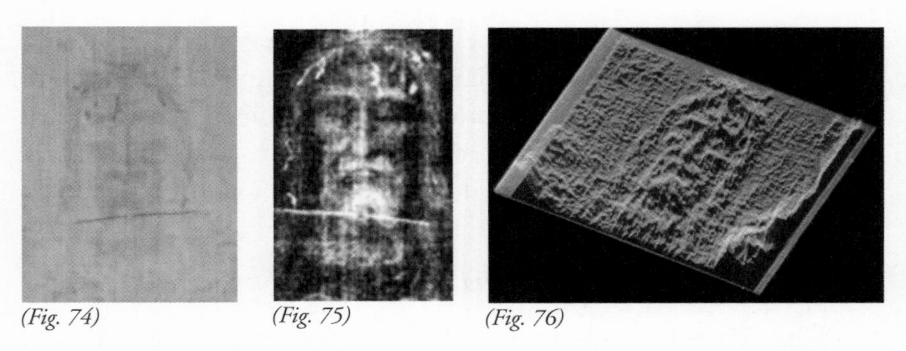

(Fig. 74) (Fig. 75) (Fig. 76)

Figs. 74 -76 show that greater quality and more information is revealed when modern technology is applied to the Shroud's image.

Photos taken under UV fluorescent lighting not only confirm these smaller features, but show that serum surrounds the blood marks throughout the man's body. Photographic enlargers and microscopes reveal the indented centers and upraised edges with serum surrounding borders on all the scourge marks. They also illustrate the superficial nature of both full-length body images. The above forms of technology, along with the infrared, X-ray and other forms of emission in the electromagnetic spectrum demonstrated that the body images did not consist of any material. Instead, they consisted only of oxidized, dehydrated cellulose.

No other physical object, let alone a burial garment, contains such unprecedented information and body images. These full-length images not only make quantum leaps in development and reveal increasing amounts of information as modern technology is applied to them, but they contain a number of unique and remarkable features that had never been seen in centuries of earlier examinations.

The above analysis does not even consider such things as the human hemoglobin, human albumin, human whole blood serum, human immunoglobulins and human DNA that have been chemically and physically identified from the man's blood marks. Nor does it consider other information acquired by normal investigative techniques that relate to the age or origin of this burial shroud, or where and when certain events occurred to this man. This information was acquired from the physical inspection or historical analysis of such things as the man's scourge marks and side wounds, Roman instruments, executioners and coins, as well as pollens, limestone and flowers. The encoding process that reveals so much additional information and makes quantum leaps in development as time and technology advances, simply cannot be forged or acquired naturally. It had to be the result of an unprecedented event. The full-length, frontal and dorsal body images, their many unique and exceptional features and their 130 blood marks, which we will continue to discuss in this book, have never been duplicated by any natural or artistic methods. The unprecedented event that encoded so many unique features must be considered in evaluating the identity of the victim in the Shroud.

Actually, only a couple of items of evidence on the Shroud appear to be forgeable. A medieval forger could have possibly sprinkled pollens primarily from Jerusalem and other geographical locations throughout this burial cloth, which were proportionate to its subsequently seen flower images, and with its origin in Jerusalem, and with its stays in Turkey and Europe. Even here, a forger would have to anticipate that centuries later these tiny pollens could one day be seen by microphotography that would reveal their different shapes and features, and that science could identify the regions of the world where these pollens were and were not found.

A medieval forger could also have placed a calcium or limestone sample at the foot of the man in the Shroud that scientists would subsequently find on a tested fiber, which matched the same rock shelf in which Jesus was reputably buried, but no other rock shelf in Israel. However it's very doubtful he would have anticipated the scientific development of a high-resolution scanning ion microprobe or similar device centuries later that could identify and compare calcium from various geographic locations.

Let's examine whether this unprecedented evidence aligns with any other historical figures or sources. The most attested sources of ancient history, the Gospels, state that Pontius Pilate had Jesus scourged by the Roman guards.[120] As a non-Roman, there would have been no limitation on the number of strokes that Jesus received. The man in the Shroud received a severe scourging by the Romans, with scourge marks covering the front and back of his body, from head to feet. Many of these scourge mark features could not be seen until the application of modern technology to the man's images and to the cloth. These scourge marks have never been duplicated artistically or naturally.

According to the Gospels, Jesus was beaten by Roman soldiers at the praetorium after having been beaten before the council, their guard, the chief priests and the high priest.[121]The man in the Shroud was also beaten about the face and the head, which is clearly indicated by his swellings and lacerations. Some of Jesus' blows were received after a crown of thorns was placed over his head to mock him as King of the Jews.[122] Such a crowning has not been found among any of the recorded tortures of the condemned prior to crucifixion. Evidence of a number of puncture wounds are seen on the man in the Shroud, however, indicating that a full crown of thorns, typical of the type used in the East during ancient times, had been thrust on his head.

Several similarities can also be noted regarding the crucifixion victim being required to carry his own cross. The Gospels tell us that Jesus had to bear his own cross.[123] An abundance of evidence shows that the man in the Shroud suffered multiple injuries indicating that he, too, carried a heavy, rough object across his shoulders. The abrasions on his shoulders as well as those on his face and legs, and the dirt on his nose and knees indicate that he fell several times as well. Clearly, both Jesus and the man in the Shroud were in a weakened condition after having been beaten and scourged, and the weight of the heavy beam would have caused them to fall. Jesus was so weak, in fact, that Simon of Cyrene was forced to carry his cross.[124]

Although some crucifixion victims were tied with rope, both the man in the Shroud and Jesus were nailed at the hands and feet.[125] Nail wounds at these locations are very visible on the man in the Shroud. The two parallel blood streams on each forearm, the expanded rib cage, the enlarged pectoral muscles drawn in toward the collarbone and arms, the upraised left leg and the taut legs and buttocks are also consistent with the man in the Shroud having been crucified. While most of the parallel blood marks on the forearms consist of real human blood, a few appear to be encoded as degraded cellulose. This, too, not only appeared after modern technology was applied to the Shroud's images, but has never been seen before.

The two thieves who were crucified with Jesus, as well as

Yehohanan, all had their legs broken by Roman executioners, in the 1st century, in Jerusalem. This was done to accommodate the wishes of the local Jewish leaders to hasten the victims' deaths. However, neither Jesus[126] nor the man in the Shroud had their legs broken. This was not necessary for Jesus and the man in the Shroud had already died while on the cross. This was confirmed by a postmortem wound in Jesus' side inflicted by a Roman soldier from which blood and water astonishingly flowed.[127] This is also confirmed to have occurred to the man in the Shroud from a wound in his side, whose size and shape match excavated examples of the Roman leaf-shaped lancea that was used by foot soldiers of the Roman militia.

The only time in history that such a flow of water and blood has been recorded was when Jesus Christ was crucified. From the time of Jesus until the fourth century, a period in which crucifixions frequently occurred until they became illegal in c. A.D. 315, Christian advocates regarded the flow of blood and water as a miracle. A spear thrust at an upward angle at the location of the side wound on the man in the Shroud aligns perfectly with the right auricle of the heart, which fills with blood upon death. This would allow blood and watery fluid from the pleural cavity to flow vertically by gravity from this postmortem wound.

Only when the Shroud was first photographed at the dawn of the 20th century did science or medicine even realize that the side wound and the blood and water flows were not located on his left side, but were actually located and inflicted upon the man's right side! It is absurd to think that a medieval forger would expertly encode these things on the left side of the man so that when photography was invented 500 years later, these items would align perfectly on the victim's right side. Besides, neither he nor anyone in medieval medicine would have had the medical knowledge to understand the correct location for or the interplay between the right auricle and postmortem blood, or between the pleural spaces and watery fluid.

No forger in medieval times or any other era has ever encoded the man's reversed postmortem side wound, blood and watery flows. No forger in any era, whether using artistic or natural methods, has ever come close to encoding the Shroud's 130 coagulated blood marks that were inflicted at different times with different instruments with different intensities, yet appear on cloth in the same shape and form as they appeared on the man's body.

The many signs of rigor mortis, which clearly set in while on the cross, confirm that the man in the Shroud also died by crucifixion. Jesus' body was given an individual burial in a new linen shroud; however, since the Sabbath and Passover were approaching, the burial was incomplete and his body was not washed.[128] The man in the Shroud was also given an individual burial, in a fine linen cloth, and his body was unwashed. Both Jesus and the man in the Shroud appear to have been buried with sudaria, or chin bands, over their heads, and those who buried them apparently possessed detailed knowledge of Jewish burial customs. Moreover, while wrapped in a linen shroud, Jesus was laid in a nearby tomb hewn from rock. Invisible traces of limestone that match samples from this same rock shelf, discovered only by microscopic examination, have been found on the Shroud.

While there is much more evidence to examine, the evidence considered thus far from both full-length body images and their extensive blood marks indicates that the same series of events comprising the passion, crucifixion, death and burial of the historical Jesus Christ also occurred under all the same circumstances of time, place, people and instruments to the dead man wrapped in the Shroud.

It is understandable that an artist from medieval or other times would want to depict these critical events in the life of Jesus Christ. It is also understandable that all artistic and naturalistic forms of depicting these events, as well as the Shroud's body images and blood marks, should be thoroughly considered and examined. Yet, naturalistic and artistic methods have consistently shown that the Shroud's body images and their blood marks cannot and have not been duplicated. The unforgeable features throughout both full-length body images and their blood marks appear to have occurred as the result of an unprecedented event.

A unique *process* happened that not only *captured* all the evidence from the man's body and blood, but all the critical events that occurred to this man were also captured by this unparalleled event. The encoding process that gave these images the ability to improve and make quantum leaps in development as time and technology advanced was also an unparalleled *process* that occurred within or during an unprecedented event.

The many features that were seen for the first time ever throughout both unforgeable, full-length images, along with the similar historical circumstances in which the various events occurred, all clearly point to the man's identity as the historical Jesus. Yet, the unique processes that occurred during an unprecedented event to the man in the Shroud are more convincing than any other individual event or circumstance that the Shroud was the burial cloth of Jesus Christ.

The resurrection of Jesus Christ was recorded as an unprecedented event in which his body disappeared. The unprecedented processes and events that occurred to the man in the Shroud rule out artistic and naturalistic methods even more so than do the unforgeable features and images on this burial cloth. Unprecedented processes occurring during an unprecedented event, following a sequence of events that also occurred under the same circumstances of time, place, people and instruments, to a real victim who already appeared from both unforgeable images and numerous unique blood marks to be the historical Jesus — now leaves us no alternative but to recognize the man in the Shroud as the historical Jesus Christ, who is described in the Gospels and New Testament.

We will see in Chapter Eleven that the leading image-forming hypothesis, which is the only one to account for the Shroud's body images, and *all* other critical features on the cloth, also involves the sudden disappearance of the body of the man in the Shroud. In the next few chapters we will examine more of these features and the question whether the unprecedented act that occurred to the man in the Shroud can be proven to have been a miraculous act.

5

ALL SIGNS POINT TO A UNIQUE FORM OF RADIATION – FROM THE BODY

There are many other attributes of the Shroud's body images that we haven't even discussed yet, and these, too, can be accounted for by radiation. In fact, *only* radiation can account for all of the unique body image attributes on the Shroud of Turin.

The only comprehensive scientific examination of the Shroud of Turin in 1978 actually provided a variety of indicators that radiation caused its full-length, frontal and dorsal body images. Some of the earliest tangible signs came from comparing the body images to the light scorch on this burial cloth.

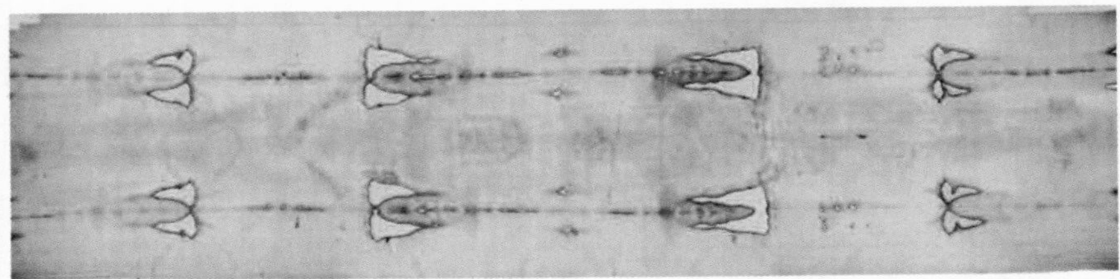

(Fig. 77) Full-length negative body images with all scorch and burn marks on the Shroud.

Unlike paint or other materials, both the body images and all the scorch marks are stable in water, neither impeding its flow nor being altered by it. Both are also stable to further heating and do not change color (up to temperatures and times that would produce equivalent scorches).[1] Neither the scorch marks nor the body images have faded with time, and neither were caused by foreign materials or particulates. The atoms within the body images and *light* scorch areas of this cloth reflect, absorb and emit radiation very similarly when irradiated under ultraviolet lighting.[2] When examined under ultraviolet lighting, the

Shroud linen fluoresces, except at its body images and its *light* scorches. Both of these areas absorb this energy without visible emission and do not fluoresce.[3] With controlled timing and heat, superficial scorches (similar to the Shroud body images) can also be produced on cloth without affecting its gross mechanical properties.[4]

Transparent fibers taken from *lightly* scorched areas closely resemble Shroud body image fibers.[5] The body image fibers are also similar to a *light* scorch in their microscopically corroded appearance and their lower tensile strength.[6] In addition, neither the chemical components of the body images nor the scorch marks could be dissolved when numerous acetic acids, oxidants, reducing agents and organic solvents were applied to them.[7] Irradiated linen can account for these and all other conditions present on the Shroud's body images.

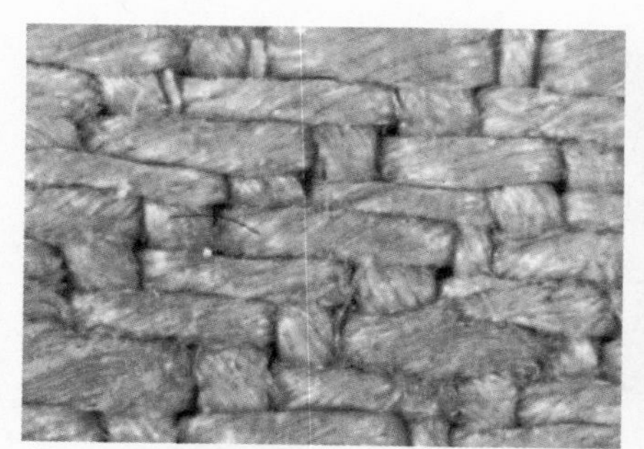

(Fig. 78) Photomicrographs taken from the Shroud of light scorch image on the left and (Fig. 79) body image on the right.

Neither artistic nor naturalistic materials such as paint, powders, plants, body fluids, liquids or other materials or combinations thereof can account for these conditions, let alone for the many extraordinary features of the Shroud's body images and their blood marks.

STURP scientists noted that "the [Shroud's] body image is due to a more *advanced* decomposition process than the normal aging rate of the background linen itself." (italics added)[8] Scientists have also established that the application of radiation (light or heat) to cellulose will artificially darken it in what amounts to a rapid simulation of the aging process.[9] By using such controlled, accelerated aging processes, scientists can produce the same overall properties as the body image and background areas on the Shroud.[10] The body images appear to have been exposed to more light or heat than the rest of the cloth and to have aged, decomposed or degraded at a faster rate than the background.

For these and many other reasons, numerous scientists and Shroud experts concluded that some form of light or heat (or radiation) caused the images on the Shroud. However, the absence of pyrolytic compounds or products expected from high-temperature cellulose degradation indicates the image-forming process took place at a fairly low temperature. This type of low-temperature radiation would not leave any residue, saturation or capillary action on cloth, all of which are absent with the Shroud's body images. Low temperature radiation could also uniformly encode the fibers on a cloth. Throughout this book, when radiation is discussed as the cause of the many features on the Shroud's body images, this term refers to non-thermal or low temperature radiation.

Some of the Shroud's image features can *only* be accounted for by radiation. As we noted earlier while referring to two of the body image features on the Shroud, the late Dr. Luigi Gonella, who served as the scientific advisor to the official Custodian of the Shroud, noted: "an agent acting at a distance with decreasing intensity is, almost by definition, radiation. The limitation of the cloth darkening to the outermost [superficial] surface pointed to a non-penetrating, non-diffusing agent, like radiant energy...."[11] Radiation seems to be the only method capable of encoding other features on the Shroud as well. For example, only vertical beams of light or radiation illustrate how the Shroud's full length, frontal body image was encoded through space in a straight-line direction from the body to the cloth. As Dr. Gonella further explained "...whatever the mechanism might be, it must be such to yield effects as if it were a burst of collimated [parallel beams of] radiant energy."[12]

Only through simulation have modern scientists been able to come close to the Shroud's vertical directionality, three-dimensionality and highly resolved, focused image. Their simulation was achieved by a mechanism in which light was attenuated in a liquid, then traveled in a vertical, straight-line direction from a plaster reference face while it was being focused in a camera.[13] All the previous scientific and medical evidence points to a very unique occurrence that caused the full-length, frontal and dorsal images to appear on this burial cloth, something that could never have occurred naturally or been created by medieval or modern techniques.

Particle Radiation

One form of radiation can account for all of the Shroud's extraordinary body image and non-image features. This form of radiation was not even proposed or tested by scientists until after the Shroud was radiocarbon dated to the Middle Ages in 1988. This form of radiation, particle radiation, can explain how an irradiated object can erroneously appear by carbon (C-14) dating to be centuries younger than its actual age. Particle radiation was first proposed to have irradiated the Shroud by physicist Thomas Phillips of the High Energy Physics Laboratory at Harvard University in 1989.[14] Particle radiation could not even be generated by scientists until the twentieth century.

Radiation is energy that is emitted in all directions in the form of waves or particles. Particle radiation consists of very small particles emitted from the basic building blocks of matter. The vast majority of these particles would consist of protons and neutrons. All matter is made of atoms. Over 99.9% of the mass or weight of all atoms is in their nuclei, which consists solely of protons and neutrons. If particle radiation was given off at the atomic or nuclear level, the vast majority would consist of proton and neutron radiation, which would easily cause the most effects throughout this burial cloth. (A small fraction of electrons, alpha particles and deuterium would also be given off and perhaps gamma rays or other electromagnetic radiation.) Neutrons (along with gamma rays and electrons) would pass through the Shroud linen cloth and anything that was on it when it was irradiated. These forms of radiation, especially neutron radiation, will cause several considerable effects throughout linen, blood and almost any other materials that they pass through and irradiate. Although these very significant effects will be discussed in subsequent chapters, this chapter focuses on the type(s) of particle radiation that could have caused the Shroud's full-length, frontal and dorsal body images. Because neutron radiation causes far more significant effects on linen, blood and other materials than do gamma rays or electrons, we can think of neutrons as the primary form of particle radiation affecting every part of the Shroud except its superficial body images. These superficial body images would be caused by protons (and a very small fraction) of other particles that act similar to them.

Proton radiation could actually explain all of the Shroud's body image features. This assertion was made most prominently by biophysicist Jean-Baptiste Rinaudo of the Center for Nuclear Medical Research in Montpellier, France in the 1990s. Dr. Kitty Little, a retired nuclear physicist from Britain's Atomic Energy Research Establishment in Harwell, also joined in the scientific debate in the 1990s regarding the cause of the Shroud's images. After summarizing previous scientific investigations of the Shroud, Dr. Little confirmed "that the source of the illumination that had formed the image came from within — that is, from the body — ... as a whole."[15] Thereafter, she stated, "Now it seemed almost certain that the image must have been caused by some sort of radiation.... However, there was one source of ionizing radiation that they did not try."[16]

Little recalled that in 1950 she had irradiated several different cellulose fibers at the nuclear reactor in Harwell with the above forms of particle radiation, which reproduced the straw yellow color that she learned subsequently is on the Shroud. The temperatures in the reactor's channels were as low as 40° C, so the radiation effects could be examined without the complication of heat degradation.[17]

(Fig. 80) Harwell Nuclear Reactor Site

Neutrons and protons behave quite differently. While neutrons are extremely penetrating, protons have very short ranges. Protons are so non-penetrating that they even absorb or attenuate in air. Protons would evenly deposit their energy to produce the uniform straw yellow color – only on the topmost fibers of the cloth.[18] Moreover, they would break many of the bonds of the molecular structure of the cellulose, but only in these topmost image fibers, thereby causing some of the single-bonded carbon atoms attached to hydrogen or oxygen to, thereafter, re-form or double bond with other carbon and oxygen atoms into conjugated carbonyl groups.[19] Comprising more than 99.9% of the mass or weight of all atomic nuclei, protons and neutrons are found in immeasurable abundance in all bodies.*

In the 1990s, Dr. Rinaudo began performing intriguing experiments with protons and neutrons.[20] Fig. 68 shows one of his proton-irradiated linen cloth samples. Dr. Rinaudo and his associates performed numerous experiments irradiating white linen cloth with proton beams of various energies with a particle accelerator at the Grenoble Nuclear Studies Center in France. As we saw, linen naturally fluoresces under ultraviolet lighting, as does the Shroud's off-image areas. However, when Rinaudo's experimental linen was irradiated with proton beams with energies of 1.4 MeV or less, the cloth's natural fluorescence disappeared, as was the case with the Shroud's body images. The protons produced uniform superficial coloration on cloth whose fibers and threads lacked any cementation, added pigments or materials of any kind. Where body image fibers crossed, underlying fibers were protected and remained white as did the inner part of the straw-yellow

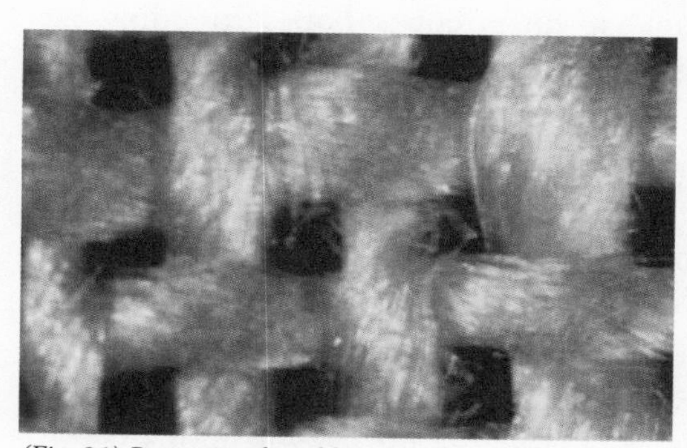

(Fig. 81) Proton irradiated linen

*A very small fraction of deuterons and alpha particles could also be emitted from a body. They contain one or two pairs of protons and neutrons; however, both behave like a proton, in that they traverse only a short distance in air and an even shorter distance in linen. Therefore, throughout this and subsequent discussions, where protons are specifically mentioned, similar results would be expected for deuterons, alpha particles or other heavy charged particles.

image fibers. The scientists were also able to duplicate the microchemistry results of dehydrated, oxidized cellulose, as is also found with the Shroud's body image.[21]

Rinaudo's straw-yellow color resulted from conjugated carbonyl (double-bonded) groups within the molecular structure of the cellulose, as did the Shroud's image fibers. These double-bonded carbon groups absorb light and reflect it as the straw-yellow color that is visible on Rinaudo's linen, as well as on the Shroud linen. Furthermore, Rinaudo demonstrated that like the Shroud's coloration, his could also develop over time if the irradiated linen was artificially aged by heating at low temperatures.[22]

Drs. Phillips, Little and Rinaudo not only indicated that particle radiation irradiated the Shroud, but the source of this particle radiation was the body wrapped within this cloth. While this is obviously extraordinary, it is indicated by a variety of evidence. We saw in Chapter Three that skeletal features such as finger bones, bones extending over the palm, part of the skull at the forehead, the left thumb, several parts of the backbone and even teeth have been identified on the man in the Shroud. Like the man's skin, blood marks and hair, these features also lie near the front and back surfaces of the reclined body of the man wrapped in the Shroud. How could a medieval forger think to encode these features when X-rays weren't invented until 1895? How could he have encoded internal skeletal and dental features that would not be visible for hundreds of years? How could these features have occurred naturally? Only radiation from the body seems to explain them.

Dr. Carter, who first recognized these internal features, thought they indicated not only that radiation came from the body, but that it resembled or had qualities analogous to X-rays. Interestingly, enlargement and diffusion of the body's bones, ligaments, and skin normally occur when X-rays are made. That is because the rays leave an *external* tube before hitting part of the person's body and recording the image on film.

The enlargement and diffusion, however, will vary with the distance of the radiation from the body. The shorter the distance between the source of X-rays and the body, the greater is the enlargement and diffusion. For the short distances that existed between the Shroud cloth and the underlying body, extensive enlargement and diffusion would have clearly been present if the source of radiation came from *outside* the body.

However, the Shroud's frontal and dorsal body images are not only highly resolved, but without any enlargements. These attributes indicate that the source of the light or radiation did not come from any source outside of the body, but came from the body *itself.*

We saw earlier that although the bottom part of the cloth (containing the dorsal image) would have received all the weight of the man's reclined body, the dorsal image is encoded with the same amount of intensity as the frontal image. Both images are encoded with the same amount of intensity, independent of any pressure or weight from the body. Radiation coming from the body would not only explain this independence of pressure, but also the left/right and light/dark reversals and the distinct body boundaries found throughout the Shroud's full-length, frontal and dorsal body images.

If the light or radiation came from any source *besides* the body wrapped within the cloth, we would not have the unique full-length, frontal and dorsal body images encoded on the inner parts of the Shroud. In photography, all objects that are illuminated naturally by artificial light in front of the camera become recorded on the film as images. The same thing also happened in the case of this unique burial cloth. The film is clearly the inside surfaces of the cloth that wrapped the body and recorded the images. Neither the outside nor the inside of the *tomb* are seen on the Shroud's images. Neither the outer side nor even the inner side of the *cloth* are found on the Shroud's extraordinary positive images. *Only* the front and back sides of the man's *body* are seen on the Shroud's incomparable images. (Fig. 82)* This means that the source of the light

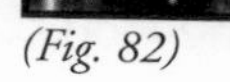

(Fig. 82)

*Along with anything else that was between the body and the film such as blood or possibly a coin or flowers.

did *not* originate *outside* of the body, but originated *exclusively with* the body itself.

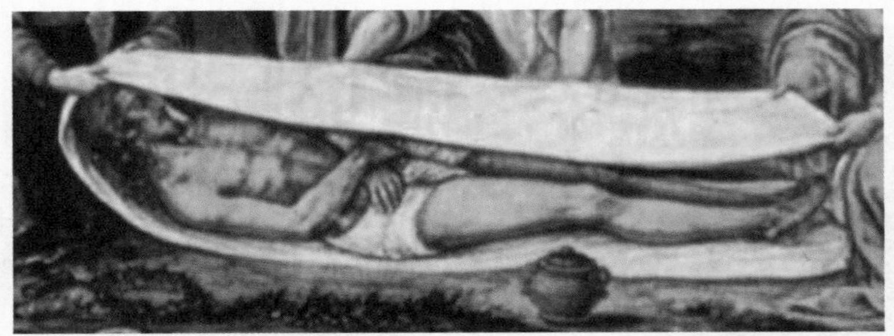

(Fig. 83) The source of the illumination could only have been the body wrapped within the cloth.

Although the three-dimensional quality of the Shroud's body image is found along the entire length of the frontal image, it is best displayed on the facial image. That is because the greatest variety of depth is found in this relatively small region. The nose, forehead, cheeks, lips, chin and hair are all at different elevations on or around the face of the supine body. The rest of the body tends to contain much less variety of relief within a concentrated area. The rest of the body image also has more blood marks, water stains, scorch marks and patches on or near it. Because these last features were not encoded like the body images, they appear distorted on the VP-8 image relief and clearly detract from the three-dimensionality of the body image. (Only the blood marks on the Shroud's facial image distort and interfere with its accurate three-dimensional relief.)

Dr. Rinaudo has worked for years with particle radiation (protons and neutrons) as has Dr. Jackson with ultraviolet (UV) radiation, and Dr. Carter with X-rays. Since proton and UV radiation, and low-energy, long-wave X-rays are very attenuating, each scientist has asserted that these particular forms of radiation could encode superficial, straw-yellow fibers and convey three-dimensional information onto the cloth.[23] However, all of these scientists agree that the source of the radiation can only be the body wrapped within this cloth.

The Shroud's truly proportional, three-dimensional frontal body image is directly correlated with distance. Its various degrees of intensity directly correlate with their original distances from the underlying body.[24] Since every point of distance information throughout the frontal body image was received *by* and is clearly contained *on* the linen cloth – it could only have come from every part of the underlying body.

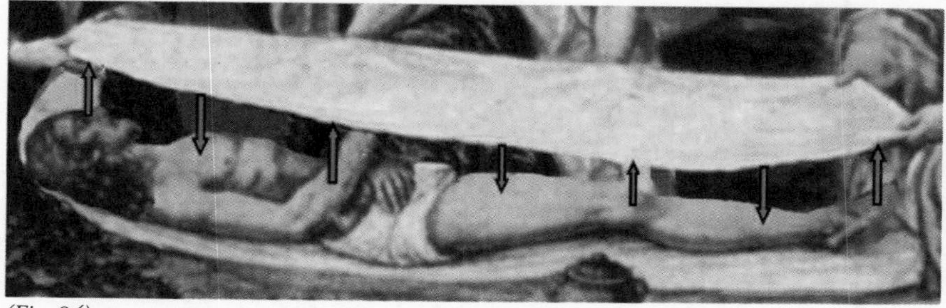

(Fig. 84)

When the two men at each end of the body let go of the cloth (Fig. 84), the top of it will conform roughly to the contours of the underlying body. Yet, regardless whether the Shroud was sloping downward, upward, or was relatively flat, all parts of its frontal body image were encoded in a vertical straight-line direction from the draped cloth to their corresponding and underlying points on the entire body.[25] Since this vertical correlation exists even where the draped cloth was not originally touching the body, the image was vertically encoded through these empty spaces.[26]

Like the three-dimensional information, since the vertically-correlated information was also received *by* and is contained *on* the draped cloth, it too, could only have come *from* every part of the underlying body.

Needless to say, vertically-correlated radiation from a body would not leave any two-dimensional (latitudinal and longitudinal) directionality on the cloth. The lack of two-dimensional directionality was a surprising discovery made by scientists who played leading roles on various NASA projects in the 1970s. This discovery helped lead to the formation of STURP and the only comprehensive, scientific investigation ever performed on this cloth in 1978. This discovery and the subsequent one that the Shroud's body image was encoded in a vertical straight-line direction are quite consistent and corroborating. Since the body image lacks latitudinal and longitudinal (width and length) directionality, the vertical direction (depth) was the only dimension left in our three-dimensional world of space in which the image could have been encoded.

Like proton or particle radiation, superficial, straw-yellow color has also been produced on cloth by ultraviolet (UV) light, which also does not fluoresce under UV illumination. It, too, consists of conjugated carbonyls (carbon double-bonded with itself and oxygen) after the irradiated fibers dehydrated and oxidized.[27] The coloring process that was initially triggered by exposure to UV light could also be accelerated and strengthened by heat. In this manner, coloring that was not visible after UV irradiation could then gradually become visible, like the coloring produced by proton radiation on linen and the body images on the Shroud.[28]

32 Features Uniquely Encoded Only Where Cloth Wrapped the Body

While you may not realize it, we have seen so far that non-thermal or low temperature radiation can cause or explain 32 unique or extraordinary features that are found on the Shroud's body images and their delicately encoded fibers. These features listed below have all been gradually documented by modern scientists who started documenting a few of the Shroud's attributes in the late 1960s and early 1970s. Some of these properties were initially observed by comparing the Shroud's body images and their fibers to the Shroud's light scorches and their fibers, or by making light scorches and conducting other laboratory tests and experiments on control linen. Most of these exceptional features have been confirmed by countless examinations and experiments performed on the Shroud itself, its body images and their fibers. Although some of these features can be duplicated by naturalistic or artistic means, only radiation can duplicate or explain all of them.

- lack of fading
- lack of foreign materials or particulates
- straw yellow coloration
- only topmost superficial fibers of threads encoded
- individual fibers encoded
- fibers colored 360° around circumference
- only outer layers of individually encoded fibers are colored

- no coloration inside of fiber
- fibers colored with similar intensity
- oxidation and dehydration of body images
- containing conjugated carbonyls (double-bonded carbon and oxygen atoms formed after single-bonded atoms within linen fibers broke apart) that
- developed over time
- accelerated aging of the body image
- stability to water and heating
- insolubility to acids, reductants and solvents
- gross mechanical properties of linen intact
- microscopically corroded appearance of fibers
- lower tensile strength of fibers
- reduction of the cloth's fluorescence at body images
- lack of residue
- highly attenuating or absorbing agent
- agent operated over skin, hair, bones, teeth (coins and flowers)
- non-diffuse image with sharp boundaries
- equal intensity for frontal and dorsal images
- lack of two-dimensional directionality
- negative images with left/right and light/dark reversals that develop into
- highly resolved, photographic quality images
- without any magnification
- with skeletal and dental features
- three-dimensionality
- encoded through the spaces between the body and the cloth
- in a straight-line vertical direction [29]

Only the Body Could Be the Source of Radiation

From a number of discussions within five chapters of this book, we have seen how radiation or radiation from a body could explain all of these remarkable features on the Shroud's body images. Several of these arguments assert the body is necessarily the source of the radiation.

However, when we consider that:

1) The 32 extraordinary, itemized features are *only* found throughout the length and width of the Shroud's frontal and dorsal body images;

2) both full-length body images are encoded *only* on the inner sides of the cloth that wrapped the crucified, dead body; and that

3) *only* radiation can duplicate or explain all 32 exceptional body image features;

we can be very confident that the source of this radiation and its effects could only have been the body wrapped within this cloth.

Based on the available evidence, you don't have to be a scientist to wonder whether a supernatural radiating event occurred to the dead crucified man in the Shroud that caused his body images to develop over time on this burial cloth. You don't have to be a theologian to realize that all of the unfakable features and events encoded into this burial shroud are similar to the very same series of events and circumstances that are recorded to have occurred to the historical Jesus Christ. Keep in mind that all the scientific and medical evidence we have discussed so far has come from the unparalleled body images and the extraordinary wounds and blood marks that are scattered throughout them. All of the evidence regarding radiation or particle radiation emanating from the body of the man in the Shroud has also derived from the body images.

In the 1980s, some scientists and researchers speculated that the Shroud's dual body images were caused by bursts of energy such as pulsed laser beams that emanated from the body of the man in the Shroud.[30] In the 1990s, scientists began suggesting that something happened to the body so that is disappeared from the Shroud almost instantaneously as it gave off radiation.[31] While we will discuss more specific hypotheses in Chapters Eleven and Twelve, the reader should understand that the instantaneous disappearance of the body from the Shroud would not necessarily have caused an explosion. In addition, we will see several more reasons why the body's instantaneous disappearance as it gave off radiation would not have caused any (or very little) energy to be released.

We will see two hypotheses that stand apart from all others in which radiation is given off uniformly from an instantaneously disappearing body. Only these two hypotheses can account for all the Shroud's primary and secondary body images. These hypotheses were developed strictly from considerations of the Shroud's body image features after observing that all naturalistic and artificial image creating models had been tested and failed. Physical objects within these hypotheses, as well as their physical environment (such as the cloth, air, gravity, radiation, chemical modification of cellulose) all behave according to scientific laws. The only thing that acts unconventionally is the disappearing radiating body within each of these hypotheses. Yet this is the primary reason these hypotheses can account for the Shroud's body images. The unconventional disappearance of the man in the Shroud is completely consistent with numerous events and circumstances that occurred to the historical Jesus Christ, described in the most attested sources of antiquity. Moreover, both body images, their wounds and blood marks reveal that neither their features nor the unique events that happened to the man in the Shroud can be forged by any known methods or techniques.

One of these hypotheses asserts that particle radiation was given off from an instantaneously disappearing body. This hypothesis will explain the Shroud's body images and their still-red blood marks, which are embedded in the cloth in the same shape and form as when they formed and coagulated on the body. This hypothesis can also explain the Shroud's excellent condition, its skeletal features, possible outer side imaging, coin and flower images, and the Shroud's aberrant medieval radiocarbon dating.

We live at a unique moment in history. Scientific technology exists that could be adapted and applied to the entire Shroud, as well as to linen, blood and other samples strategically removed from this burial garment, that could test this leading hypothesis and all other image forming hypotheses. If a miraculous event occurred to the dead body of the man wrapped in the Shroud it could be proven! Sophisticated scientific technology could not only prove whether radiation emanated from the body wrapped within the cloth, but whether this radiation was particle radiation. Scientific testing could also indicate if this event occurred as the body instantaneously disappeared. Among other things, modern technology could not only demonstrate that such a miraculous event occurred but when it happened, where it happened, whether the Shroud's medieval radiocarbon dating is valid, the actual age of the cloth and the identity of the victim.

Let us next look at this scientific technology, the leading hypotheses, and the historical evidence that could confirm whether a miraculous event occurred to this man after he incurred a series of wounds, was crucified and killed by Roman executioners, in the first century in Jerusalem, and buried according to detailed Jewish burial customs, in the same rock shelf in which the historical Jesus Christ was reputed to have been buried.

6

UNIQUE EFFECTS OF NEUTRON RADIATION

—∾—

Keep in mind that all the scientific and medical evidence we have discussed so far has come from the unprecedented body images and the remarkable wounds and blood marks scattered throughout these images. All of the evidence indicating that radiation or particle radiation emanated from the body of the man in the Shroud was also derived from the body images or their features. Yet, if the Shroud of Turin was irradiated with particle radiation from the dead body wrapped within it, unique and unfakable evidence of this event would be found in samples taken from all over this cloth. This evidence would be found at the *atomic* level throughout both body images, all of the blood marks, the non-image areas and the entire outer side of the Shroud of Turin. Moreover, these atomic signatures would be found in varying amounts that reflected their closeness to and their locations over the body at the time of this unprecedented event.

This atomic evidence will comprise an entire new line of evidence and proofs that we have not discussed before. This evidence will be even more abundant than the evidence discussed in the previous chapters. Whereas the Shroud's unique blood marks number well over a hundred and its many body image features are encoded throughout tens of thousands of fibers on both full-length body images, these numbers are quite small in comparison. We will see that the number of items of evidence that would be available from testing the Shroud at the atomic level would easily number in the *billions*.

These unique signatures could even reflect the historical time in which they were deposited into the atomic levels of the Shroud. Neither this radiating event, the types of atomic evidence, their quantities nor the historical period in which they were implanted can be forged. This and the next chapter will show how scientific tests and experiments could be conducted at the atomic level on this burial cloth that may very well prove:

1. the entire Shroud was irradiated with particle radiation;
2. the amount of particle radiation each part of it received;
3. the Shroud's radiocarbon dating is erroneous;
4. the Shroud linen and its blood are from the 1st century;
5. the source of the radiation was the length, width and depth of the dead body wrapped within it;
6. whether this event occurred in the 1st century; and
7. where this radiating event happened.

As we saw in earlier chapters, particle radiation consists of very small particles of matter that can be emitted from atoms. Of particular interest for research on the Shroud are protons and neutrons, which are tiny energetic particles that reside within the nuclei of atoms. The Big Bang Theory asserts that light or energy was created before matter or mass. Protons and neutrons were created at the very beginnings of matter and comprise the vast majority of it. Even though protons and neutrons behave very differently, they *both* have had unparalleled effects on the Shroud. Protons have extremely short ranges and would only have affected the Shroud's superficial body images. While the effects of protons on linen and the Shroud's body images have been discussed previously, they will be examined in more detail in Chapter Eleven. Neutrons are very penetrating and would easily pass through the Shroud linen, its blood marks and any materials that were on it when it was irradiated. The extraordinary effects of neutrons on the Shroud will be a primary point of focus in this chapter.

Neutron radiation could not be generated by humans until the 20th century. After English physicist James Chadwick (Fig. 85) discovered the neutron in 1932 (for which he was awarded the Nobel Prize), scientists began to learn of its critical role in almost every aspect of matter, and thus, life itself.

Within a decade, neutron radiation would make a very dramatic impact upon the world when it was used to bombard and split the

(Fig. 85) Nobel laureate James Chadwick

uranium atom and achieve the first controlled release of nuclear energy in 1942. Comprising more than 99% of the mass of atoms, neutrons and protons are not only at the heart of all scientific matter, but at the heart of future scientific research on the body images, blood marks, age and condition of this famous burial cloth — and the critical events that happened to the dead man wrapped within it.

Neutron Radiation Creates Two New Radioactive Atoms in Linen and Blood

If the dead body of the man wrapped within the Shroud only gave off 0.000000015% of the neutrons within his body, then 3.0×10^{18} neutrons would have been emitted from the body.[1] This number consists of a three with 18 zeroes after it. This number is so large that we don't have a common name for it. A trillion only has 12 zeroes. While only a very small percentage of these neutrons (and protons) would have been acquired by the Shroud, *trillions* of these neutrons would have been acquired by and left unfakable signatures of this radiating event throughout both body images, their many blood marks and the non-image areas of this famous cloth.

When neutrons are initially released they usually travel at extremely high speeds. If they were released from a body within a burial tomb, they would have ricocheted throughout the interior of the tomb. Each time the neutrons collided with atoms in the rock, cloth, blood or other matter, they would have lost energy and speed. When neutrons slow to a certain energy and speed (called thermal energy),[2] the nuclei of atoms within the irradiated material (the burial cloth and its blood marks) have a better chance to react to the neutron as it passes through them. Occasionally, some nuclei react by "capturing" an individual neutron, where it then joins other protons and neutrons *within* the nucleus.* When this capture occurs within the nuclei of common chlorine or calcium atoms (contained in linen and blood), it produces unique results that are not fully appreciated by most of us.

When a neutron particle is added within the nuclei of the most common chlorine or calcium atom, a new, distinct atom is *created* that is so *rare*, it virtually does not exist in nature. The most common chlorine atom is called chlorine-35 (Cl-35).[3] (This simply means its nucleus consists of a total of 35 protons and neutrons.) The most common calcium atom is called calcium-40 (Ca-40), (which means its nucleus consists of a total of 40 protons and neutrons.)[4] While you wouldn't think that adding one neutron to the nucleus of an atom that already has 35 or 40 protons and neutrons would make that much of a difference — it makes all the difference in the world. The new neutron causes a *new* chlorine or calcium atom to be created that is now unstable and radioactive (and will decay). These newly-created atoms are distinct and identifiable, and are simply called chlorine-36 (Cl-36) and calcium-41 (Ca-41). However, these infinitesimally rare atoms that virtually do not exist in nature can be identified and measured with instruments called accelerator mass spectrometers (AMS).

(Fig. 86) Accelerator Mass Spectrometer

*When this chapter refers to a "nuclear reaction," it merely means the addition or replacement of one of the many neutrons or protons within the nuclei of an atom.

Chlorine and calcium atoms normally reside within molecules. When new Cl-36 and Ca-41 atoms* are created by neutron capture, these extremely rare atoms will also reside within the molecular structures of the neutron irradiated material. Like the additional neutrons that only occur within the nuclei of these new radioactive atoms because of neutron radiation, the new radioactive atoms only reside within the molecular structure of the irradiated material as a result of neutron radiation.

These distinct, newly-created Cl-36 and Ca-41 atoms can not only be measured, but can be compared to the Cl-35 and Ca-40 atoms to which they were added. If Cl-36 or Ca-41 atoms were detected in the Shroud's linen, blood marks or charred material above their natural infinitesimal limits, they could *only* have resulted from neutron radiation. Their presence well above their natural infinitesimal levels within the molecular structures of linen, blood or charred material would prove these materials were irradiated by neutrons. This event could not possibly have been performed by humans before the 20th century and could not possibly have occurred naturally.

All radioactive atoms are unstable and undergo nuclear decay at known rates. Some radioactive atoms will decay within microseconds or minutes. Fortunately, Cl-36 and Ca-41 have long half-lives of 301,000 and 102,000 years, respectively. If the Shroud linen or its blood marks were irradiated by neutrons from the body of Jesus Christ, almost *all* of the Cl-36 and Ca-41 atoms that would have been created two thousand years ago throughout this burial cloth would still be present today.**

***When the name or abbreviation of a chemical element is given along with its total number of protons and neutrons (or its atomic mass), it is called an isotope, such as Cl-35, Cl-36, Ca-40 or Ca-41. However, an isotope is just another form of an atom. An isotope is an element or atom that contains both a particular number of protons and neutrons within its nucleus. Since most people are unfamiliar with the term isotope, but understand what an atom is, I will use the latter term throughout. However, I will supply the name or abbreviation of the element and its total number of protons and neutrons (or its atomic mass) wherever it is useful or required.**

****Only about 0.46% of the Cl-36 and 1.3% of the Ca-41 created within the Shroud linen cloth and its blood marks (or other material present) during the radiation event would have disappeared naturally since 30 A.D.**

After the discovery of neutrons in the 1930s and their dramatic application in the field of nuclear energy in the 1940s, scientific experiments with neutron bombardment and its resulting nuclear reactions became a major focus of research in many parts of the world for the next several decades. As a result, the very precise rates in which newly-created Cl-36 and Ca-41 occurs within neutron irradiated material, as well as the rates for other specific nuclear reactions, have been well-known and long established by scientists throughout the world for many decades.

Modern scientists have demonstrated that if a known number of neutrons irradiate material containing known amounts of chlorine or calcium, then *known* numbers of new Cl-36 and Ca-41 will be produced within the irradiated material. Both the presence of the newly-created Cl-36 and Ca-41 atoms *and* their specific amounts can be determined by measuring the Cl-36 to Cl-35 and Ca-41 to Ca-40 ratios of the neutron irradiated material. As stated earlier, modern science can measure these atomic ratios with an AMS. This instrument is also used for measuring atomic ratios within material during C-14 or radiocarbon dating.

I've been explaining the basic nuclear reactions and the scientific techniques to detect them so the reader can understand what could be proven by testing the Shroud of Turin at the atomic level. Calcium exists naturally in blood and limestone and has been identified throughout the Shroud linen cloth. Calcium would also have remained in the charred samples that were removed from the Shroud in 2002. Chlorine occurs naturally in blood and should also be present throughout the Shroud linen cloth.

If scientists were to measure the Cl-36 to Cl-35 and Ca-41 to Ca-40 ratios in cloth, blood and charred samples removed from a variety of strategic locations throughout the Shroud of Turin — these measurements would not only prove this famous linen and its blood marks were irradiated with neutron or particle radiation — but they would reveal the *amount* of neutron or particle radiation that each location received. (It should be noted that charred material was removed in 2002 from behind various patches running along the sides of both full-length body images and kept in 42 small bottles in Turin.)[5] Five corroborating *sets* of Cl-36 to Cl-35 and Ca-41 to Ca-40 ratios could be obtained confirming that this famous burial cloth was irradiated with particle or neutron radiation throughout its length and width. The ratios within these samples could also demonstrate a direct correlation between the amount of neutron radiation they received and their proximity to, as well as their positions over the body, at the time of this neutron radiating event.

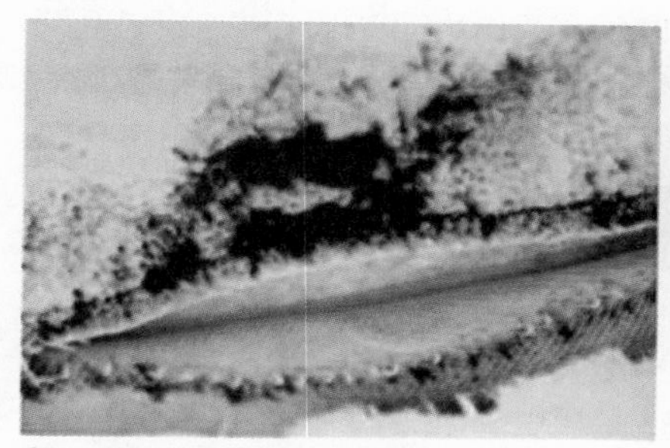

(Fig. 87) Charred material from one location behind Shroud's patches.

(Fig. 88) Charred material from behind Shroud patches now stored in small bottles with their exact provenance noted.

We will see shortly that neutron radiation also creates C-14 atoms. Ironically, C-14 atoms only exist on our planet because of a nuclear reaction caused by neutron radiation. Like Cl-36 and Ca-41 atoms, C-14 atoms are also created at very explicit rates within neutron irradiated matter. We will see that if the Shroud was irradiated by neutrons, this event would necessarily invalidate the cloth's C-14 dating and would explain why the Shroud erroneously carbon dated to a period much younger than its actual age.

Neutron Radiation Creates New C-14 Atoms in Linen and Blood

It is difficult to imagine a greater scientific disservice to humanity than the Shroud's C-14 dating conducted in 1988. The medieval age that it attributes to the cloth perpetuates the lack of public interest and knowledge about the Shroud that exists throughout the world. Unfortunately, this medieval dating has an exclusive effect. If it is accepted, it is impossible for the cloth to have been Jesus' burial garment. Consequently, neither the Shroud's body images, its blood marks, nor most of its other unique features could have been caused by the body of the historical Jesus. This dating eliminates the incentive to study the technical and complex scientific and medical evidence contained on this unique burial garment. This evidence happens to relate to the ultimate philosophical, historical and religious issues of life that are also avoided if the C-14 dating is accepted.

Yet the present state of affairs on this subject is much worse. The vast worldwide public not only knows very little about the Shroud of Turin, but what little it does know (that it derives from the Middle Ages) is completely erroneous. Although most people in the world have not heard of the Shroud, the ones who have think it is a medieval forgery. Unfortunately, this erroneous conclusion was believed for 600 years before the Shroud was carbon dated in 1988. In 1389, the Shroud was declared to be a medieval painting by Bishop Pierre d' Arcis of Troyes in a rambling, ranting memorandum, whose conclusion was accepted by the vast majority of the public. However, when the Shroud was comprehensively examined for the first and only time in 1978, scientists were easily able to prove that the Shroud was not painted in the medieval ages or at any other time.

If the Shroud was examined at the atomic level, the results could easily prove that the cloth's medieval C-14 dating is erroneous. Testing the Shroud linen cloth and its blood marks at the atomic level could also establish the actual age of the cloth, utilizing the same technology that is employed by C-14 or radiocarbon dating. Cl-36, Ca-41 and C-14 measurements can not only repudiate the Shroud's C-14 dating, but play important roles in confirming that a unique neutron radiating event happened to the body of the man wrapped in the Shroud. We will see these measurements could also help determine when this unprecedented event occurred, as well as where it occurred.

The worldwide public should understand that C-14 is just another rare, unstable radioactive atom that is also *created* by neutron radiation. In fact, C-14 can *only* be created by neutron radiation. While C-14 exists naturally, it exists only in trace amounts. And, even where it is found naturally, it was created by neutron radiation.

After modern scientists discovered the neutron in 1932, they demonstrated that new C-14 atoms could be created within material irradiated by neutron particles. In fact, they demonstrated that C-14 is also created at very precise rates within any irradiated material containing nitrogen atoms, such as linen or blood. If the Shroud was irradiated by neutrons, new C-14 atoms would also have been created throughout this unique burial cloth. Neutron radiation would have substantially increased the C-14 content within the cloth making it appear much younger than its actual age. Neutron radiation would have created even more C-14 atoms in the Shroud's blood marks.

The process that creates radioactive C-14 atoms within neutron irradiated linen, blood or other matter is called neutron conversion. Like neutron capture, this process also takes one atom and changes it into a different atom. While Cl-36 and Ca-41 atoms are created from the most common chlorine and calcium atoms, C-14 atoms are created from the most common nitrogen atoms called nitrogen-14 (N-14).[6] The N14 atom has 7 protons and 7 neutrons and does not decay. When the ricocheting neutrons within the burial tomb slowed to thermal energy, nuclear reactions would have occurred within some of the nuclei of the nitrogen atoms distributed throughout the Shroud linen cloth and its blood marks.

This nuclear reaction occurs after a neutron collides with and enters the N-14 nucleus, creating instability within the nucleus and causing the ejection of a proton.[7] The resulting nucleus is a C-14 nucleus containing 6 protons and 8 neutrons.* This nuclear reaction releases considerable energy, which is carried away by the kinetic energy of the C-14 nucleus and the proton as they rapidly fly apart in opposite directions.

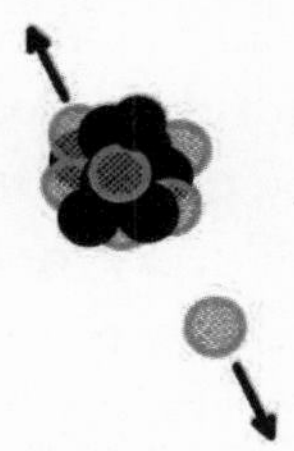

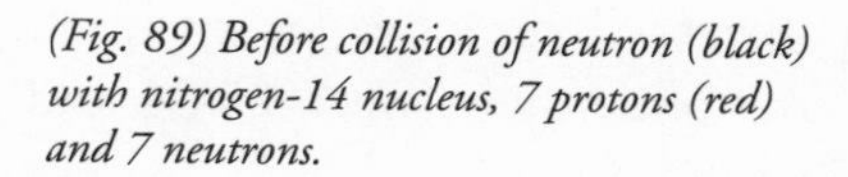

(Fig. 89) Before collision of neutron (black) with nitrogen-14 nucleus, 7 protons (red) and 7 neutrons.

(Fig. 90) After collision, neutron is captured and proton is ejected, resulting in carbon-14 with 6 protons and 8 neutrons.

***When new C-14 atoms are created from N-14 atoms, they also become new elements. Elements are comprised of specific types of atoms containing the same number of protons within their nuclei (or have the same atomic number).**

Only after modern scientists learned to create C-14 by neutron bombardment in laboratories, did they realize that new C-14 is also created in our atmosphere by neutron radiation by the *same* process. This occurs regularly because cosmic rays from outer space (consisting of highly energetic particles) constantly bombard the earth's atmosphere, thereby causing neutrons to fly around and circulate in the atmosphere.[8] These neutrons cause nuclear reactions to occur within the nuclei of nitrogen atoms in the atmosphere,[9] in the same way they're caused by neutrons within the nuclei of nitrogen atoms in linen, blood or other irradiated material.

The neutrons and protons comprising the newly-created C-14, Cl-36 and Ca-41 atoms all reside within their nuclei. The only way that new, additional neutrons could have penetrated into or been added to the nuclei of the chlorine, calcium and nitrogen atoms of any material is by neutron radiation. Like Cl-36 and Ca-41, the newly-created C-14 atoms will also reside within the molecular structure of the Shroud samples, which could only have resulted from neutron radiation. The creation of these new radioactive atoms could have unparalleled importance for us.

C-14, of course, is the key atom that is measured in C-14 or radiocarbon dating. Like Cl-36 and Ca-41, these new radioactive C-14 atoms would also be found throughout both frontal and dorsal body images, all the blood marks and the non-image areas of both sides of the cloth. Like Cl-36 and Ca-41, the number of newly created C-14 atoms within Shroud linen and blood samples could also reveal a direct correlation with their closeness to and their position upon (or under) the body at the time of the radiating event.

The rate of decay for radioactive C-14 is much faster than for Cl-36 and Ca-41. Whereas the half-lives of Cl-36 and Ca-41 are 301,000 and 102,000 years, the half-life of C-14 is 5,730 years. (This means that half of the C-14 atoms within an object will disappear every 5,730 years.) We will see that the different rates of decay between Cl-36 and Ca-41, on the one hand, and C-14 on the other hand, will be very critical to scientists. These different decay rates will play contributing roles in determining that these radioactive atoms could not have been embedded into the Shroud by a forger and will allow scientists to calculate when this unique neutron radiating event occurred and the actual age of the incomparable Shroud of Turin.

Basic Principles and Assumptions of Carbon Dating

While the next few pages may be a little (more) technical in nature, they will provide a basic understanding of the principles and limitations inherent with C-14 dating. These principles and assumptions are provided so the reader can understand that if the Shroud was irradiated with neutrons, this event would *necessarily* cause an erroneous C-14 dating result. If it was just stated that neutron radiation would make the irradiated object appear to be younger, the reader would not understand why this is so and why the effect would essentially be permanent.

C-14 is not only radioactive or unstable, but it only comprises approximately one part in a trillion of the overall carbon content (1/1,000,000,000,000). Carbon-12 (C-12) accounts for 98.9% of the earth's naturally occurring carbon. Carbon-13 (C-13) accounts for the other 1.1 percent. C-12 and C-13 are stable atoms that were formed when the earth's other atoms were formed. Yet, despite C-14's extremely minute quantity, which naturally decays, the overall amount of C-14 on earth remains nearly the same because new C-14 is created in the atmosphere at essentially the same rate it is decaying on the earth's surface. Thus C-14 is said to be in balance.

This very tiny amount of C-14 formed in the atmosphere, along with the much larger amounts of C-12 and C-13, is taken up in atmospheric carbon dioxide by photosynthesizing plants and is, thereby, spread throughout the biosphere, thus allowing all living things to have a similar ratio of C-14 to C-12. Since the carbon atoms have the same chemical behavior, this ratio is maintained while the organism lives. However, upon its death, the organism's C-14 gradually disappears according to its radioactive half-life, which is approximately 5,730 years, while its C-12 amount remains stable. By measuring its C-14 to C-12 ratio, scientists can calculate the date of the organism's death.[10]

In the case of the Shroud linen, its organism is flax, an annual plant that produces long, slender stems up to four feet in height that are used in the manufacture of linen. When scientists carbon date a linen cloth, they are actually measuring how long ago the flax plants that comprise it were harvested and woven into that particular cloth. Carbon or C-14 dating is a valid, scientific dating technique *only if all* the C-14 within the measured object was acquired from the atmosphere while the organism was alive. If any outside or foreign C-14 was acquired by the linen cloth and measured by the radiocarbon dating laboratories, it would necessarily invalidate the cloth's age. Radiocarbon laboratories do not thoroughly date objects. They only measure the C-14 to C-12 ratios within these objects and then assign ages to them based upon their C-14 to C-12 ratios. Inherent within their assignment of ages is the *critical* assumption that the C-14 to C-12 ratios measured within their samples do not contain any C-14 atoms from any outside sources.

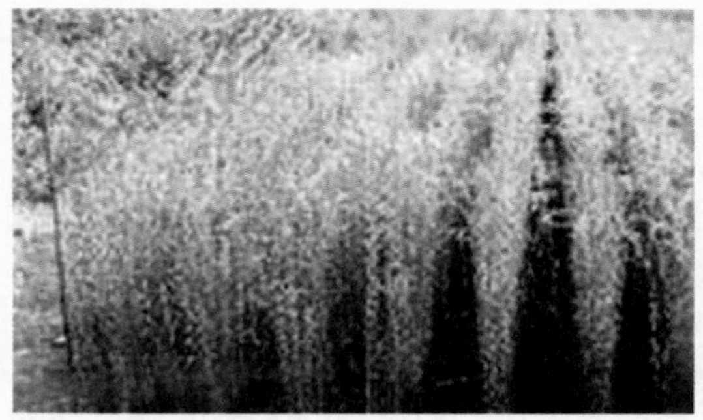

(Fig. 91) Flax plants before harvest

As we saw, the Shroud linen cloth not only contains indigenous amounts of calcium and chlorine, but it also contains indigenous amounts of nitrogen (N-14) within it.[11] If this linen cloth was irradiated by neutrons, the N-14 in the cloth would react to neutron radiation in the same way it reacts in the atmosphere or during neutron bombardment. A known number would convert to new C-14 atoms. If neutrons were emitted from the body of Jesus Christ while he was wrapped in the Shroud, not only would radioactive Cl-36 and Ca-41 atoms be created at known rates throughout this burial cloth, but so would C-14 atoms.

These C-14 atoms, however, would be acquired from an *outside* source, *after* the flax plants were harvested. Their presence within this linen cloth would *necessarily* invalidate the cloth's C-14 dating as the

AMS that measures the C-14 atoms has no way of distinguishing between the C-14 acquired by the plant while it was alive and the C-14 created within the linen cloth by neutron radiation. The C-14 atoms that were created within the cloth after it was woven would make it appear to date much younger than its actual age.

Scientists who participated in the Shroud's radiocarbon dating in 1988 readily acknowledge the effect that neutron radiation would have had on the validity of their results. When asked by a journalist soon after their results were announced whether such a process could have caused an incorrect dating of the Shroud, Michael Tite, who coordinated the carbon dating of the Shroud for the British Museum, commented: "It is certainly possible if one gave the Shroud a large dose of neutrons to *produce C-14 from the nitrogen in the cloth.*" (emphasis added)[12] Robert Hedges, one of the scientists who participated in the carbon dating of the Shroud at the Oxford laboratory, also acknowledged to the journalist that a "sufficient level of neutrons from radiation on the Shroud would *invalidate the radiocarbon date which we obtained.*" (emphasis added)[13]

While C-14 dating is a valid scientific technique, it has always been subject to enormous error. Even most scientists are unaware how error-prone carbon dating can be. While our major focus will be how particle or neutron radiation would have invalidated the dating of the Shroud, we would be remiss if we did not briefly note how other sources of contaminating carbon have also caused C-14 dating to be grievously inaccurate. Radiocarbon dating has dated mammoth fur more than 20,000 years younger than its actual age and living snail shells to be 26,000 years old. It has dated a newly-killed seal to be 1,300 years old, one-year old leaves as 400 years old and a Viking horn to the early 21st century. C-14 dating has dated bone tools to 27,000 years old, while a sample from the innermost portion of the bone dated to 1,350 years.[14]

While many more examples could be given in which dates have been erroneously ascribed to samples, rarely have the C-14 to C-12 ratios within the samples been incorrectly counted or measured. For various reasons, however, carbon from outside sources has been exchanged with the natural or innate carbon within the dated sample causing the C-14 to C-12 ratios to be altered. The outside carbon could be from a source that is younger or older than the natural carbon remaining within the measured object, but it will necessarily cause a different overall C-14 to C-12 ratio within the measured material.

Sometimes the reasons for and the extent of carbon contamination are never fully understood. Sometimes the best and only method of evaluating the extent or effects of carbon contamination on an object or its site is to observe the divergence of the radiocarbon date from the site's historically datable context. For these and other reasons, many radiocarbon dates have been rejected by archaeologists and geologists as being anomalous or in conflict with other C-14 dates or more reliable data.[15] Good scientists do not rely on carbon dating in isolation when there is other evidence available to help confirm an accurate date.

One of the ways that radiocarbon laboratories combat the very real problem of extraneous carbon contamination is to routinely pretreat or clean the sample before dating it. This is accomplished by rinsing the samples in various combinations of acids, bases and water in order to clean or rid it of any foreign carbons or materials that are not indigenous or native to the sample. For example, if you spilled coffee on your cloth sample before it was carbon dated, the coffee would spread throughout the cloth by capillary action and contaminate your sample. However, the radiocarbon laboratory could easily pretreat and clean this sample so that all the coffee was rinsed or removed from the cloth before it was dated.

Radiocarbon laboratories make another *critical* assumption when ascribing an age or a date to a sample whose C-14 to C-12 ratio has been measured. They assume that if any extraneous or outside carbon has contaminated the sample that this new carbon has been removed during the standard pretreatment cleaning of the sample. Yet, history has shown that carbon from outside sources can be acquired and trapped, especially within porous materials such as bones, wood or cloth. Removing contaminants from the pore spaces and fissures of some samples has been described as almost impossible.[16] Unfortunately, even with specialized pretreatment, contamination cannot always be detected, and, if detected or identified, cannot always be eliminated.

Newly Created C-14 Atoms Remain in Linen and Blood

As we saw, if neutrons irradiated the Shroud of Turin, they would have irradiated the nitrogen atoms within the cloth and its numerous blood marks. The neutrons would have acted within the nuclei of these nitrogen atoms in the same manner that they act in the atmosphere and by neutron bombardment in laboratories. A *certain* fraction of N-14 atoms would have converted to C-14 atoms at scientifically established rates.

Recent tests and experiments with neutron radiation by American and Italian scientists have demonstrated additional critical findings.[17] These tests, led by physicist Arthur C. Lind, confirmed that if linen is irradiated with neutrons, C-14 is created within the irradiated linen in two related ways. The first is by nuclear reactions with the nitrogen in the air that surrounds the linen. This new C-14 will be created in the same way that C-14 is naturally created by neutrons from N-14 in the atmosphere. These newly-created C-14 atoms combine with oxygen in the air and rapidly diffuse into the linen as carbon-14 dioxide. However, this carbon-14 dioxide will also slowly diffuse out of the cloth naturally in several years at room temperature. (The inward diffusion is more rapid than the outward diffusion because linen fibers have a crystalline structure that acts as a labyrinth.) This speed of outward diffusion can be greatly increased by heat or by standard pretreatment cleaning methods,[18] so these particular C-14 atoms would have no detrimental effect on radiocarbon dating.

The second way that C-14 is created within the irradiated linen is of far more significance. Nitrogen (N-14) is also indigenous to and found within linen. As we saw, when linen is directly irradiated by neutrons, nuclear reactions occur *within* the atomic nuclei of the nitrogen atoms within the linen itself. Like chlorine and calcium atoms, these nitrogen (N-14) atoms also reside within molecules inside the linen itself. When the newly-created C-14 atoms are created within the nuclei of atoms, they necessarily become part of the molecular structure of the irradiated linen.[19] These C-14 atoms do not diffuse away in several years or at any time. They are not removed by the application of heat beyond temperatures incurred by the Shroud in 1532, or by standard stringent pretreatment cleaning methods that the Shroud's samples received in 1988. These new atoms will not even be removed by combinations of time, heat and standard stringent pretreatment cleaning processes.[20]

These tests and experiments demonstrate that if the Shroud was irradiated with neutrons, its newly-created C-14 atoms would *not* have been removed by the stringent standard pretreatment cleaning processes that were applied to its samples in 1988. Nor would these newly-created C-14 atoms have been removed by any other conditions that the cloth or its samples have incurred throughout its history. As part of the molecular structures of the neutron irradiated linen, these newly-created C-14 atoms will only disappear if the linen is completely destroyed or when they decay naturally according to their half-life of 5,730 years.

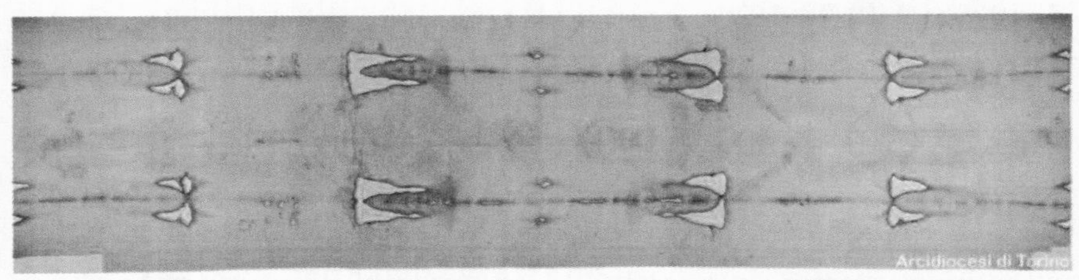

(Fig. 92) Neither pretreatment cleaning nor heat nor natural aging would have removed any C-14 atoms acquired throughout the Shroud by neutron radiation.

The Stage is Set

Now that we have some familiarity with the principles and assumptions behind C-14 dating and the permanent reactions that occur at the nuclear, atomic and molecular levels in cloth and blood from neutron radiation — the hardest parts are behind us. Now comes the interesting and innovative science that could alter the course of world affairs and history. We can now appreciate that the neutrons and protons comprising the newly-created, radioactive C-14, Cl-36 and C-41 atoms all reside within their nuclei and that the only way new neutrons could have penetrated into or been added to these nuclei is by neutron radiation. We can also understand that nitrogen, chlorine and calcium atoms are found in shroud linen, blood, charred material and limestone. Thus, we can comprehend the significance of the proposed tests and all their possible implications.

Since we also understand that all three of these radioactive atoms are formed by neutron radiation at established rates, but disappear (or decay) at different known rates, we will be able to appreciate in the next chapter how scientists could calculate not only *if*, but *when* such a neutron radiating event occurred to this burial cloth and its blood marks. Since we can also understand that the amounts of all three radioactive atoms within these samples could vary according to their closeness to and their positions near the center of the mass of the crucified body wrapped within the Shroud — we can also appreciate that the *source* of this neutron radiation can be determined. Please keep in mind that the atomic evidence that would prove the body was necessarily the source of the neutron radiation — is completely different from other extensive and unfakable evidence acquired from the body images indicating that only the body wrapped within the Shroud could have been the source of the radiation that caused all of its full-length body image features.

We will be able to understand how scientific testing at the atomic level could be conducted on linen, blood and charred samples from the Shroud of Turin, and on limestone samples from Jesus' reputed burial tomb(s) that could prove:

1. The entire Shroud was irradiated with particle radiation;
2. The Shroud's radiocarbon dating is erroneous;
3. The amount of particle radiation each part of it received;
4. The Shroud cloth and its blood are from the 1st century;
5. The radiating event happened in the 1st century;
6. The source of radiation was the length, width and depth of the dead body wrapped within it;
7. The identity of the victim; and
8. Where this neutron radiating event occurred.

7

TESTING THE SHROUD AT THE ATOMIC LEVEL

Numerous Ways to Disprove Shroud's C-14 Dating

Just finding Cl-36 and Ca-41 well above their natural infinitesimal levels in the Shroud would invalidate its medieval radiocarbon dating of 1988. The presence of Cl-36 and Ca-41 at these levels within its cloth and blood samples could only have resulted from neutron radiation. Of course, if these materials had been irradiated with neutrons, C-14 atoms would also have been created within them that would *necessarily* have made the Shroud appear much younger than its natural age.

The cloth's 1988 radiocarbon dating would also be disproven if you carbon dated the man's blood from *any* location on the Shroud of Turin. Nitrogen-14 (N-14) is estimated to comprise more than 10% of dried human blood.[1] This is 100 times more N-14 than is found in linen. If you measured the C-14 to C-12 ratio within the Shroud's blood it could date tens of thousands of years into the *future*, possibly on the order of 100,000 years or more. This, of course, would be an *impossible* age for the blood or any known material. There's no manner of contamination on earth *except* neutron radiation that could cause material to date so far into the future. This date, of course, would not only confirm the Shroud was irradiated with neutron particles, but would glaringly expose the invalidity of the cloth's 1988 medieval C-14 dating.

In most instances where outside or extraneous carbon contaminates a sample, the C-14 *and* the C-12 from the contaminating source become mixed with the sample's remaining C-14 and its original C-12.

Thus a new combined C-14 to C-12 ratio results that makes the sample appear younger or older than its original C-14 to C-12 ratio indicated. Even if the contaminating source was modern and comprised 99% of the new C-14 to C-12 ratio, the new ratio would not carbon date to the future; at worst it could only date to the present. However, when neutrons irradiate material containing N-14, *only* new C-14 atoms (but not C-12) are created within the material. Neutron radiation could easily create enough new C-14 atoms within the irradiated blood or linen to make it appear to date in the future.

If scientists carbon dated a cloth sample from a part of the Shroud linen that laid under or draped over the central mass of the man's body — such as the middle of his back or his rib cage — you could also *easily* disprove the Shroud's 1988 radiocarbon dating. These parts of the cloth would have received the largest amounts of direct neutron radiation, which would have traveled uniformly in all directions from within the body.[2] These areas of the cloth (especially, the middle of the man's back) would also have acquired the largest number of neutrons ricocheting back onto them from the limestone bottom on which the Shroud rested. Neutrons would also have ricocheted from the walls and ceiling of the burial tomb. More neutrons would have ricocheted onto the Shroud's dorsal body image regions because they would have lain immediately underneath the body and directly on the limestone.[3] These parts of the linen cloth would date significantly younger than the 1988 radiocarbon sample, which was removed from a non-image part of the cloth that was almost 18" from a very narrow part of the top of the man's right foot. Off-image parts of the Shroud closer to the body images should date younger than the 1988 site.[4]

If the man in the Shroud was the historical Jesus Christ, his dead body probably would have been placed upon a bench hewn into the limestone walls. The bench usually ran along all three sides of new or single chamber burial tombs of the Second Temple period.[5] Although Jesus' corpse could have been placed on a bench to the right or to the left of the opening of the tomb, the Gospels suggest that it was laid on a bench across from the opening,[6] which also would have been easier for his buriers. If neutrons were given off in 360° directions from the length, width and depth of a disappearing body, they would have ricocheted throughout the tomb and become distributed throughout every location on the cloth.

(Fig.93) Drawing of typical newly-hewn tomb of Second Temple Period

Nuclear engineer Robert Rucker, who has four decades of experience in advanced nuclear reactor design and criticality safety, has about 12 years' experience with a computer code called MCNP (Monte Carlo Neutron Particle). This code, developed at the Los Alamos National Laboratory, is used in extremely complex geometries involved in fissile material operations. In 2014 he began adapting this extremely sophisticated code to perform nuclear calculations for the first time ever in a manner related to the Shroud of Turin. His codes can calculate with an acceptable uncertainty the effects of *billions upon billions* of neutrons (this is the number with 18 zeroes) that would have been released, if only 0.000000015% of the body's neutrons were emitted in a fraction of a second from the disappearing corpse once wrapped in the Shroud. These codes can calculate the various amounts of Cl-36, Ca-41 and C-14 that would have been created throughout all parts of the length and width of the Shroud of Turin, including both body images and its 130 blood marks. These fascinating results have been calculated on the assumption that the body was placed on the bench across from the opening to the tomb.[7]

Rucker's calculations are discussed because they can predict the quantities and distribution patterns of Cl-36, Ca-41 and C-14 atoms that would be produced by neutron radiation at any location on the Shroud. Among other things, these predictions could provide a guide for sampling sites that could easily and independently confirm that the source of the neutron or particle radiation was the body wrapped within this cloth.

Shroud cloth samples removed from parts of the body image such as the back could carbon date several thousand years or more into the future, despite its linen having 100-200 times less N-14 than blood.

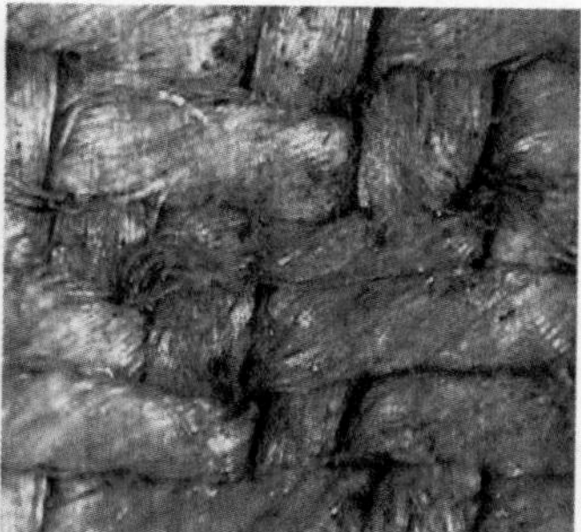

(Figs. 94 and 95) Blood mark images taken by photomicroscopy.

Many other parts of the Shroud's body images, and all of its blood marks, would carbon date to the future. While these C-14 dates would be even younger and more erroneous than the Shroud's 1988 radiocarbon dating, all of them would be consistent with and corroborate a number of critical points already indicated by the overall evidence in the previous six chapters: The burial cloth and its blood marks were irradiated with varying amounts of particle or neutron radiation thereby making every part of it appear much younger than its actual age. A correlation exists between the number of neutrons received by the non-image parts of the Shroud and their closeness to the body wrapped within it. A correlation also exists between the number of neutrons that each part of the cloth and its bloodstains received and their locations *over or under* the length or width of the body. Furthermore, a correlation exists between the amount of neutrons that each part of the Shroud's blood marks and its body images received and their proximity to the *center* of the body's mass. These dating patterns would confirm that the length, width and depth of the dead human body wrapped within this cloth was the *source* of the neutron radiating event.

A neutron radiating event of this kind could not possibly have been performed by scientists in the 21st or any century and could not have occurred naturally. Although scientists have been able to generate particle radiation since the 20th century, they can only do so with a nuclear generator. Modern scientists can't begin to make neutrons or protons radiate from *any* part of a body, let alone the length or width or depth of a dead body. Only the resurrection of the historical Jesus Christ, reported to have occurred in the 1st century, bears any resemblance to such a miraculous event involving a dead body. Extensive scientific and medical evidence in the previous chapters not only indicates that a unique radiating event happened to the body of the man wrapped in the Shroud, but this event also appears to have *encoded* unfakable, full-length body images and still-red blood marks of a man who recently incurred a series of wounds, was crucified and killed, and buried within a linen shroud under all the same circumstances as the historical Jesus Christ.

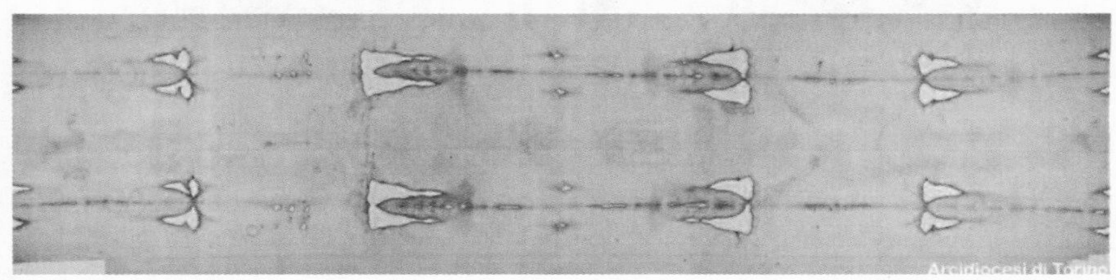

(Fig. 96)

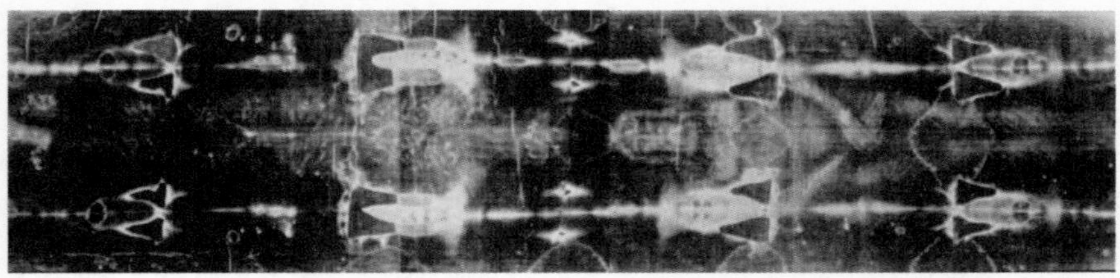

(Fig. 97)

Calculating When the Miraculous Event Occurred, the Age of the Linen and its Blood Marks

While the discussion in the next couple of pages may have to be re-read slowly a couple of times, it is provided so the reader can understand that the same AMS technology used in radiocarbon dating could do far more than just confirm that the Shroud's C-14 dating is erroneous. This technology can confirm what the previous independent scientific investigations have indicated — that an unprecedented, neutron radiating event happened in the first century to this contemporary first century burial cloth.

Once strategically located samples were removed from the Shroud, scientists would thoroughly wash or rinse all inorganic elements or material from them and non-destructively, or with very minimal invasion, measure the different amounts of organic chlorine, calcium and nitrogen contained within each cloth and blood sample.[8] These samples could then be submitted for AMS analysis to measure their organic Cl-36 to Cl-35 (or Ca-41 to Ca-40) ratios. Since Cl-36 to Cl-35 measurements can be detected in smaller samples than Ca-41 to Ca-40 measurements, they are best suited for testing precious, limited cloth and blood samples removed from the Shroud. After the Cl-36 to Cl-35 ratios have been measured from these samples, their C-14 to C-12 ratios should then be measured. Because AMS testing is destructive, each of these testing methods and procedures must be carefully thought out, thoroughly developed and refined, and repeatedly proven with neutron irradiated control cloth and blood samples before being performed on Shroud samples.[9] (See Appendix A for a discussion of laboratory procedures.)

Layman's Formula

While new radioactive Cl-36, Ca-41 and C-14 atoms are created by neutron radiation at very precise rates — they *decay* at different rates. The number of these radioactive atoms remaining in the Shroud's linen and blood marks today would allow scientists to calculate whether this radiating event occurred in the first century or in modern times, and to calculate the age of the samples.

As we saw earlier, the Cl-36 to Cl-35 ratios contained within each cloth and blood sample removed from the Shroud would reveal whether they were irradiated with neutrons.[10] These ratios would also reveal the *number* of neutrons that each cloth and blood sample received — regardless where they were located on this burial linen.[11] Knowing the amounts of N-14 and Cl-36 within each sample, scientists could then calculate the amount of new or *additional* C-14 atoms that were *necessarily* created at very precise rates within each Shroud sample by the neutron radiating event.[12] These new C-14 atoms, of course, would have been added within cloth and blood samples that already contained C-14 atoms remaining from the time when both the flax plant and the man's blood were alive.

Since the additional C-14 atoms are in the molecular structures of the Shroud samples and cannot be removed by pretreatment cleaning methods, their C-14 to C-12 ratios cannot possibly date the Shroud with any accuracy. These C-14 to C-12 ratios would make the samples appear to be much younger than their actual age. However, these C-14 to C-12 ratios would provide scientists with the *total* number of C-14 atoms *remaining* within each of these irradiated samples today.

Because scientists would have performed all the above calculations and measurements on every cloth and blood sample removed from the Shroud, they would know how many new C-14 atoms were added to *every* sample by neutron radiation. With this knowledge, scientists could also calculate how many of these additional C-14 atoms would have *disappeared* from each of these neutron irradiated linen and blood samples. Of course, practically none of the newly-created C-14 atoms would have disappeared if the Shroud had been irradiated in the 20th century. However, a significant and calculable number of the new C-14 atoms would have disappeared if the neutron radiating event occurred in the 1st century. Scientists could calculate how many of the newly-created C-14 atoms would have disappeared (and thus how many would remain today) — if the neutron radiating event occurred in the 1st century, in the 20th century, or any century in between.

Similarly, scientists would also know the original C-14 to C-12 ratios that existed within the Shroud's samples when its flax was first harvested or when its blood originally flowed from the victim's body. This ratio would be approximately the same whether the samples originated in the 1st century (as indicated by the overall evidence) or in the 14th century (as indicated by the Shroud's medieval C-14 dating). Both ratios would be approximately the same as the C-14 to C-12 ratio in all living organisms. Scientists could then accurately calculate how many of these original C-14 atoms would also *remain* in the Shroud's cloth and blood samples today — if they originated in the 1st century, in medieval times, modern times or any century in between.

The above two calculations of original and additional C-14 atoms that would have remained within each sample from every century could be added together. The totals for every century could be compared to the total amount of C-14 atoms remaining today within each one of the cloth and blood samples from the Shroud. If only the two C-14 calculations from the 1st century match the total amount of C-14 presently remaining within every one of the Shroud samples — it means that every cloth and blood sample not only originated from the 1st century, but that an unprecedented radiating event also occurred to each one of them in the first century.* (Please also read Appendix B for a scientific calculation to determine when this hypothetical neutron radiating event occurred and the age of the neutron irradiated Shroud.)

(Fig. 98) Accelerator Mass Spectrometer

A forger could not get away with irradiating a 1st century or a 14th century Shroud in modern times because he would leave too many C-14 atoms in the cloth and the blood. He would have had to irradiate the Shroud's cloth and blood samples with the precise amounts of neutrons that are indicated by the specific Cl-36 to Cl-35 and Ca-41 to Ca-40 ratios found within each sample. (Because the half-lives of Cl-36 and Ca-41 are 301,000 and 102,000 years, the amounts of Cl-36 and Ca-41 remaining

---

***Please keep in mind that the chlorine, calcium and nitrogen contents will vary between the cloth samples. Please also keep in mind the cloth and blood samples should be taken from off-image areas of the cloth and from different locations upon the frontal and dorsal body images. Thus, the samples would also have received various amounts of neutron radiation. Although Cl-36, Ca-41 and C-14 are all created by neutron radiation at very precise rates, significant differences in the amounts of these three radioactive atoms will be found throughout all of the various samples. The various amounts of all three radioactive atoms in the variety of samples would provide extensive corroboration for all of these independent calculations.**

within each sample would be virtually the same whether they were irradiated in the 1st century or in modern times.) While it would not have been *possible* for a forger to have neutron irradiated the Shroud before the 20th century, even if a medieval forger had somehow managed to irradiate a 1st or 14th century sample with the correct amount of neutrons, he would still *leave* too many C-14 atoms in every irradiated sample, which could be detected and measured with today's technology.

If the Shroud was irradiated with neutrons in the 1st century, 21% of the additional C-14 atoms that were created within it would have decayed or disappeared naturally by today. (Similarly, if the flax comprising the Shroud was harvested in the 1st century, 21% of its original C-14 atoms would have disappeared or decayed by today.) Regardless of the number of neutrons that irradiated a particular location on the Shroud in the 1st century, 79% of the newly-created atoms would still remain on the cloth today. While this percentage would not change, the *number* of newly-created C-14 atoms remaining within the various samples today would be noticeably and measurably different.

Since the 1988 radiocarbon site lies almost 18" from a narrow part of the right foot, it only dated to medieval times, and not to any younger or future dates. The 21% difference at this location, however, would consist of hundreds of millions of C-14 atoms, whose presence or absence are measurable with today's technology. The mid-back region on the dorsal side of the cloth would have received about ten times (10Xs) the amount of neutron radiation that the 1988 radiocarbon site received. The 21% difference in the amount of C-14 atoms remaining at this location from a neutron radiating event in the first century or in modern times, could translate to a difference of *billions* of C-14 atoms. The presence or absence of these neutrons could be even more easily measured by today's technology. The differences in the number of C-14 atoms remaining at this location from Jesus' time, medieval times or modern times could all be calculated by modern scientists and measured by accelerator mass spectrometers (AMS).

As stated earlier, blood contains approximately *100 times* (100 Xs) more nitrogen than linen. Almost all of the blood marks are found on the body images with a majority of them located on the dorsal image. If the Shroud received roughly the amount of neutron radiation calculated by Rucker, enormous differences would exist in the amounts of C-14 atoms that would remain in the man's blood from a 1st century neutron radiation and from a modern day neutron radiation. If blood samples were tested from the man's side wound, the back of his right foot or from the scourge marks in the middle of his back, these differences would range from *hundreds of billions* to *trillions* of C-14 atoms. These very large differences in the number of C-14 atoms remaining in these blood samples from a neutron radiating event during Jesus' time, or medieval times, or in modern times, could all be calculated by modern scientists and measured by AMS.[13]

Possible Sampling Sites

Depending upon the minimum size that future research determines is necessary to measure the Cl-36 to Cl-35 or Ca-41 to Ca-40, and the C-14 to C-12 ratios in each sample, our foundation recommends removing cloth samples from the middle of the man's back, as well as other body image locations that would not have received as much neutron radiation, such as his chest or abdomen, or the front or back of his thigh, leg or the hair from his head. We also recommend that off-image cloth samples immediately next to the body images (such as next to the thigh on the frontal side and next to the buttocks on the dorsal side) be similarly tested along with samples still remaining from the 1988 radiocarbon site. Since the C-14 to C-12 ratios were previously measured from the Shroud samples, the 1988 Cl-36 to Cl-35 or Ca-41 to Ca-40 ratios should also be measured from contiguous portions still remaining at the Oxford, Zurich and Arizona laboratories. We also recommend that the minimum size blood sample necessary to measure the Cl-36 to Cl-35 or Ca-41 to Ca-40 and the C-14 to C-12 ratios be tested from the middle of the man's back, the small of his back, the back of his head or foot, or his side wound. Blood from two non-images areas near the back of the man's right foot and near the front of the right elbow would also make excellent candidates for Cl-36 to Cl-35 or Ca-41 to Ca-40 and C-14 to C-12 testing. Hopefully, these measurements can be perfected so well that very small amounts of cloth and blood can be tested.

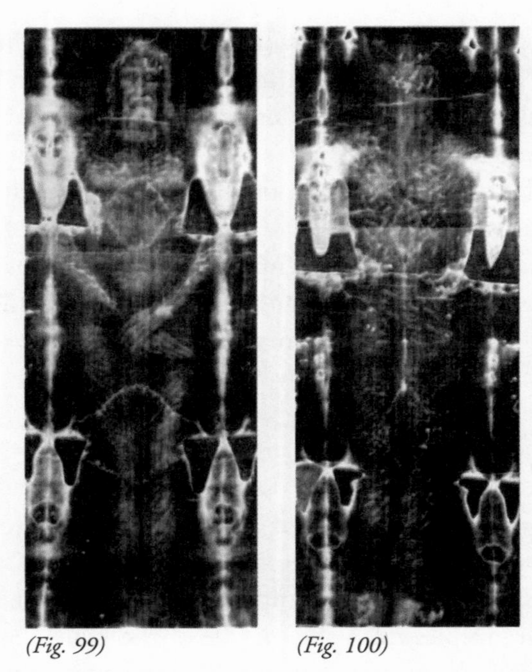

(Fig. 99) *(Fig. 100)*

The largest amounts of radioactive Cl-36, Ca-41 and C-14 atoms would be found on the dorsal body image and its blood marks. The frontal image and its blood marks would contain many more of these radioactive atoms than their surrounding non-image locations.

While there are cloths that are older than 2,000 years, I'm not aware of any that are as large as the Shroud and in as excellent condition. A major reason for its excellent condition was pointed out by Dr. Kitty Little who stressed that penetrating neutrons and perhaps gamma rays released from the body would have strengthened the cloth by passing through it and causing some of the molecular bonds to break and reform in the non-crystalline regions of the cellulose that comprises the linen.[14] (Crystalline regions have specific internal and symmetrically arranged structures.)

As we saw, neutrons would have ricocheted off the bench, walls and ceiling of the tomb, and many would have passed through the cloth again especially from the dorsal side. Although gamma rays would not ricochet within the tomb like neutrons, they would have easily passed through the cloth once. Such repeated breaking and reforming in the non-crystalline areas would cause these molecules to cross-link thus giving this burial cloth greater resistance to solubility, oxygenation and chemical reactions. This type of cross-linking combined with the high crystallinity of good quality linen would account for the Shroud's lack of degradation and contribute in several ways to its excellent condition.[15]

If the Shroud was irradiated with neutron or particle radiation, it could not possibly have originated in medieval times and would necessarily be older than the date indicated by its C-14 dating. Yet, even if a medieval or a 1st century cloth was irradiated in modern times, it would be far too late for such an older, more degraded cloth to have acquired the long-lasting benefits that a new cloth would have acquired from neutron radiation.[16]

Furthermore, if a forger duplicated the amount of neutron radiation that is on the Shroud, he would had to have neutron irradiated the cloth before 1988. Yet, our forger would not have known the organic or indigenous amounts of Cl-35, Ca-40 or N-14 that were found at every location on the Shroud. These amounts would not only vary from location to location throughout the cloth, but will vary from spot to spot on the thumbnail size linen sample. (That is why an average amount for each tested sample must be acquired.) These organic amounts have never been measured from the Shroud (and the techniques to capture and measure the ratios of these organic atoms with their radioactive atoms have never been developed for linen). The forger would not have known how many Cl-36, Ca-41 or C-14 atoms to leave at every location on the Shroud, regardless of its age. Even if he somehow managed to get the Cl-36 and Ca-41 amounts correct in one location, he could not have gotten them correct at all locations, nor would he have left the correct amounts of C-14 atoms.

Cloth and blood samples could be removed from the Shroud of Turin and proven to have been irradiated with neutrons. If these neutrons were distributed in any pattern roughly similar to the calculations of nuclear engineer Robert Rucker, no forger could have reproduced the amounts of Cl-36, Ca-41 and C-14, which would be found in various amounts throughout and within the off-image, body images and blood mark regions of the Shroud of Turin.[17] However, it could be shown that only a neutron radiating event that occurred in the 1st century to a new contemporary cloth could have duplicated these various amounts. Additionally, it could be shown that this neutron radiation miraculously emanated from the length, width and depth of the bloodied, crucified body of the man who was wrapped in this burial shroud.

These conclusions would be demonstrated from the distribution of unfakable radioactive atoms throughout the length and width of the entire Shroud. These conclusions would be the same as those that were reached by extensively studying the Shroud's many unique body image features at the microscopic and macroscopic levels throughout the lengths and widths of both body images. We will see further in Chapter Eleven that this miraculous radiating event can also explain both unprecedented, full-length body images, the still-red, pristine blood marks and many other unique features that are only found on this particular burial cloth.

Testing Limestone and Charred Material

As we saw in Chapter Six, if neutrons emanated from the length, width and depth of the dead body of the man in the Shroud, approximately three quintillion (one billion times one billion)[18] neutrons would have easily passed into and through the cloth and its bloodstains, and ricocheted off the limestone bench, walls, ceiling and floor of the burial tomb. (This is the number with eighteen zeroes in it.) Most of the neutrons released from the body would have penetrated and landed within the limestone structure inside the burial tomb.[19] Most of these neutrons would have landed approximately a foot within the limestone interior; however, they are so penetrating that some could have traveled as much as a meter (39.37") within the interior structures of the limestone tomb.[20]

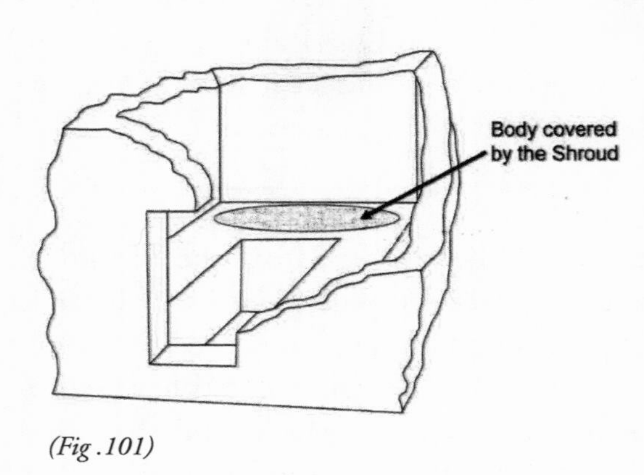

(Fig.101)

Although chlorine (Cl-35) and nitrogen (N-14) are absent in limestone, calcium (Ca-40) comprises about 28% of the weight of limestone in Jerusalem. If such a neutron-radiating event occurred inside a limestone burial tomb, an abundance of radioactive calcium-41 (Ca-41) atoms would have been created within and throughout the benches, walls, ceiling and floor of the tomb. These Ca-41 atoms would exist well above their natural infinitesimal levels. No other burial tomb in the world would possibly contain such natural levels of Ca-41 within it.

Interestingly, carbon-14 (C-14) atoms would also have been created within this tomb. In all of our discussions about creating new C-14 atoms by neutron radiation, we spoke of its creation from N-14 atoms. Without a doubt, this is the most efficient way to create new C-14 atoms. Even though limestone does not contain nitrogen, C-14 can still be created within it by neutron radiation in the same manner that Cl-36 and Ca-41 is created — by neutron capture. C-14 can be created in limestone when a carbon-13 (C-13) atom captures a neutron passing through it. (Although this occurs after the neutron particle has lost some of its energy and speed within the tomb, like all earlier reactions, it happens within a fraction of a second.) Like Ca-41, these C-14 conversions would also occur at known rates within the molecular structure of the limestone and at the same depths within the limestone.

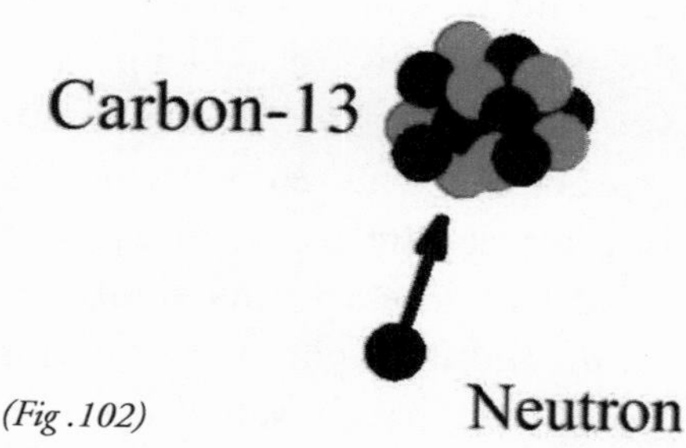

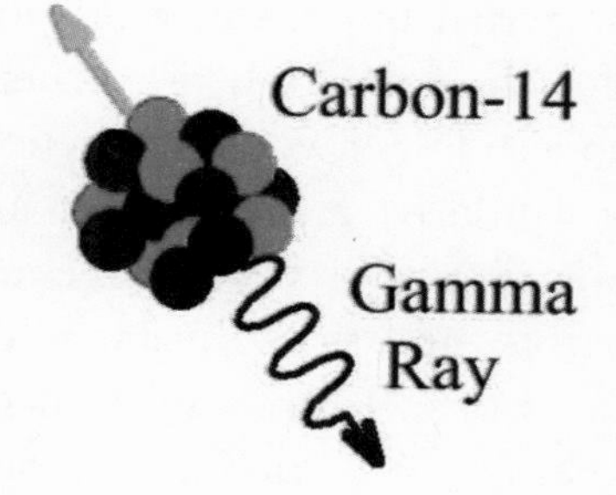

(Fig .102)

Before collision of neutron (black) with carbon-13 nucleus, 6 protons (red) and 7 neutrons.

After collision, neutron is captured, resulting in C-14 with 6 protons and 8 neutrons.

Carbon comprises about 10% of the weight of limestone in Jerusalem. Although C-13 only comprises 1.1% of all carbon, this would amount to about one-tenth of a percent (.10%) of the limestone, which is the approximate amount of nitrogen (N-14) contained in linen. If a neutron radiating event occurred within a limestone burial tomb, C-14, like Ca-41, would also have been left within the benches, walls, ceiling and floor of the burial tomb. No other limestone tomb or rock shelf in the world contains C-14. The only other time C-14 could have been found naturally within a limestone rock shelf was when the limestone rock first formed on the earth millions of years or more ago. All traces of this original C-14 would have long ago disappeared from the limestone's formation.

The only way a forger could possibly have caused Ca-41 and C-14 atoms to appear within the atomic and molecular structures of limestone, let alone deposit them as much as a meter within the tomb, would be if he used neutron radiation, but neutrons weren't even discovered until 1932. In addition, if a forger neutron-irradiated limestone samples within a tomb, scientists today could easily ascertain that the irradiation occurred in the 20th or 21st century. For the reasons stated in the previous chapter, the Ca-41 to Ca-40 and C-14 to C-12 ratios found within each limestone sample would tell scientists whether this unique neutron-radiating event occurred in the first century, or the 20th century, or any century in between.

Although more than one location in Jerusalem claims to contain the actual burial tomb of Jesus, the Tomb of Christ located in the Church of the Holy Sepulcher has the strongest historical claim. If a modern forger would have encoded Ca-41 and C-14 within this tomb, he would have encountered another problem. The remnants of this limestone tomb have been encased for centuries under several layers of marble, and the forger would not have had access to the tomb. If Ca-41 or C-14 is found embedded within and throughout the interior limestone structure of this tomb, it could only have resulted from an unprecedented event.

We strongly recommend that the marble encasing over the Tomb of Christ at the Holy Sepulcher be removed. If parts of the original limestone floor, walls, bench or other interior structures remain, their samples should be tested for the presence of Ca-41 and C-14. Limestone samples from the Garden Tomb, another location considered by many to be the burial tomb of Jesus, should also be tested, as should other limestone rock formations alleged to be the site of Jesus' burial. Detecting Ca-41 and C-14 within the limestone interior of these reputed burial tomb(s) could not only confirm the location where a miraculous neutron radiating event occurred, but could also identify Jesus' actual burial tomb.

(Fig. 103) Two thousand-year-old tomb with circular stone.

The Garden Tomb was not proposed as a possible location until the 19th century. Much of the original rock around this tomb has been carved out or displaced. The Tomb of Christ located within the Holy Sepulcher was also severely damaged in 1009 when Arabs attempted to destroy every trace of it.[21] Professor of Medieval Archaeology Martin Biddle thinks the entire rock-cut roof and much, if not all, of the east and west walls of this tomb were removed in 1009. However, he thinks that much of the south wall and a burial bench to the right of the tomb's entrance and, possibly, part of the original north wall might remain.[22] (Parts of a wall above the bench were also removed and hidden from the crusaders in 1099.)[23] If the Arabs did destroy this tomb, it is difficult to think they destroyed all of the surrounding walls to the depth of a meter; even if they could have destroyed the floor to this depth and completely destroyed the ceiling and the typical three-sided benches within the tomb. There is a reasonable chance that some of the original limestone still remains from the tomb's interior walls, or from a meter within their original interior surfaces, from which Ca-41 and C-14 could be detected and measured. These radioactive atoms could not only demonstrate that a neutron radiating event occurred within this tomb, but when this radiating event occurred.*

Limestone samples should be removed from all the reputed burial tombs of Jesus and tested for the presence of Ca-41 and C-14. If these extremely rare atoms were not detected within limestone samples from the remnants of these tombs, it would only mean that these particular samples were not irradiated with neutrons. However, if Ca-41 or C-14 were found within one of these limestone samples, it would indicate that an unprecedented neutron radiating event occurred within the particular burial tomb from which it was extracted. If neutron radiation was also found to have irradiated the Shroud's linen and its blood samples in a manner that correlated with their distances from and their positions upon or under the body wrapped within the cloth, then it would confirm that this dead body was the source of this miraculous radiating event. Such an event would further confirm that the dead body of the man in the Shroud was Jesus Christ and that this particular location was his original burial tomb. The occurrence of such a miraculous event within the reputed burial tomb of Jesus Christ would also be consistent with the Gospel accounts of his resurrection.

The above testing of limestone samples would be very similar to the types of testing that could be performed on the charred materials removed from the Shroud in 2002. While chlorine would be found in the Shroud linen, it would not have survived in the parts of the cloth that were charred in the fire of 1532. However, calcium and carbon would survive such a burn and remain within charred material. If the Shroud was irradiated with neutrons, Ca-41 and C-14 would both be found within its charred material.

Radiocarbon laboratories burn linen and other material in order to

***Even though these remaining radioactive atoms were created as much as a meter within the limestone walls, the Ca-41 to Ca-40 ratios that are above their natural infinitesimal levels would reveal the number of neutrons that each sample received. The remaining C-14 atoms within these irradiated samples (revealed by their C-14 to C-12 ratios) would reveal when the neutron radiating event occurred.**

measure their carbon. (This burned linen product is similar to the charred linen material caused by the fire of 1532.) Most, if not all, of the Shroud's C-12, C-13, C-14, Ca-40 and Ca-41 would still remain within the charred material taken in 2002 from behind the patches that covered its burn holes. If accurate, Ca-41 to Ca-40 and C-14 to C-12 ratios can be measured from irradiated control linen samples* reduced to charred material, then such tests could also be performed on the Shroud's charred samples. Like the limestone from Jesus' reputed burial tomb, the charred material could only be tested for Ca-41 to Ca-40 and C-14 to C-12 ratios. Since Cl-36 to Cl-35 measurements can be detected in smaller samples than Ca-41 to Ca-40 measurements, they are best suited for testing precious and limited linen and blood samples removed from the Shroud. Fortunately, large quantities of charred material from the Shroud exist and relatively large amounts of limestone could still be present from Jesus' reputed burial tomb(s).

(Fig.104) 42 small vials of charred material have been kept in Turin since they were removed from the Shroud in 2002.

As we saw in the last chapter, small cloth and blood samples could be tested from off-image locations and from both the frontal and, especially, the dorsal body images on the Shroud. The various amounts

***Many tests initially performed on irradiated control linen samples reduced to charred material in a closed container (as occurred to the Shroud in 1532) could measure the Ca-41 to Ca-40 and C-14 to C-12 ratios in the remaining char. These ratios could be compared to the Ca-41 to Ca-40 and C-14 to C-12 ratios in the same original irradiated control linen.**

of Cl-36 atoms contained within each sample could demonstrate, among other things, that the length, width and depth of the corpse wrapped within the Shroud was the source of the neutron or particle radiation. While some of this atomic evidence may be contained within the charred material and within one of Jesus' reputed limestone tombs, they would not contain as much direct information regarding the source of the radiation as would the above cloth and blood samples.

The charred material came from behind eight pairs of patches seen in Figs. 105 and 106 that covered burn holes on the cloth.* Four pairs of patches were located near the front and back of the man's knees and four more pairs were located next to the front and back of the man's torso. The latter four sets of patches were near the areas where the top part of the man's arms, his elbows, and, perhaps, part of his torso were encoded on the Shroud before 1532. Although the charred material would have moved around under these patches, this material could contain higher amounts of Ca-41 and C-14 than the charred material removed from behind the patches at the sides of the knees. If charred material taken from the back of the man's arms (on the dorsal side of the cloth) could be tested, it could contain more Ca-41 and C-14 than the other charred materials. That is because more neutrons would have ricocheted onto the dorsal body image and its adjoining areas of the cloth.

An analysis of neutron radiation on the Shroud could provide valuable information concerning Christ's tomb. The remaining parts of the north and south walls within the Tomb of Christ in the Holy Sepulcher may not contain any of their original interior surfaces. If Ca-41 and C-14 were found within the limestone interior of the remnants of this tomb, it may be difficult to determine the original width of the wall or the precise location of the source of the neutron radiation. However, the amount of neutron radiation emitted from the body and distributed throughout the entire Shroud can be established from its various cloth, blood (and charred) samples. From this information, the various locations of the body on any

***Although the Shroud's patches were removed in 2002, we have continued to use the full-length images taken by STURP photographer Vernon Miller in 1978, and not the very professional-quality photos taken after the patches' removal in 2002. STURP's 1978 photographs still seem to provide the best contrast and context. This may be due, at least in part, to the presence of the patches, which appear to provide more of a framework for the torso and the lower limbs of the body.**

of the three benches within a typical Second Temple tomb, as well as the corresponding distribution of neutrons and their depths of penetration within the interior, can be calculated and estimated. These amounts can be compared to the amounts and locations of Ca-41 and C-14 from the remnants of the Tomb of Christ. Not only would the location of Jesus' burial tomb be known, but the position of the body and the amounts and patterns of neutron distribution within this original tomb might also be indirectly confirmed.

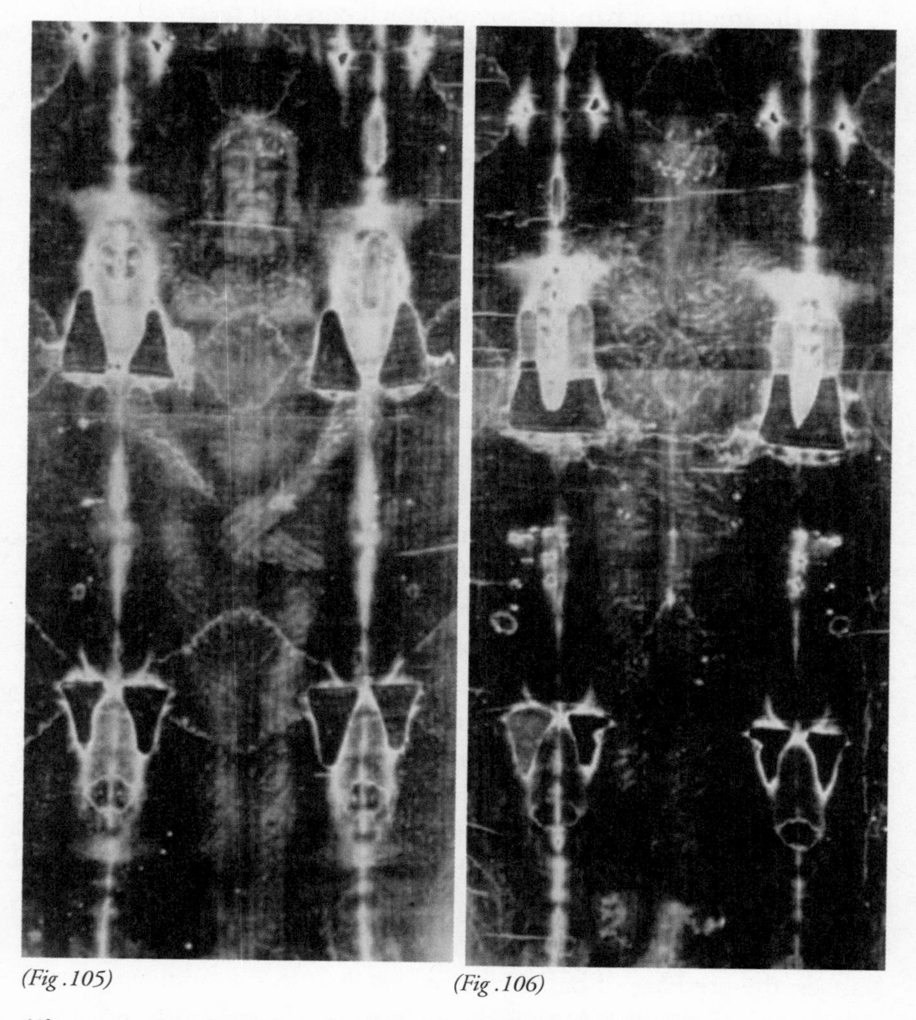

(Fig .105) *(Fig .106)*

(Above images show Shroud frontal and dorsal images framed by eight sets of patches.

When the results from the atomic testing performed on samples from the Shroud linen cloth, its blood marks, its charred material (and on limestone from the reputed tomb(s) of the historical Jesus) were combined with all the results from the previous investigations, this extensive and objective evidence could prove that:

1. the entire Shroud was irradiated with particle radiation;
2. the Shroud's radiocarbon dating is erroneous;
3. the amount of particle radiation each part of it received;
4. the Shroud is from the 1st century;
5. the radiating event happened in the 1st century;
6. the source of the radiation was the length, width and depth of the dead body wrapped within it;
7. all of the events that occurred to the man happened in the 1st century;
8. the events occurred in Jerusalem;
9. under all of the surrounding circumstances described in the Gospels; and
10. this man was the historical Jesus Christ.

These tests, and those discussed in the next chapter, are just part of a new series of scientific tests and examinations that should be conducted on the Shroud of Turin. A growing number of scientists contend the Shroud's body images, its blood marks, off-image and outer side should be examined at the molecular and atomic levels. These new test results could not only provide more extensive and definitive evidence that this cloth is the burial garment of Jesus Christ, but that an unprecedented and miraculous event occurred to his dead body. This miraculous event occurred *after* he incurred a series of wounds, was crucified, killed and wrapped in a burial shroud under all the same circumstances of time, place, people and instruments as attributed to the historical Jesus in the Gospels. The reader should keep in mind that neither the frontal and dorsal body images; the 130 blood marks; all of their unique features; the unprecedented radiating event; the surrounding circumstances; the presence of radioactive atoms that do not occur in nature; their proportional distribution throughout this burial cloth; nor most of its extraordinary evidence can be faked or encoded in any kind of artistic or naturalistic manner.

All of this evidence appears to have been encoded as a result of this miraculous event. While the next chapter will discuss how even more extensive evidence can be acquired from the Shroud at the molecular and other levels, Chapters Eleven and Twelve will provide the best explanations to date as to how all of this unique evidence was encoded by this unprecedented and miraculous event.

8

MOLECULAR AND OTHER INNOVATIVE EXAMINATIONS OF THE SHROUD

Since (**a**) *trillions* of unique radioactive atoms would be distributed throughout the Shroud's linen, blood and charred material; (**b**) that can *only* be created from neutron or particle radiation; and (**c**) their source was the crucified dead body wrapped within it — extensive scientific evidence would exist for the first time in history for the occurrence of a miraculous event. However, even more scientific evidence can be acquired from the Shroud to confirm the occurrence of this unprecedented event.

New technology exists, which could be further developed and adapted, that could also examine the entire Shroud and strategic samples removed from it at the molecular and elemental or chemical levels. This new technology could conceivably scan the entire Shroud in a matter of hours, thereby allowing scientists to spend years analyzing all of its data. This non-destructive technology could map the entire cloth and its samples, identifying not just every fiber of every thread, but also what is on every fiber. It could also extensively examine materials from and on the Shroud such as the blood marks, charred material and the scorched or water stained areas.

When humans see objects, we perceive only the colors reflected from them in the visible spectrum from red to violet. Below red lies the infrared part of the spectrum and above violet lies the ultraviolet. These energies consist of electromagnetic waves with frequencies and wave lengths that are higher and lower than visible light. Although these forms of radiation are invisible, they allow us to identify and determine the chemical composition of unknown materials that they irradiate. We obtain this information because every chemical compound reflects, absorbs or emits infrared and ultraviolet radiation in its own unique spectrum that is routinely used to determine the chemical composition of unknown materials. Previously, one of two procedures was used by scientists to obtain spectral information from an object under infrared radiation. They could obtain a complete spectrum at only one point on the object and repeat for all points of interest or obtain a complete image of the object at only one wavelength within the infrared spectrum and repeat for all special wavelengths of interest.

New technologies such as multi-spectral imaging and molecular microscopy continue to develop that allow objects to be viewed under the entire visible, ultraviolet and infrared light spectra simultaneously, allowing their composite images to become visible from all spectra simultaneously. Cary 620 FTIR microscopes claim to provide focal plane array imaging that enables the collection of hundreds to thousands of spectra simultaneously. Its measurement modes also include transmission (back lighting), reflection, attenuated total reflectance and grazing angle. Agilent Technologies, which makes these instruments, claim they deliver unmatched imaging for biomedical and materials research including applications to polymers, chemicals and forensics.[1] (Cellulose, which comprises linen, is a naturally occurring polymer.) In these ways, the Shroud linen, its numerous blood marks, and the various limestone and other materials it has collected over the centuries can be examined at the molecular level, allowing their individual chemical or elemental compounds to be identified.

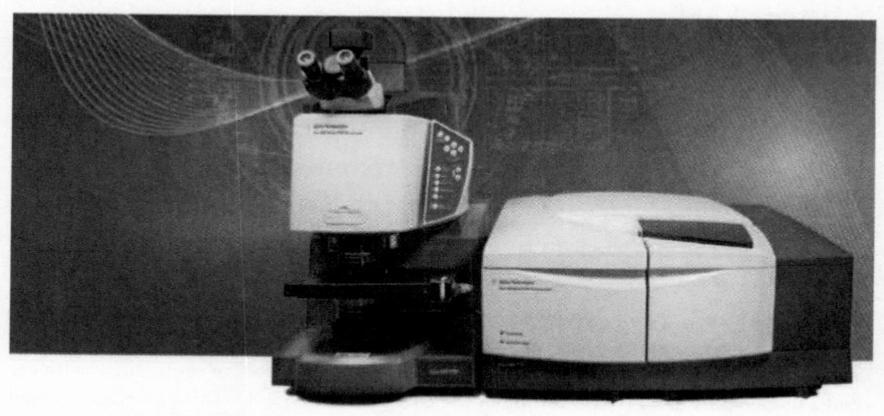

(Fig.107) Cary 620 FTIR microscope and imaging system

A Cary 620 FTIR microscope and imaging system's viewing mechanism could possibly be adapted by placing it on an arm or in a stable position that moves over the length and width of the entire Shroud. If this is not feasible, perhaps the entire length and width of the Shroud could be slowly slid or moved through the viewing aperture while it lays outstretched (or on a thin, lightweight surface). If so, the entire linen, all of its threads and fibers, its many blood marks, pollen, limestone and any of its other materials could be examined at the molecular, chemical and/or elemental levels.

This non-destructive technology should also be adapted and applied to the entire outer surface of the Shroud, especially behind the facial and hand regions of the frontal image.* As we saw, images at these locations were identified by computer enhancement of the outer surface of the Shroud after its 16th century backing cloth was removed in 2002.[2] Cloth discolorations at these locations are possible with the two radiant cloth-collapse image forming methods involving particle and ultraviolet radiation, as well as a corona discharge method. These will be discussed in Chapter Eleven. Only these radiant hypotheses have accounted for these features. If broken chemical bonds or conjugated carbonyls are more prevalent at these locations on the outer side of the cloth, this could confirm the presence of radiation at the time of image formation.

Multi-spectral imaging techniques might be able to confirm the oxidized, dehydrated state of the Shroud's cellulose and the extent of its double-bonded or conjugated carbonyl groups on the cloth's inner side. While these features occur naturally with aging, they should be far more extensive on the superficial body images, where they were accelerated by proton or other radiation from the body, than on the non-body images. Furthermore, these techniques might also ascertain whether the molecular bonds of the cellulose had been broken and reformed in the non-crystalline regions. This last feature is accounted

*** X-ray fluorescence, X-radiography, transmission and all appropriate forms of nondestructive photography and examination of the entire outer side of the Shroud should also be undertaken with its backing cloth removed. The backing cloth interfered to some extent with the results obtained from STURP's testing of the entire cloth in 1978. This interference could not only be eliminated, but new information could be obtained from the outer side.**

for by neutrons, electrons and gamma rays within the radiation. This feature should be most prevalent in the mid-back regions of the dorsal body image than in other locations on the Shroud. (This location is not only near the center of the body's mass, but more neutrons would have ricocheted onto the dorsal side of the cloth.) These assessments from the Shroud's body image and non-image regions should also be compared to existing 650 and 2,000-year old linen samples.

Modern control linen samples could be irradiated with various amounts of protons, long and short wave X-rays, ultraviolet light and electrical or corona discharge and examined under multi-spectral imaging to ascertain whether the last three forms of radiation also tend to deposit their energy superficially and uniformly like protons. The control samples could also be irradiated in various combinations and amounts of neutrons, protons, alpha particles and gamma rays. All of these irradiated samples should also be artificially aged and compared to the Shroud's body image and non-body regions under multi-spectral imaging and by all the other scientific Shroud measurements acquired in 1978 and thereafter. All naturalistic and artificial encoding techniques could also be applied to control linen, which is then artificially aged and examined under multi-spectral imaging and compared to the Shroud in all the above manners.

Other Contaminants on the Shroud

Multi-spectral imaging or molecular microscopy could also identify substances such as wax deposited on the Shroud. Dr. James Chickos, Professor of Chemistry at the University of Missouri, St. Louis, found that tallow, which candle wax is made from, can become chemically bound to cellulose by a process known as transesterification.[3] Tallow contains a long carbon chain. If it becomes attached to the glucose molecular structure of the linen, it would naturally add carbons and alter the appearance of the cloth's actual age.

(Fig.108) Wax visible on the Shroud.

Wax has been identified by STURP scientists on samples

they removed from the Shroud,[4] and can also be seen with the naked eye (Fig. 108) at some locations on the cloth itself. Photomicrographic investigation discovered wax, loosely adhered and not visible to the naked eye, on one of the remaining Shroud samples that were carbon dated in 1988 at the University of Arizona.[5]

STURP chemist, Dr. Alan Adler, positively identified wax from the edge of one of the relatively small round burn holes located off both sides of the back of the man's thighs. This pattern of burn holes was first observed in a painting of the Shroud in 1516. While this pattern of "hot poker" burn holes will be discussed further in Chapter Ten, many people think they could be from cinders falling onto the folded Shroud from an incense container swung during a religious ceremony. Candle wax would not only have been convenient, but would also have been an excellent substance with which to stiffen cloth and support it. This would have kept the cloth from fraying or tearing at the delicate edges of these burn holes and could have been applied intentionally. Tallow or candle wax that is chemically bound to the molecular structure of the linen would not be visible to the naked eye. Furthermore, according to Dr. Chickos, a significant amount of it would survive the normal pretreatment cleaning processes used on samples prior to the dating.

Starch, like wax, could also stiffen and support cloth and has been identified on a thread from the Raes sample located immediately next to the 1988 radiocarbon site.[6] The molecular structure of starch is very similar to that of cellulose. Starch could chemically bind to cellulose, not be detected with the naked eye, and not be removed by standard pretreatment cleaning processes. Starch could also make an ancient cloth appear much younger. Even a small amount of chemically bound starch or wax could affect the dating results of a sample by many centuries, especially if either was applied in the cloth's recent past. For instance, STURP sent a thread from the Raes sample to be radiocarbon dated in 1982. This was undertaken without permission from the cloth's custodians or knowledge at the time that the thread contained starch. Intriguingly, one end of this thread dated to A.D. 200 while the opposite end – containing starch – dated to A.D. 1000.[7]

Although the facility that dated this thread was not a dedicated laboratory that regularly carbon dated samples, its results demand attention. If one end of the Shroud thread dated to A.D. 200, the error range of the dating could place the Shroud in the first century. The presence of starch could explain why the other end of the Shroud thread dated to A.D. 1000. Since starch was located near the 1988 Shroud sample site, this and all future sampling locations on the Shroud should be thoroughly investigated. The application of multi-spectral imaging to this and all sites on the Shroud could also tell us what other locations may have such foreign substances physically or chemically attached to the linen, and thus, what sampling area or areas to avoid.

As discussed previously, traces of cotton were found by Professor Raes in one of his samples. Cotton has also been observed on other threads and fibers taken from the Shroud's radiocarbon site and other samples taken immediately next to this location or the side seam on the cloth.[8] This cotton could have derived from the Shroud having been woven on a loom that previously wove cotton fabrics.[9] (STURP scientists examined the entire Shroud while wearing white cotton gloves and this, too could be responsible for the trace presence of cotton anywhere on the Shroud.)

Textile expert, Mechthild Flury-Lemberg, who has examined the Shroud's texture and construction more than anyone, concludes that "the narrow added strip has been proven by the weave analysis to come from the same bolt of fabric as the large section [of the Shroud] and would not have been available at a later point in time."[10] She states that ancient weaving mills wove cloth as much as three times wider than the Shroud, dividing it in thirds and adding strips, if necessary, to make cloths a certain width. Since the Shroud lacks woven borders that are different in structure or material that would indicate the start or end of a woven fabric, and its lengthwise seam and the hems at its short edges have been sewn with the same professional competence, Flury-Lemberg concludes the Shroud was completed in one working process.[11]

The side seam was added to the cloth with the same professional competence as the rest of the finely woven linen. The seam and stitches were designed to be as unnoticeable as possible on the face of the cloth. The fabric was not only folded onto the back, but was sewn from the back, causing the seam to be flat on the face with the stitches hardly noticeable, as seen in Fig. 109.

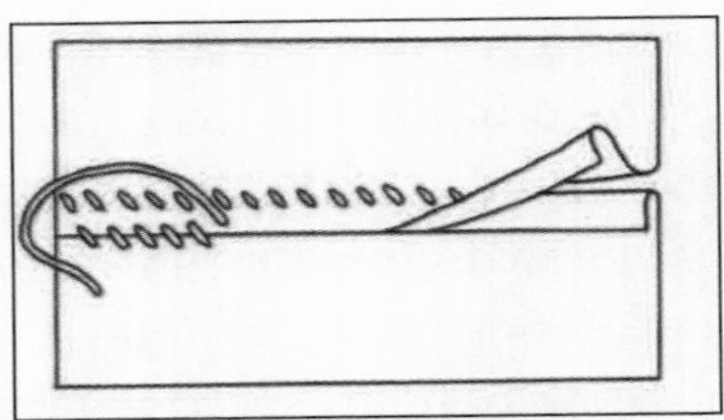

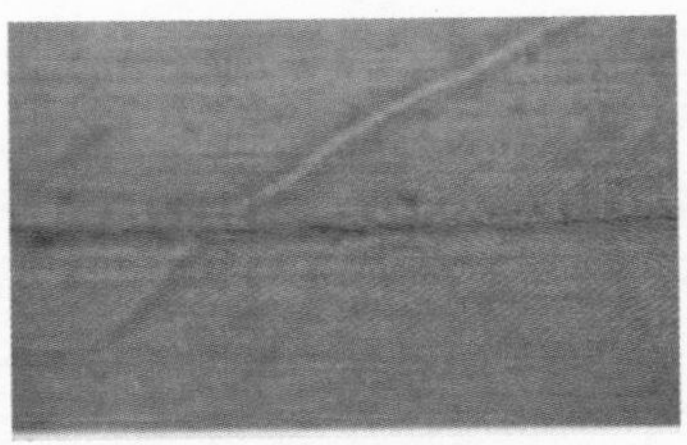

(Fig. 109, left and Fig. 110, right) Cl-36 to Cl-35 and C-14 to C-12 measurements should also be made on the sewing threads at the Shroud's side seam.

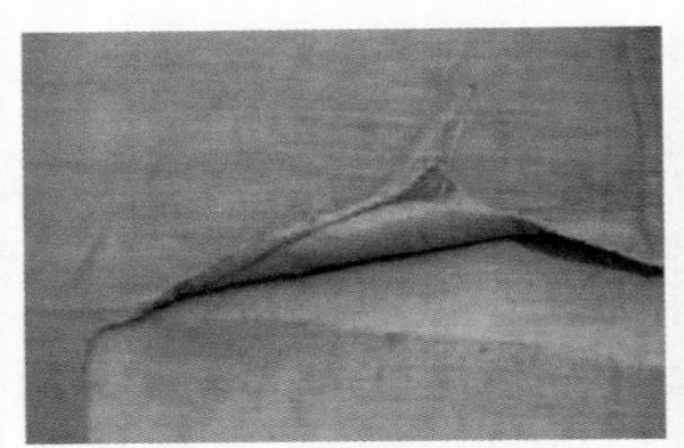

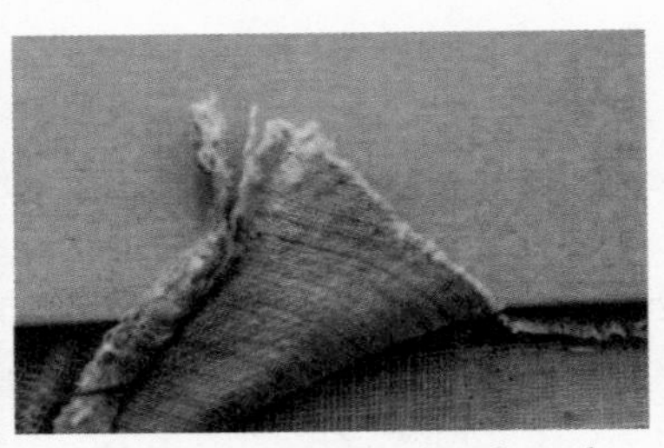

(Figs. 111, left and 112, right) The actual age of the Shroud's sewing threads on the left could be similar to those on right recovered from the fall at Masada in 73 A.D. Both textile parts have the same raised or rolled effect on the reverse side.

Although this would cause the reverse or outer side of the Shroud to have a raised or rolled effect, this same technique has been found among textile fragments from the fall of Masada in 73 A.D. Flury-Lemberg concludes that the Shroud of Turin displays numerous indicators of its original high quality production by textile workers of the first century.[12]

Multi-spectral imaging of the Shroud could provide valuable information on the extent and location of starch and cotton on the Shroud and whether their composition is modern. Radiocarbon dating the threads sewn at the side seam from the back of the cloth could also provide insight as to when these particular threads may have been sewn onto the Shroud. In case these threads have been present since the Shroud was originally manufactured, they, too, should first be non-destructively tested for their indigenous amounts of chlorine (Cl-35), calcium (Ca-40) and nitrogen (N-14).* Their Cl-36 to Cl-35, Ca-41

***Chlorine and calcium can now be measured in small samples nondestructively by X-ray fluorescence. N-14 cannot be measured in this manner. However, a small part of a thread (aliquot) can yield the N-14 content by giving it an extremely high dose of neutron bombardment, whose extremely elevated C-14 to C-12 ratios would reveal its N-14 content.**

to Ca-40, as well as their C-14 to C-12 ratios should all be measured.* If these threads were present when the Shroud was originally manufactured, they could also have been present if and when an unprecedented event involving particle radiation occurred while the cloth wrapped the crucified, dead man. If so, these threads could contain elevated amounts of radioactive Cl-36, Ca-41 and C-14 atoms. If these threads were present when such an event occurred, they should have received similar amounts of neutron radiation along the entire seam as the original Shroud linen adjacent to the seam received. Since the sewing thread is plentiful, Ca-41, Cl-36, C-14 and N-14 testing should be performed on it. If it was on the cloth since its original manufacture, it, too, could reveal its age, whether the neutron radiation came from the length of the body within the cloth and when this miraculous event occurred.

Recent Shroud Dating

Perhaps the most scientifically-based and accurate testing of the age of the famous Shroud of Turin was conducted at the University of Padua by scientists led by Professor Giulio Fanti. Based on chemical and mechanical changes that take place in linen due to aging, they applied three scientific tests that dated various Shroud samples to the first century.[13] These scientists tested control linen samples that ranged from 3250 B.C. to 2000 A.D. and originated from quite dry environments. Using FT-IR/ATR spectra and FT-Raman Spectra, scientists were able to find a correlation between the spectroscopic properties and the molecular structures within these samples. Using an IR spectrometer and laser lighting they were able to demonstrate a correlation with the age of linen material and its ability to reflect light.[14]

Testing samples by infrared radiation was briefly discussed earlier in this chapter and reference was made to ATR (attenuated total reflectance), whereby an infrared beam can be sent at a certain angle through a highly refractive crystal, creating an evanescent wave that

---

***Unlike the ancient manufacturing process for fine woven linen, the plant material comprising the sewing threads may not have been retted. Thus, X-ray fluorescence measurements may confirm that an insufficient amount of organic Cl-35 is present in the sewing threads. In this case, Ca-41 to Ca-40 testing, which is more practical for testing charred material and limestone should be considered for the sewing threads.**

penetrates a few microns beyond the crystal into the tested sample. In regions of the IR spectra where the sample absorbs energy, the evanescent wave will be attenuated or altered. (This altered energy will again pass through the highly refractive crystal to the IR beam, which then exits the opposite end of the crystal and passes on to the detector in the IR spectrometer.) This system can generate single or several reflections as seen in Fig. 113 and improves traditional spectral acquisition and reproducibility with more precise material verification and identification.

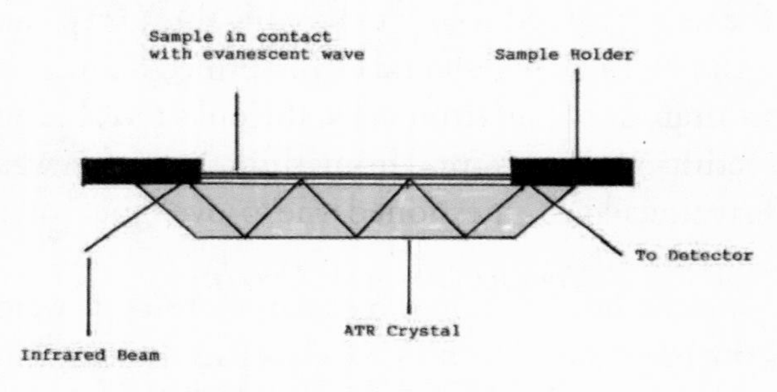

(Fig. 113) Diagram of an ATR reflection system

While FT-IR/ATR techniques provided the best correlations and results, FT-Raman spectra and mechanical testing also yielded similar results. Although these techniques are new to the field of dating, their application and accuracy continue to be refined and improved. Machines to measure the weight bearing load for threads are readily available; however, Fanti had to construct his own equipment to mechanically measure the load bearing strength of fibers. With this new equipment, Fanti was also able to demonstrate a connection between the age of fibers and their strength.[15] When the non-destructive data and spectral information was acquired from the Shroud's samples and compared to those from the control samples, the Shroud samples dated to 33 B.C. +/- 250 years. This date, of course, is well within the range of the life of Christ and is consistent with all other scientific test results, as well as an overwhelming amount of medical, archaeological and historical data derived from this burial cloth.

Although some aging effects on the Shroud of Turin are accelerated on its frontal and dorsal body images, probably due to radiation, the Italian scientists also tested samples from off-image areas on the Shroud.

It must be noted, however, that although radiocarbon dating has been utilized much longer than these new techniques, it has been scientifically demonstrated that if the Shroud was irradiated with neutrons, any C-14 or radiocarbon dating results would be off by many centuries. If neutrons irradiated the Shroud, the accurate determination of its age would not be affected by these new dating methods. Even though these new techniques and the age they ascribed to the Shroud have been published since 2013, I'm not aware of any criticism that they have received regarding their scientific merits. Scientists that I interviewed were impressed with the consistency in the data, especially the results obtained by infrared testing. Ironically, the only public discussion that the dating results have received from sindonologists was from two Shroud institutions that questioned the provenance of the Shroud samples that were dated.[16]

If multi-spectral imaging and molecular microscopy were performed on the cloth regions from which these samples were removed, as well as the entire Shroud linen itself, and compared to the above Shroud samples, this data would readily confirm whether these samples were removed from the Shroud of Turin. The elemental, molecular and chemical composition of these various samples should easily resolve any questions as to whether they originate from the famous Shroud of Turin.

If multi-spectral imaging was also performed on the remaining Shroud samples at Arizona, Oxford and Zurich and compared to the above imaging results, they could easily resolve the question whether any invisible reweaves or other repairs occurred at or near the radiocarbon site. This non-destructive analysis could easily resolve whether scientists at the University of Padua and those at the AMS laboratories at Arizona, Oxford and Zurich have actual Shroud samples in their possession. I am quite confident that the scientists at each of these institutions all examined samples that originated from the Shroud of Turin.

If coin and flower images were encoded on the Shroud, they would provide a far more specific indication of when they were encoded and where the encoding event took place, than could most other scientific tests. Particle radiation emanating from the body wrapped in the Shroud is the only hypothesis that can account for these images on the cloth. When neutrons hit the nucleus of copper, the primary component of ancient bronze coins, or hit heavier elements in flowers such as iron, calcium and potassium, they can absorb the neutron and give off protons, alpha particles, deuterium or low energy gamma rays. Each of these first three particles would encode superficial images of coin or flower features. Gamma rays from all four heavier elements could, in turn, cause long-wave X-rays to be given off that could encode superficial images, or cause short-wave X-rays or visible light to be given off that could encode coin or flower imaging more than two or three fibers deep.

(Fig. 114) Possible coin features on Shroud.

(Fig. 115) Possible flower image.

Multi-spectral imaging could help determine whether oxidized dehydrated cellulose comprised of conjugated carbonyl groups are found at the possible coin features and flower images on the Shroud of Turin, and whether this discoloring is superficial or extends further into the cloth. This discoloration could be compared to the molecular composition of other Shroud body images or irradiated control linen samples. Multi-spectral imaging could determine whether discoloration is even found at the possible coin and flower image sites.

Other hypotheses to explain the Shroud's radiocarbon dating have garnered some attention over the years such as the biofractionization hypothesis [17] and the bioplastic coating hypothesis; [18] however, both of these hypotheses have been investigated, tested and rejected by scientists. The former explanation hypothesizes that when the Shroud was folded in its reliquary during the fire of 1532, the C-14 and C-13 content within the cloth not only greatly increased, but each reached their respective peaks after approximately 1-1/2 – 2 hours. The latter hypothesis claims that a coating of bacteria and fungi built up throughout the entire Shroud that not only explained its radiocarbon dating, but also the Shroud's unique images.

While both of these hypotheses were easily refuted by various experiments, previous examinations of the Shroud and by recent and past examinations of its samples (see Chapter Five, Appendices E and F of *The Resurrection of the Shroud*), multi-spectral imaging and molecular microscopy could also shed further light on these hypotheses. Multi-spectral imaging and atomic testing could test every hypothesis ever proposed to explain the Shroud's medieval radiocarbon dating and its unique body images. Multi-spectral imaging and atomic testing performed on the Shroud of Turin and its various samples could answer all of the outstanding issues regarding this famous burial cloth.

Development of Multi-Spectral Imaging Technology and Techniques for Atomic Testing

In the last few years multi-spectral imaging and molecular microscopy have made great advances. Market forces have driven this development because there are many areas of science where the discerning and identification capabilities of this new technology are of value and importance. As we discussed earlier in this chapter, it must be adapted so that it can examine every part of the large Shroud.

Because testing at the atomic level could prove that a miraculous, radiating event actually occurred to the dead body of the crucified man in the Shroud, and even indicate when, where and to whom it happened, it is clearly the most exciting and critical area of scientific testing. While atomic testing has been performed on relatively solid materials such as limestone, it needs to be adapted to less solid materials such as linen, blood and charred material. One of the keys for advancement in this area of research lies in capturing all the radioactive chlorine (Cl-36) and calcium (Ca-41) atoms within the neutron irradiated material, along with the natural chlorine (Cl-35) and calcium (Ca-40) atoms from which they convert. Once they're all captured, their ratios can be measured at an AMS facility. Since only the Shroud's linen cloth was C-14 dated in 1988, and one of the primary objectives of atomic testing is to verify or refute this particular dating, we will discuss the techniques that need to initially be developed for linen. In addition, since Cl-36 to Cl-35 ratios can be detected in smaller amounts of material than can Ca-41 to Ca-40 ratios, we will discuss developing new techniques to measure Cl-36 to Cl-35 ratios in linen.

The development of atomic testing techniques has not occurred as rapidly as multi-spectral imaging. Part of the reason is that identifying radioactive isotopes within neutron irradiated material is not a common scientific problem to which market forces have needed to respond. Since the amounts of radioactive atoms (or isotopes) that will be created within neutron irradiated material have long ago been established, there is also less scientific intrigue. One of the primary reasons to develop these techniques will be to undertake the challenge of testing the Shroud of Turin. In order to reach chemists, donors and interested members of the public, I have sketched out the problem in layman's terms in Appendix A so that interested parties can take up the challenge.

There is no need for chemists with the proper facilities to be afraid to undertake these experiments. The work is not that sophisticated. Diligence and persistence may be the most important ingredients next to making sure the experiments take place under the right conditions in the proper facilities. The chemists will know if they are capturing all of the normal and radioactive chlorine and if the AMS laboratories are accurately measuring these ratios within the irradiated control linen samples. This is because the rates of conversion to radioactive atoms are well-known scientifically. Scientists can first non-destructively measure the organic amounts of chlorine by X-ray fluorescence analysis, which will vary in each control linen sample.[19] Since scientists will also know the amounts of neutron flux or radiation that will be given to each of their control samples, they will also know the small fraction of radioactive atoms (Cl-36 and Ca-41) that will be created in each irradiated sample.

(Fig. 116) AMS facilities can measure C-14 to C-12, Cl-36 to Cl-35 and/or Ca-41 to Ca-40 ratios in linen, blood, charred material and limestone.

Too often in the last several years I have been unable to find chemists who are willing to undertake such work, especially if they realize the work could have important implications with the Shroud of Turin. Too often, they know only of the Shroud's radiocarbon dating.

These scientists are unaware that the Shroud's C-14 dating is the only scientific test result that is inconsistent with its authenticity as Jesus' burial cloth, and that this dating result is probably incorrect. They are unaware that radiation is the only agent that can explain the more than 30 unique or extraordinary features found solely throughout both body images located on the inner sides of the cloth that wrapped this body. They are also unaware that the source of this radiation could only have been the severely wounded, crucified corpse that was wrapped within this cloth. Neither scientists nor the worldwide public have ever had to deal with this kind of objective, independent and unforgeable evidence before. Yet, if such unforgeable evidence is located throughout the man's unique body images, his unparalleled blood marks, the rest of this burial cloth, the charred material and Jesus' reputed burial tomb, then people throughout the world would have an inherent human right to know of this information.

As we saw in the last chapter and in this one, the amount of neutron radiation dispersed throughout the cloth will vary in direct proportion with its closeness to and with its positions over the body at the time of this event. The cloth's C-14 (along with its Cl-36 and Ca-41) content will also vary accordingly throughout the cloth. C-14 or radiocarbon "dating" the Shroud again without concurrent Cl-36 and Ca-41 testing, would yield a vast range of meaningless and inaccurate dates. Until atomic testing techniques for Cl-36 and Ca-41 are first developed and perfected with control linen, blood and charred samples and then applied to the Shroud, any additional C-14 dates obtained from this cloth will *necessarily* be varied and erroneous.[20] However, if atomic testing has been developed, perfected and applied to the Shroud's linen, blood and charred material, then the samples' various C-14 to C-12 ratios combined with its Cl-36 to Cl-35 or Ca-41 to Ca-40 ratios will have extraordinary meaning and relevance. The variety of C-14 to C-12 ratios (as well as Cl-36 to Cl-35 and Ca-41 to Ca-40 ratios) will confirm that this burial cloth has been irradiated with particle radiation in direct proportion with its closeness to and with its position under and over the body at the time of this miraculous event.

Protocol Summary for Testing the Shroud at the Atomic Level

Multi-spectral imaging and molecular microscopy, which will reveal the most extensive information ever acquired from the Shroud, along with X-ray fluorescence testing (that can non-destructively measure chlorine and calcium) should initially be performed on the entire Shroud with its backing cloth removed. This extensive information should be studied in combination with Robert Rucker's MCNP codes indicating the distribution of neutrons throughout this burial cloth. These detailed studies will help scientists choose the best locations from which to remove cloth and blood samples for Cl-36 to Cl-35 and C-14 to C-12 measurements.

The first two procedures will also identify limestone remnants between the threads of either side of the cloth. Since chlorine is not present in charred material or limestone, Ca-41 to Ca-40 and C-14 to C-12 ratios should be measured from the Shroud's charred material (which has already been removed) and from any limestone on the Shroud itself and from Jesus' reputed burial tombs.

Once these cloth and blood samples have been selected and removed from the Shroud, the cloth samples should be thoroughly rinsed and cleaned so that all inorganic elements or materials have been removed from the cloth samples. The remaining organic amount of chlorine on the small, cleaned and rinsed cloth sample should be non-destructively measured by X-ray fluorescence (even though this amount should also be revealed in the Cl-36 to Cl-35 ratio). The amount of organic N-14 in the cloth sample must also be measured in order to calculate the age of the samples and when the neutron radiating event occurred.[21] The organic or indigenous amounts of chlorine and nitrogen in dried human blood should be the same for all people.[22]

Numerous small lengths of the *individual* warp and weft threads from throughout this entire cleaned sample[23] can then be thoroughly mixed together for Cl-36 to Cl-35 and C-14 to C-12 testing. In this way, the *same* sample can reveal how much neutron radiation it received and how many additional C-14 atoms were created within it, both of which are key in determining when the event occurred and the age of the sample. This is a better method than using two samples side by side.[24] For one thing, it may conserve material and further, the amounts of neutron radiation and the chlorine and nitrogen contents will vary from sample to sample.

Additional Testing

Extensive testing conducted on the entire Shroud linen and its various samples, along with physical, mathematical, medical and chemical examination of the frontal and dorsal body images and their corresponding blood marks, consistently show that a real human victim, who died in the vertical position from a crucifixion, was once wrapped in this burial cloth. Some of these tests included DNA and chromosome studies in 1997 on two blood samples from the back of the man's head.[25] These studies merely confirmed that this victim was a human male, which was actually first apparent when the Shroud was photographed in 1898. DNA and chromosome analysis have advanced considerably since this limited study in 1997. So have other testing techniques that were performed in the 1980s by American and Italian scientists that identified human blood on the Shroud. If non-destructive DNA and other forms of testing could be conducted on Shroud blood samples prior to and without interfering with atomic testing (Cl-36 to Cl-35, Ca-41 to Ca-40 and C-14 to C-12 analysis) of these blood samples, then such testing should also be undertaken.

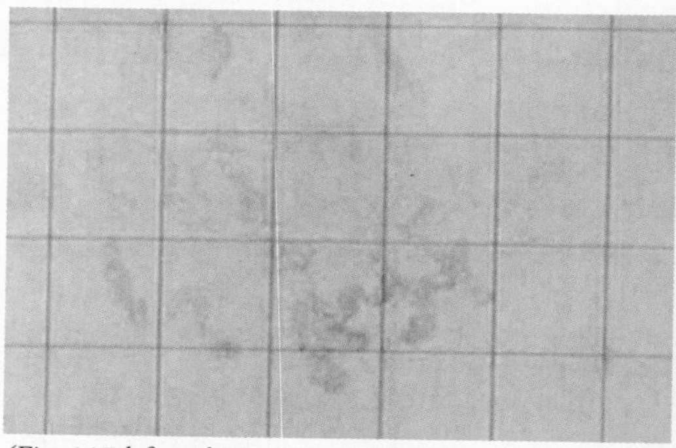

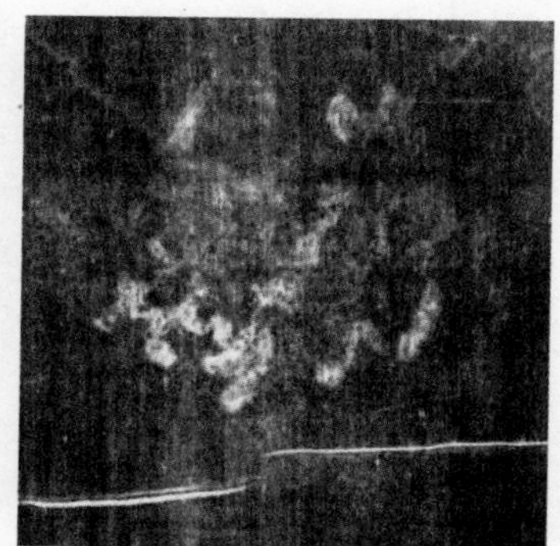

(Fig. 117 left and Fig. 118, right) Blood mark images taken at back of head.

The blood on the Shroud still retains a reddish coloration. Even if the Shroud is medieval, its blood should have long ago turned dark brown or black, like other centuries-old blood. Blood will actually start turning dark within days or weeks of leaving the body and being exposed to air. When the Shroud is exposed to sunlight, it appears to be even redder.[26] The late Dr. Carlo Goldoni undertook experiments to explain these remarkable attributes and concluded that when blood marks are first exposed to neutron radiation and then to ultraviolet light (such as the Shroud would naturally receive from sunlight during exhibitions) it resulted in the blood marks having a bright red coloration. Dr. Goldoni found this reddish coloration existed regardless of the blood's bilirubin content,[27] which has been the traditional explanation for the Shroud's centuries-old red coloration.[28]

Further experiments with particle and other forms of radiation should be undertaken to build upon these initial experiments. Particle radiation emanating from the man in the Shroud is the only hypothesis to claim it could account for the still-reddish color of the man's blood marks after all these centuries.

Particle radiation emanating from the man in the Shroud is at the heart of the most intriguing hypothesis ever proposed. It claims to explain all of the Shroud's features. This hypothesis not only claims that radiation can account for the more than 30 unique or extraordinary features on the Shroud's images, but also that the radiation is particle radiation and its source was the dead body wrapped within this burial cloth. The Historically Consistent Hypothesis can not only account for all of the primary but even the secondary body image features throughout both body images, and all of their 130 corresponding blood marks. It can also account for the cloth's excellent condition, its reverse side imaging at the hands and face, its possible coin and flower images, the still-red color of its centuries-old blood marks, the cloth's 1988 radiocarbon dating and more. No other hypothesis begins to account for all of these features.

This hypothesis will be discussed in more detail in Chapters Eleven and Twelve. If multi-spectral imaging and atomic testing is performed on the Shroud, these techniques could prove whether such a miraculous event occurred to the dead body of the man in the Shroud, when and where this miraculous event occurred, the age of this burial cloth, and the identity of the victim whose images were miraculously encoded on this cloth. If this miraculous event can be proven to have occurred to the man wrapped in the Shroud, this hypothesis would explain how all of the unprecedented features discussed above are then encoded onto the Shroud, all in accordance with natural scientific laws.

While molecular and atomic testing are easily the most exciting and revealing forms of testing that could be applied to the Shroud, they are not the only forms of tests that should be undertaken on this cloth. Professor Bruno Barberis and Professor Paolo Di Lazzaro have both proposed series of tests.[29] While STURP last formally presented a testing proposal in 1984, I'm sure some of its surviving members would also be interested in making a new proposal, as would other qualified scientists and experts.

Let us now turn our attention to the question of whether there is any scientific or other tangible evidence on the Shroud of Turin to support the assertion that this cloth was invisibly repaired or rewoven at the 1988 radiocarbon site.

9

WAS THE SHROUD INVISIBLY REPAIRED?

—\~—

The hypothesis that the Shroud was invisibly repaired initially attracted a number of followers after it was first presented by Joseph Marino and M. Sue Benford at the International Shroud Conference held in Orvieto, Italy in 2000. This hypothesis was also thought to have acquired scientific credibility when STURP chemist Raymond Rogers published his research and conclusions from 2001 to 2005 following his examination of 14 thread segments from the Raes sample and from warp and weft threads from the original radiocarbon material seen below. The lettered portions on the right side of the rather large radiocarbon sample below were removed in 1988 and subdivided among the three radiocarbon laboratories who dated them that same year. See Fig. 119 below for a diagram of the various samples removed from this area of the Shroud.

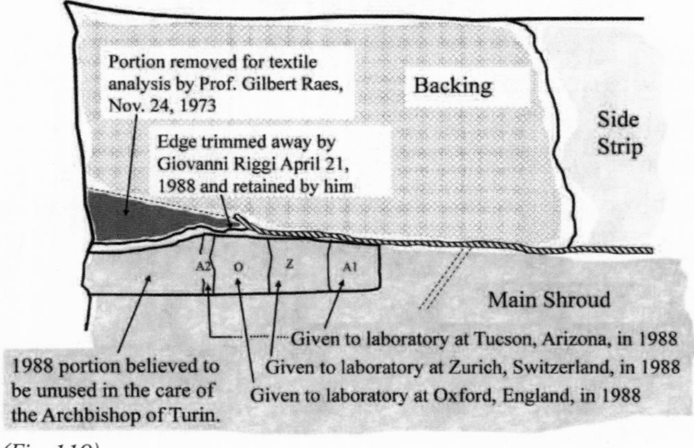

(Fig. 119)

The invisible repair hypothesis and Rogers' research caused the entire Shroud itself to be thoroughly inspected, as well as many scientific techniques and observations to be performed and re-examined at this part of the cloth. These new examinations and re-examinations were not only undertaken by several STURP scientists and shroud experts, but also by one of the scientists who carbon dated the Shroud in 1988 and re-examined part of a remaining sample in 2010. These examinations and re-examinations have consistently shown that, at best, this hypothesis is extremely unlikely.

Rogers claimed that his examination and procedures "prove that the radiocarbon sample was not part of the original cloth of the Shroud of Turin,"[1] He states "all threads from the Raes sample and yarn segments from the radiocarbon sample show colored encrustations (or coatings) on their surfaces. . . . suggesting the color and its vehicle were added by wiping a viscous liquid on the outside of the yarn."[2] He asserts that the "coating is easy to observe on Raes and radiocarbon yarns."[3] He further observes ". . . the Raes and radiocarbon samples indicate that the color has been manipulated. Specifically, the color and distribution of the coatings implies that repairs were made at an unknown time with foreign linen dyed to match the older original material. Such repairs were suggested by Benford and Marino."[4]

Rogers concludes, "The radiocarbon sample had been dyed. Dyeing was probably done intentionally on pristine replacement material to match the color of the older, sepia-colored cloth."[5] He adds the dye "is probably the same age as the Raes and radiocarbon [replacement] yarn."[6] He further specifies: "The dye found on the radiocarbon sample was not used in Europe before about A.D. 1291 and was not common until more than 100 years later."[7]

Benford and Marino initially claimed in 2000 that the Shroud was invisibly repaired and that the new material ran across all three laboratories' samples comprising most, but not all of them.[8] In subsequent articles, they no longer designated precise borders or edges for the new material. Based on a report and research by Raymond Rogers, they asserted in 2002 that the Shroud was invisibly rewoven with another linen sample to match the Shroud material.[9] They reported that Rogers observed a Raes thread with an overlap of another thread spliced into the middle of it.[10] In a subsequent article that year, Rogers claimed to have clearly identified a Raes thread that "is obviously an end-to-end splice of two different batches of yarn."[11] This suggests the ends of the Shroud threads were separated and then rewoven by hand with the ends of the threads from another medieval piece of linen at all the surrounding edges where the imitation piece was spliced into the Shroud. Raymond Rogers even claimed from his examination of Shroud samples that he "found a *medieval* splice in the sampling area."[12] (italics added) Yet, Rogers' examinations and research do not begin to prove any of his above claims.

Among Rogers' fundamental mistakes is that he frequently failed to perform elemental analysis of his samples and ignored or misunderstood the elemental analysis that had been performed on the Shroud's samples and on the entire cloth. Rogers had most of his above samples for twenty years and was a chemist at Los Alamos National Laboratory where extensive elemental analysis could have been conducted on them. Unfortunately, he was away from Shroud studies for 15-20 years, and in the last few years of his life he essentially looked at the samples through a microscope and made subjective interpretations of what he saw. From these subjective interpretations he then made enormous leaps in logic to arrive at his unsupported and erroneous conclusions. Frequently, he failed to understand or ignored the basic facts that his samples came from a scorched area and were at the edge of a water stain.

One of the fundamental mistakes in Mr. Rogers' analysis is that no evidence has been found anywhere on the Shroud itself that this burial cloth has been rewoven or patched at the radiocarbon site. Although the nearby side seam is skillfully sewn to the main body of the Shroud throughout its entire length, it is clearly visible at a distance. Eight sets of patches and other mending and threads were also sewn onto the Shroud in several locations before and after 1534; all of these repairs have always been clearly visible to the naked eye.

(Fig. 120) Typical mending visible on the Shroud.

The Shroud was examined by scores of scientists and various experts in 1969, 1973, 1978, 1988 and 1997, both before and after the Raes and radiocarbon samples were removed and before the repair/reweave hypothesis was first presented in 2000 — but no repairs were ever discovered at this site. How could a repair or reweave be made that could fool so many experts and scientists on six occasions in the last half of the 20th century? How could a medieval (1260-1532?)[13] repairer fool photo microscopy that was performed on the Shroud? In 1978 between 5,000-7,000 photographs of the Shroud were taken in various wavelengths and magnifications, but no photographs and microphotographs have indicated such a repair. Since medieval restorers would not have had the extensive magnification abilities or techniques available today, it is extremely unlikely they would be able to repair cloth in a manner that would be undetected by subsequent magnification, as the Shroud has clearly been subjected to on several occasions.

(Fig. 121) Scientist Sam Pellicori examining Shroud with a photographic microscope.

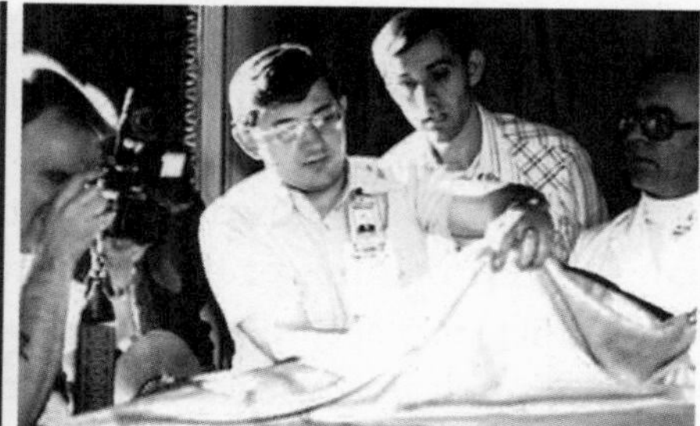

(Fig. 122) Vernon Miller, Eric Jumper, John Jackson, and Giovanni Riggi examining the Shroud.

When the backing cloth was removed from the Shroud during its restoration work in 2002, this area of the cloth was again specifically inspected on the image side, as well as on the reverse or outer side of this entire garment. All of the inspectors and observers confirmed what the many previous examinations, photographs and photomicrographs of the front of the Shroud revealed — that the Shroud had not been patched or rewoven at the radiocarbon or any nearby site. In addition, there were no threads or stitching found on either side of the cloth at the radiocarbon site on the Shroud. After weeks of painstakingly examining and restoring the Shroud, Dr. Mechthild Flury-Lemberg, one of the world's leading textile experts on ancient textiles and the Shroud of Turin, who specifically led and authored the report of its 2002 examination and restoration, stated:

> I would like to add here a note on the hypothetical "reweaving done in the 16th century." There is no doubt that the Shroud does not contain any reweaving. . . Reweaving in the literal sense does not exist. Once the piece of fabric is taken off the loom the weaving process is finished. Afterwards one can only alter a fabric by using needle and thread. An example would be a hole which has been mended by imitating its weave structure. This process will always be recognizable as mending and in any case visible on the reverse of fabric.[14]

Flury-Lemberg has published many photos in other publications with excellent repairs or "reweaves" that have been performed on textiles that are "invisible" to an undiscerning eye; however, close examination with the naked eye will invariably reveal the mending.[15] The most success has been obtained on tapestries, which are heavy, hand-worked fabrics with pictures or designs formed by threads inserted over and under the warp according to the requirements of color. They are not worked in selvage to selvage as in weaving. Flury-Lemberg notes if the lengthwise (warp) threads are undamaged and not in need of replacement, and only the less numerous crossing (weft) threads need to be replaced, this situation stands the best chance of being undetected. However, even with heavy tapestries, slight ridges, threads or the ends of the repair thread will always be visible to even the naked eye on the front or the back of the tapestry upon close or expert inspection.[16] Flury-Lemberg states, "This kind of technical proceeding applicable exclusively to coarse weaving structure of a tapestry, seems to be the basis of the argumentation of Benford/Marino."[17]

Flury-Lemberg notes that a completely different repair process would be required for a delicate fabric. Remember, the Shroud has the approximate thickness of a man's undershirt. While such material can be repaired by imitating the weaving structure, she states unequivocally, this repair or reweave on a delicate fabric will always be visible.[18] No historical examples exist from the 16th or any previous century where rewoven linen has fooled close inspection by scores of experts, photographs in all wavelengths and photomicroscopy! Such undetected reweaves or splicing on delicate fabrics are simply hypothetical and unknown to this time of history.[19] We will see later that for the size of area that would had to have been replaced on the Shroud, this type of reweaving could not be performed invisibly on delicate fabrics today, even if state-of-the-art microscopes are used *during* the reweave.

Threads that were rewoven together would also be fuller or thicker at the hypothetical splices, or more numerous at areas surrounding the damaged part of the cloth. In addition, unlike almost all other textiles, the Shroud was kept rolled on a large spool from 1578 to 2002 — a period of 424 years. This would necessarily have caused stretching and pulling that would have been revealed with material whose woven threads were supposedly unraveled and spliced or rewoven by hand into the cloth, but not attached by sewing threads. New material could not have been invisibly blended and maintained in this cloth without some type of mending or permanent attachment, which would also be revealed by inspection of either or both sides of the cloth. Unraveling and hand splicing the threads of a repair piece into threads on the Shroud, which would then be rolled on a spool for 424 years, simply could not survive detection by detailed visual inspection on one or both sides of the cloth by textile experts, or by scores of scientists and other Shroud experts at six other scientific examinations, or by photomicroscopy, or by thousands of other photographs.

Flury-Lemberg succinctly states, "In any case, neither on the front nor the back of the whole cloth is the slightest hint of a mending operation, a patch or some kind of reinforcing darning to be found."[20] She provides the pictures of both sides of the Shroud (Figs. 123 and 124) for us to observe ourselves.

The various scientific testing and imaging of the Shroud not only fail to reveal a reweave or repair, but they also indicate that the cloth was not repaired or rewoven with replacement material. X-ray fluorescence analysis on thirteen threads from the Raes sample indicated they had roughly the same relative concentrations of calcium, strontium and

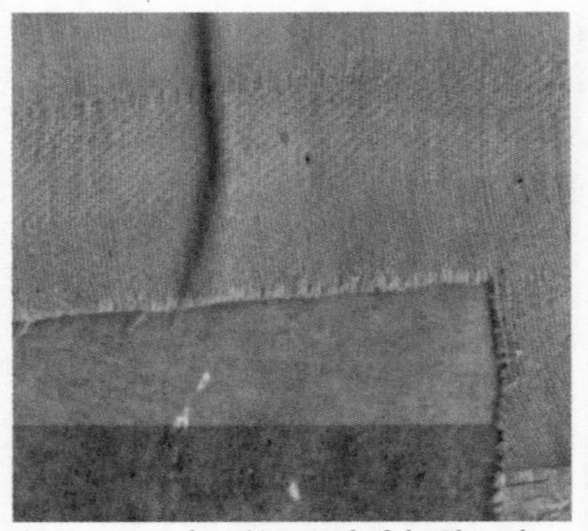

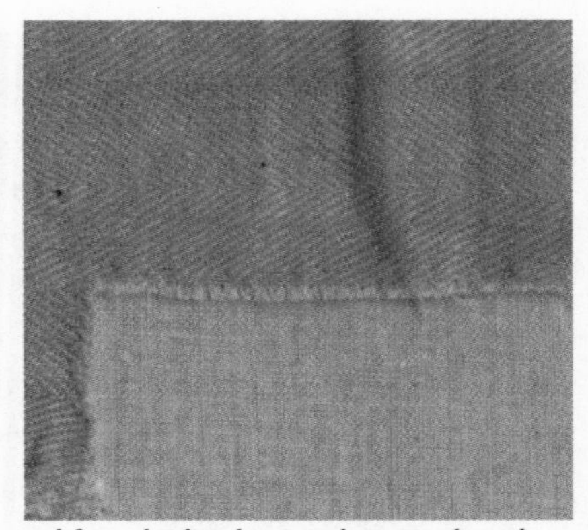

(Figs. 123 and 124) Detail of the Shroud, reverse and frontal sides, showing the area where the radiocarbon sample was taken. The woven material displays the irregularly spun threads of the warp and the weft commonly found on ancient textiles, but does not display any evidence of a reweave.

iron that was found on the rest of the Shroud.[21] Neither the existence of, let alone the concentrations of calcium, strontium and iron, or any other elements in nature, or in the Shroud, could possibly have been known to a medieval or a 16th century repairer. STURP founder John Jackson noted, "the density ratios of the calcium, iron and strontium in both the Shroud and the Raes samples . . . is a compelling argument that the fabric of the radiocarbon site is very likely not due to a fabric that is alien to the Shroud."[22]

Dr. Jackson prefaced his above comment with the statement, "The thesis that the radiocarbon sample site was a medieval reweave would be an excellent solution as to why the radiocarbon date yielded a medieval date. However, this thesis is, in my opinion, profoundly incorrect and this can be demonstrated to be so using available data that was collected by STURP in 1978."[23] Jackson noted a radiograph also taken in 1978 of a fairly large area of the Shroud, which clearly included what ten years later would become the site of the radiocarbon sampling in 1988. The authors of the publication in which this particular radiograph appears suggested that the side strip consists of the same material as the main portion of the Shroud. This is because alternating high- and low-density bands attributed to the Shroud's horizontal weave (weft) are continuous through this area and the side seam.[24] If the Shroud had been rewoven in this area, the continuity bands would have been disrupted at the reweave intersections; however, they were not. For this and other reasons, Jackson states ". . . we must conclude unambiguously that there has been no reweave whatsoever surrounding the radiocarbon site.[25]

Ironically, the X-ray fluorescence information and the above radiograph were both contained and discussed in a 1982 article by STURP scientists Larry Schwalbe and Raymond Rogers titled "Physics and Chemistry of the Shroud of Turin," which was a summary of STURP's findings from its investigation. (The X-ray fluorescence information was in an endnote.) Rogers seems to have forgotten about or misunderstood the elemental analysis of these samples and the cloth itself, or the other numerous photographs and photomicrographs taken of the Shroud.

Other examples of Rogers' unconcern with or misunderstanding of the Shroud's photographic evidence can be found with the ultraviolet fluorescent photographs and reflected light imagery, which allowed details to be seen that were not visible with the naked eye. They revealed that Rogers' samples from the radiocarbon site were in the midst of a scorch mark. This observation was made in 1989 by Vernon Miller, chief photographer of the 1978 Shroud investigation, after examining the above imaging.[26] It was also confirmed by STURP chemist, Dr. Alan Adler, who compared 15 threads from the radiocarbon area with 19 fibers from non-image, image, water stain, scorch, backing cloth and serum-coated locations on the Shroud by Fourier Transform Infrared (FTIR) micro spectrophotometry and by scanning electron microprobe.[27]

Rogers claims that he utilized more than microscopic or microchemical *observations*, for his overall claims. He claims that pyrolysis mass spectrometry performed by STURP also "prove(s) that the radiocarbon sample was not part of the original cloth of the Shroud of Turin."[28] Yet, Rogers only compared STURP's pyrolysis mass spectrometry studies of samples from the rest of the Shroud to those from the Raes sample to make a tertiary point that the coating on the Raes and radiocarbon samples was a pentosan. (The alleged existence of the coating was found through Rogers' microscopic examination, which he illustrated by photomicroscopy.) Even here, this work was not very thorough or convincing and was incomplete. Of all the other tests that STURP performed on the rest of the Shroud or its samples, Rogers chose only those mentioned, which were never published by STURP. Furthermore, he did not perform (or did not give) the mass spectrometry results for the radiocarbon sample. Lastly, Rogers stated that "Cellulose pyrolyzes to produce hydroxymethylfurfural (mass 126) . . . . the Raes fibers showed a signal for furfural at mass 96 . . . . These results prove that the gum coating on the Raes and radiocarbon samples is a pentosan."[29] Yet, if the Raes samples (the *only* non-image area from which Rogers used a sample) were in a lightly scorched area, as the radiocarbon samples were, bonds broken during the scorching of the cellulose may have allowed furfural to be released at lower temperatures. Thus, Rogers' subjective observations and interpretations are the primary support for his erroneous conclusions.

Rogers also fails to understand that his radiocarbon samples were on the edge of a water stain. This, too, was not only stated by Vernon Miller from his above examination in 1989, and supported by STURP chemist, Alan Adler, by his above examinations in 1996 and 1997, but the normal photograph of this region taken in 2002 during the Shroud's restoration clearly shows this in Fig.125.

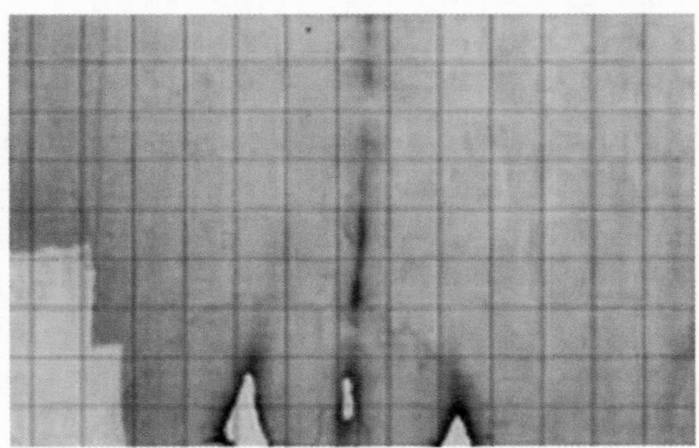

(Fig. 125) A relatively small water stain near the lower left-hand corner of the Shroud can be seen to flow into the 1988 radiocarbon site.

The edge of a water stain is where much of the debris acquired by or contained in the flow of water is going to be deposited. Adler's FTIR and scanning electron microprobe data show gross enrichment of the inorganic mineral elements in the radiocarbon samples. Adler stated, "In fact, the radiocarbon fibers appear to be an exaggerated composite of the water stain and scorch fibers."[30] Both the water stain's edge and the scorch at this location would better explain the different chemical composition of the samples than an undocumented, invisible reweave never seen by the eyes of scores of experts or photomicroscopy.

The edge of the water stain at this location also means that the cloth material at the radiocarbon site had to have been present when the water flowed over this and its adjacent areas. (Making an invisible reweave or splices without permanent attachment is difficult enough, but even reweave advocates do not contend the original Shroud material had a water stain that was replaced with new material also containing a perfectly matching water stain.) However, as seen in Chapter Four, evidence and analysis presented by photographer Aldo Guerreschi and Michele Salcito at international conferences held on the Shroud in Paris in 2002 and in Dallas in 2005 indicate that these particular water stains are not from the fire of 1532.[31] The authors show that when the Shroud is folded so that its burn holes and scorch marks from the fire of 1532 line up over each other within its folds, many water stains on the cloth also line up within this folding pattern.

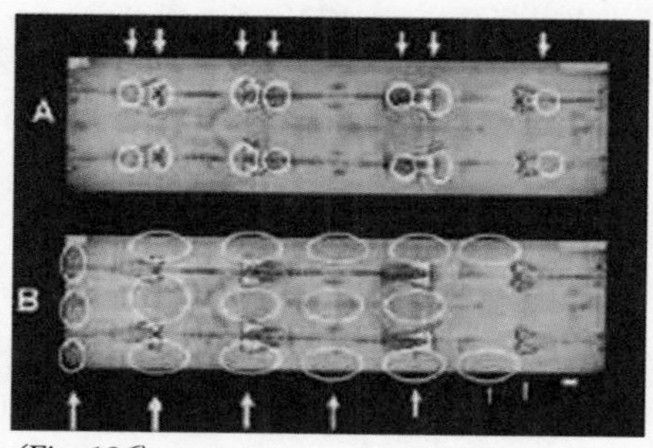

(Fig. 126)

However, the water stains along the bottom row of the Shroud, including the one just off the letter A in Fig. 126, and other water stains on other parts of the Shroud, do not line up with this fold configuration. According to the authors, these water stains line up in an accordion style fold pattern while the long cloth was kept inside a container, which acquired a small amount of water on the bottom. See Figs.127-130.

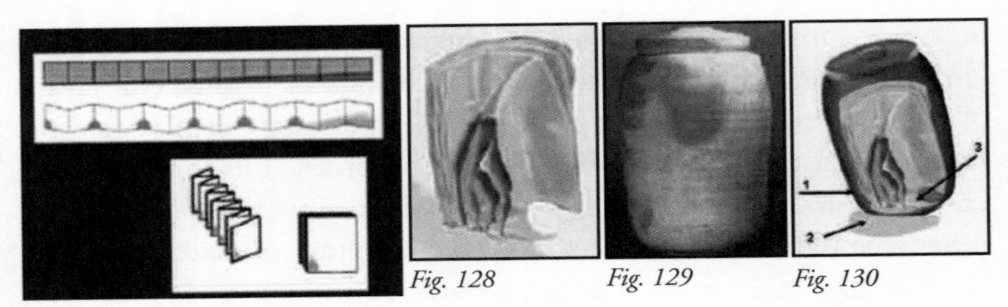

Fig. 128 Fig. 129 Fig. 130

Fig. 127

This indicates the Shroud was in two different folding configurations at the time it received the two different sets of water stains. The Shroud's history in Europe since the 14th century is well documented and no staining incidents other than the 16th century one in 1532 are recorded. Unlike the water stains from 1532, the stains near the radiocarbon site are not associated with any fire or other damage requiring the Shroud to be repaired. These water stains could have occurred centuries earlier. Interestingly, earthenware of the size in which the 52 segmented cloth would snugly fit was very commonly used in ancient times, such as the kind that were discovered at the archaeological site of Qumran.[32]

Rogers' assertions that the Raes and radiocarbon samples were dyed and were part of the replacement material that was radiocarbon dated in 1988 fails on several other important grounds. In 2010, a photomicrographic investigation was conducted on a Shroud sample that was removed from one that was used in the radiocarbon dating study at the University of Arizona in 1988. This article appears in the journal *Radiocarbon* and one of its two authors is Timothy Jull, who participated in the Shroud's 1988 radiocarbon dating. The authors state, "Under UV fluorescence, the fibers fluoresce uniformly and do not show any indication of an overall coating."[33] They continue, "In addition, we find no evidence for any coatings or dying of the linen . . . . Linen does not readily accept dye, and any surface 'coating' would be loosely adhered. We viewed a textile fragment dyed using traditional methods under UV light, and observed absolutely no similarity in UV fluorescence consistent with such a dye."[34]

Furthermore, if a "reweave" of foreign linen had been dyed centuries ago to resemble the undyed older Shroud, then natural aging over the succeeding centuries would have caused a difference in color between the two areas. The reweave would have become lighter as the color of the dye slowly disappeared, but the Shroud linen would have yellowed and gotten darker as it oxidized and dehydrated over the centuries. Similarly, textile expert Mechthild Flury-Lemberg has also examined a UV-fluorescent picture of woven material comparable to the Shroud that has been darned, yet whose foreign threads were easily visible.[35]

Perhaps the most subjective observation with the greatest leap of logic is Rogers' claim that a Raes thread "is obviously an end-to-end splice of two different batches of yarn."[36] I assume this is the same thread that he refers to when he states in the *Skeptical Inquirer*, "I found a medieval splice in the sampling area."[37]

He devotes only five lines of text in the first publication and two in the second to build his case. Rogers' photomicrograph of the Raes thread (Fig. 131), however, shows no sign of a splice.

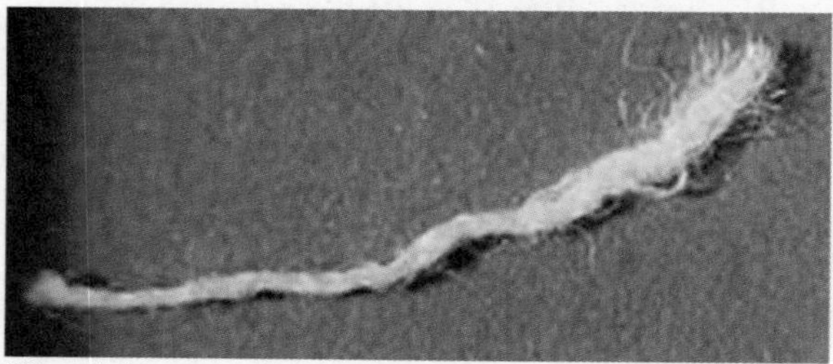

(Fig. 131)

The right third of the thread is white because the fibers on the inside of the thread have been clearly exposed or popped out of the yellowed thread. The left two-thirds of the regular thread are straw yellow in color. The fibers toward the end of the exposed white inner region are more loose or frayed. They simply point toward the end of the thread in the direction they ran. They show no signs of any foreign threads or fibers or of any splicing or reweaving. Since the fibers at both ends point to their respective opposite ends in which they ran, it is puzzling why Rogers concludes the entire thread is obviously an end to end splice.

Moreover, when both ends of this thread were examined with X-ray photoelectron spectroscopy (XPS) producing high resolution spectra, chemist Robert Villareal of the Los Alamos National Laboratory reported in 2008 that, "The two ends are chemically similar."[38] This clearly indicates that the thread is not an end to end splice. In addition, the spectra of both ends were comparable to the spectrum of cotton, which was also indicated by FTIR.[39] An obvious objection to Rogers' curious interpretation is raised by A.A.M. van der Hoeven when she asks, "Why splice a cotton thread to a cotton thread in order to repair a linen Shroud?"[40] Furthermore, if threads were spliced onto the cloth, the threads would also appear thicker or fuller at the splices. Yet, that is not the case with Shroud threads at or near the radiocarbon site.

Rogers' failure to realize or acknowledge that the radiocarbon samples that he studied were in the midst of or near a scorch also contributes to another faulty argument that he makes. He claims that the presence or absence of vanillin in the lignin of the Shroud indicates its age is between 1000 B.C. and 700 A.D., a 1700 year age range. (Such an age range is hardly a precise scientific calculation.) He also implies that the vanillin content on the cloth supports his medieval repair argument.[41] Yet, his reasoning appears very inconsistent. According to Rogers, "No samples from any location on the Shroud gave the vanillin test" and that "The lignin on Shroud samples . . . does not give the test." He further states, "The lignin at growth nodes on the Shroud's flax fibers did not give the usual chemical spot test for lignin (i.e. . . . for vanillin)."[42] If vanillin is not present in the Shroud's samples *and* the larger cloth, then it indicates all the material is ancient and the same. It does not indicate that medieval or 16th century material is found with 1000 B.C. to 700 A.D. material.

Yet, the vanillin argument cannot even be applied to the Shroud. The presence of vanillin in lignin is greatly affected by heat. The effect is so great that it operates exponentially. Rogers uses a constant temperature gradient of between 20-25° C (68-77° F) for the Shroud to state that its ". . . vanillin loss suggests that the Shroud is between 1300- and 3000- years old."[43]

Yet, the effect of heat is so great that STURP physicist John Jackson and chemist Keith Propp state if the temperature incurred by the Shroud or its samples during the fire of 1532 was just 200° C that it would lose 95% of its vanillin in a mere 6.4 seconds.[44] The Shroud could have lost much or all of its vanillin while folded in a reliquary during the Chambery fire of 1532. The samples that Rogers examined were in the midst of or next to a scorch. Furthermore, the Shroud has a likely 1200 year history in the Middle East (Jerusalem, Edessa and Constantinople) and a known 650 year history in France and Italy, where the temperatures are known to frequently exceed 20-25° C. The vanillin test expounded by Rogers simply cannot be used to estimate the age of the Shroud.

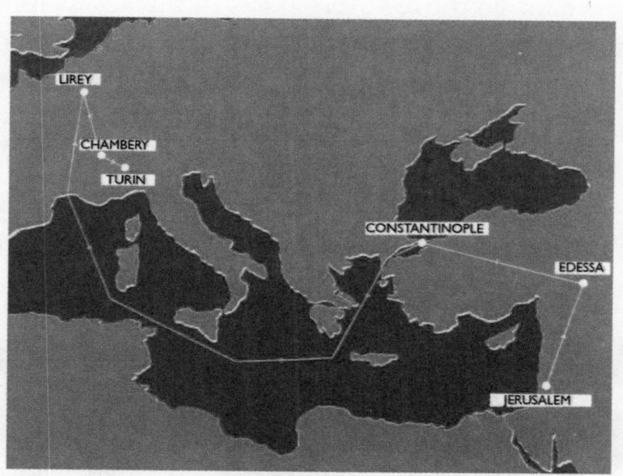

(Fig. 132) Known and probable locations of Shroud during its history.

It's not that I don't think the Shroud is from the first century. Almost all of the evidence is consistent with this age. I think the scientific testing at the atomic and molecular levels discussed in Chapters Six, Seven and Eight and the dating conducted by scientists at the University of Padua (also in Chapter Eight) are far more applicable and accurate than the vanillin test proposed by Rogers or the radiocarbon dating conducted on the Shroud in 1988.

As I state in earlier chapters, if a naturalistic, artistic or human method is proposed to have occurred to the Shroud then its proponents have the burden of proving their claim. If Rogers' undocumented findings and conclusions are accepted, a medieval restorer would had to have separated and spliced by hand or rewoven foreign material many centuries younger than the Shroud into the entire Raes/radiocarbon area; then added coloring that continued to match the rest of the Shroud even though the dye would have lightened and the Shroud would have yellowed or darkened with age over succeeding centuries; repaired the cloth before it was ever stained with water at this location; used material with approximately the same relative concentrations of calcium, strontium and iron as the rest of the Shroud; repaired or rewoven the radiocarbon site so invisibly that it could not be detected by X-radiographs at this location or the surrounding area; permanently attached the new material without any stitching or mending on either side of the cloth; not allowed the rewoven splices or threads to appear any fuller or more numerous; yet withstood all the stress or pull they would have incurred from having been repeatedly rolled and unrolled during public and private displays over centuries and kept on a spool for 424 years.

Furthermore, neither photomicrographs, nor any other photographs taken of the cloth in any other wavelengths could show this repair; nor would any direct examinations by anyone including countless textile experts, scientists, or other professionals be able to detect the repair.

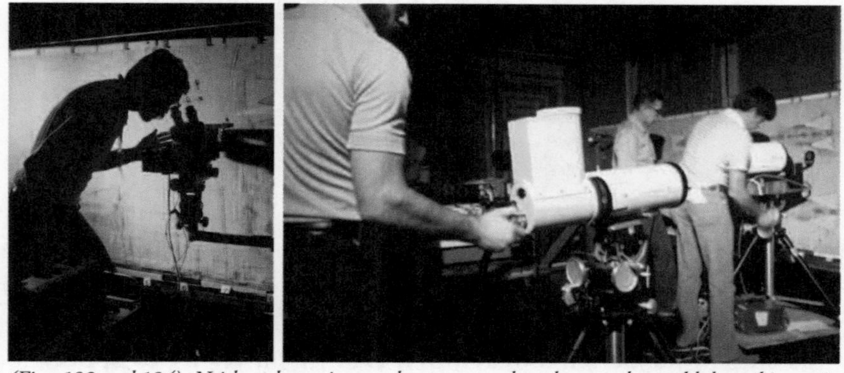

(Figs. 133 and 134) Neither photomicrographs, nor any other photographs, could show this repair.

Lastly, the reweaver must remain anonymous throughout history, and, unlike other repairs and events with the Shroud, there can be no historical record whatsoever among the numerous Savoy family records or by any other owners, custodians, priests, nuns, or textile specialists or restorers; all without any reasons for such secrecy. The alleged repair would be somewhat like the alleged medieval painting of the Shroud: its truly masterful artist is unknown; its brilliant technique remains undetected even today; there is no historical record anywhere; without any reason for such secrecy, and without any actual evidence on the cloth for the claims.

No one point or item of evidence absolutely precludes such a repair, but when these items are taken collectively, such a hypothesis is obviously extremely unlikely. As I have stated many times, I do not object to future testing related to this hypothesis, as long as it is only a part of the testing that also includes other hypotheses and testing included in Chapters Six, Seven, Eight, Eleven and Twelve of this book. As seen in my 2000 book, I once had some small concerns about known repairs in this area. However, the photomicrographic evidence, the evidence derived from expert examination of both sides of the Shroud itself and the scientific evidence, all of which I have seen or studied in this context subsequently (combined with the lack of any historical evidence) have shown me how extremely unlikely it is that any repair hypothesis, especially an invisible reweave could explain the Shroud's medieval C-14 dating results.

I have concentrated primarily on Raymond Rogers' arguments because some people think that he provides scientific evidence in support of the repair or reweave hypothesis. The X-ray fluorescent analysis, X-radiographs, UV fluorescence, elemental analysis, the photomicrographs, the many other photographs in numerous wavelengths and techniques, numerous direct examinations by a great number of people with a wide variety of professional backgrounds and numerous possible historical records — not only fail to provide any palpable evidence of a repair — but they refute the invisible repair hypothesis or are quite inconsistent with it.

A somewhat better reweaving method than trying to splice fine linen threads can be found with the reweaving performed with needle and thread on heavier coarser material like tapestries that Benford and Marino discuss. Joe Marino led me to a video in which a hole in coarse material that looks very loosely woven is repaired with threads that probably came from a hem or other location on the same material, for it is very similar. By overlapping the hole extensively with the weft and warp threads, the hole is covered without attaching the new threads. The needle can be pushed and pulled through the material surrounding the small hole, as well as through the hole and the replacement material that has been woven over the hole to match the surrounding weave. The ends that stick out from the new threads can be settled in or cut with scissors. However, the material surrounding the hole would now be fuller or denser, having more threads than the material over the hole and elsewhere on the cloth.

The fact the new material is not attached would work for a tapestry hanging high on a wall where it would not feel any pull or stress, unlike the countless times the Shroud has been rolled up and unrolled, as well as hung and displayed at private and public exhibitions. Some of the public exhibitions lasted for weeks at a time when the Shroud would be exhibited outside, so it would have been taken in and probably folded or rolled at night (and during the day if rain or inclement weather was present) and rehung again that day or the next day. The hypothetical reweave would also have been near the weight-bearing, upper left-hand corner of the Shroud where it was invariably displayed horizontally as seen in numerous paintings, drawings and photographs of the famous cloth on display during exhibitions over seven centuries in Europe.

A tapestry hanging high on a wall would not only avoid the daily stresses and tension from being rolled, unrolled and hung during centuries of exhibitions, but also the stress that came from being wound on the spool itself for 424 years. Nor would any first-hand inspections occur, in this position, let alone close inspections by scores of experts, radiography, photomicroscopy and photography in all other wavelengths and magnifications, all of which would detect the excellent reweave.

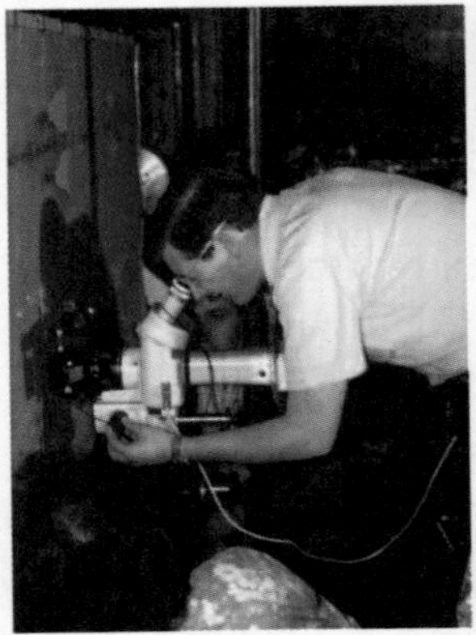

(Figs. 135 and 136) Examining the Shroud.

It must also be remembered that no completely invisible 16th century examples, even of such coarse tapestries, survive. Even the experts cited by Benford and Marino can't think of any documented examples.[45] They contend that such reweaving could have occurred on tapestries in the 16th century, but even with such coarse material, the reweave would only be "invisible" to a non-discerning eye. It would not be invisible to a discerning eye that looked up close and certainly not to modern technology.

Benford and Marino refer to French weavers or reweavers of the 16th century. Today those that employ this technique use state-of-the-art microscopes that allow reweavers to view the threads of finely woven fabrics up close, which a reweaver would not have had access to in 1532 or earlier. Yet, even these modern companies state that this type of reweaving of finely woven fabrics with state-of-the-art microscopes will only work on holes as small or smaller than a dime.[46] If the repair or reweave on the Shroud only covered the area first claimed by Marino and Bedford, it would be much larger than the size of a dime, which even modern repairers with state-of-the-art microscopes could not reweave undiscernibly. In that case they would have to use a patch and attempt to blend the patch in with the surrounding material. Yet, the hypothetical 16th century reweaver could not have used surrounding material from the Shroud for it, too, would have dated the same age as the Shroud. If he used a patch from foreign material, he would have as many or more problems to overcome than was discussed earlier.

Furthermore, Monsignor Giuseppe Ghiberti, a participant at the 2002 examination and restoration, succinctly stated in regard to their and Dr. Flury-Lemberg's detailed inspections: "The truth is that there is no patch and no darn . . . . there are no added threads." He also added the common sense observation that "you apply a patch or a darn where there is a hole,"[47] but that no such hole, tear or worn area could be seen at or near the radiocarbon site on either the inner or the outer side of the Shroud of Turin.

We saw that under UV fluorescence, the fibers on the Shroud's radiocarbon sample at the University of Arizona fluoresced uniformly and did not show any indication of an overall coating. When textile fragments that had been traditionally dyed were viewed under UV light, no similarity was found with the Shroud's radiocarbon sample. (When Dr. Flury-Lemberg examined a UV fluorescent picture of woven material comparable to the Shroud that had been darned, the foreign threads were also easily visible.)

It is seemingly impossible to have rewoven new material into the Shroud at the radiocarbon site and to have dyed it as claimed by Rogers and others, so that the dye visibly and by photomicroscopy matched the rest of the Shroud for centuries until its removal near the end of the 20th century. Moreover, it is completely impossible to have done so and to also have the above fluorescence properties that were documented on the radiocarbon sample at the University of Arizona. These results on the radiocarbon sample are mutually exclusive from those that would have appeared on a rewoven, dyed sample that was allegedly rewoven at the radiocarbon site.

A 16th century reweave on an area of the Shroud that included all three 1988 C-14 sample sites and/or adjacent areas according to Rogers, Benford and Marino and others simply cannot be made on delicate fabrics such as fine linen. Furthermore, it would not have been possible to have dyed the rewoven foreign piece to visibly match the Shroud for centuries and to not display any properties of a dye under fluorescent lighting. In addition, all forms of up-close detection with the visible eye and all of the scientific instruments and forms of photography that we have discussed earlier would have detected such a reweave or repair. All of the arguments that were applied to the earlier forms of splicing or reweaving can also be applied to the needle and thread reweave method. Even though this method is a little better, at best, it is still extremely unlikely to have occurred on the Shroud or to its radiocarbon samples.

Unfortunately, there appears to be little, if any, evidence that the Shroud was invisibly repaired or rewoven in medieval times or any century. However, let us turn our attention to the question of whether this cloth has a plausible history prior to medieval times or its first 14th century appearance in Europe.

10

THE HISTORY OF THE SHROUD

The Shroud has been publicly exhibited on fifteen different occasions since it first arrived in Turin in 1578. During its previous 225 years in Europe, it was displayed publicly and privately at least twenty or more times in France and what is now Western Europe. The Shroud was actually first exhibited in Europe by Geoffrey de Charny at a Catholic church in the mid-1350s in Lirey, France, and again in 1389 by Geoffrey's widow, Jeanne de Vergy and his son Geoffrey II. During these exhibitions the Shroud was displayed and billed as the actual burial cloth of Jesus Christ. Tragically, when Geoffrey de Charny died a hero's death at the Battle of Poitiers in 1356, the history of his acquisition of the Shroud died with him. From these first exhibits in France through today, the Shroud has a known and documented history. Records show that while the Shroud was in Chambery, France in 1532, it incurred fire and water damage when the church in which it was kept caught fire. Throughout its stay in Europe, the Shroud's locations and residences have been a matter of record.

Many theories exist as to the whereabouts of the Shroud between 1204 and its first European exhibit in Lirey in the 1350s. These theories could easily explain how the cloth arrived in Europe from its documented existence in the Byzantine Empire in the 1200s. Two references exist recording the existence of the Shroud in Constantinople, a shroud containing the full-length figure of Jesus' dead and naked body.

The first documented record of the Shroud was written by Nicholas Mesarites, a Greek and overseer of the emperor's relics in the Pharos Chapel of the Boucoleon Palace in Constantinople. Mesarites had to defend the chapel against a mob during a palace revolution, which took place in 1201. In that connection, he wrote . . .

> In this chapel Christ rises again, and the sindon [Shroud] with the burial linens is the clear proof....still smelling fragrant of myrrh, defying decay, because it wrapped the mysterious, naked dead body after the Passion....[1]

The next reference to the Shroud occurs during the Fourth Crusade. From 1203-1204, crusaders besieged Constantinople, then the wealthy and powerful capital of the Byzantine Empire and today's city of Istanbul, Turkey. Led by Marquis Boniface de Monteferrat, this army was demanding payment for having deposed Alexius III (who had usurped the Byzantine throne), and reinstating his nephew, Alexius IV, back upon the throne. Soldiers, biding their time, both inside and out of the city walls, waited for this payment, while noting the vast and unparalleled riches of the empire. In one of the darkest hours of Christianity, the soldiers of the Fourth Crusade attacked Constantinople, pillaging and looting any and everything of value they could carry. Most of the capital's priceless relics and artifacts disappeared, including the Shroud.

In a first-hand chronicle, Robert de Clari, describes his daily visits to Constantinople while the army camped outside the walls. In this chronicle he describes the Shroud and its disappearance:

> . . .there was another of the churches which they call My Lady St. Mary of Blachernae, where was kept the shroud [sydoines] in which Our Lord had been wrapped, which stood up straight every Friday so that the figure of Our Lord could be plainly seen there, and no one, either Greek or French, ever knew what became of this shroud when the city was taken.[2]

The Shroud, the Mandylion And The Image of Edessa

Whether stolen or destroyed, the Shroud had disappeared from Constantinople. Interestingly, another sacred and revered cloth had also disappeared and that cloth was the Mandylion. While the Shroud would appear again in Europe, the Mandylion would never appear again in history. Said to bear the divine and miraculous imprint of the face of Jesus, the Mandylion was labeled *acheiropoietos*, or "not made by hand." Nor was it the product of a painter's art or his pigments. Artists so revered this cloth that they came from great distances to copy it. Arriving in Constantinople in 944, the Mandylion, too, disappeared in 1204, during the looting of the Fourth Crusade. Prior to arriving in Constantinople in 944, it had been kept in Edessa, now Urfa, Turkey, since at least the sixth century. While in Turkey, it was equally revered, albeit differently labeled, as the Image of Edessa. Several sources contend the cloth was brought in the first century to Edessa from Jerusalem, but differ as to whether the cloth was taken to Edessa by the disciple Thaddaeus or a messenger of Edessa's King Abgar V.[3] Shortly after its arrival in Edesaa, this cloth disappeared and did not reappear again until the sixth century. Its sixth century reappearance, however, had a profound effect on the development of the traditional image of Christ in art. According to historian Ian Wilson, the emergence of this traditional likeness of Christ was due to the influence of the Mandylion:

> From the point of view of the tradition of the Eastern Orthodox Church, there is absolutely no mystery about this. The universally recognized source of the true likeness of Jesus in art was an apparently miraculously imprinted image of Jesus on cloth, the so-called Image of Edessa, or Mandylion, so highly venerated that a representation of it is to be found in virtually every Orthodox Church even to this day.[4]

In the early part of the 20th century, Paul Vignon began a pioneering study to examine the artistic depictions of Christ through the centuries.[5] Other scholars would follow his lead, which not only broadened in the field of art, but would include Christ's depictions on coins. These studies in the field of art not only influenced, but were later expanded upon by Wilson in his work *The Shroud of Turin*, published in 1978.

The image of Christ that is so familiar to us today actually did not begin to appear until the sixth century. Researchers found that prior to this time, the artistic depictions of Jesus varied greatly from the standard and conventional likeness of Christ we know today. Furthermore, the

New Testament does not describe Christ's physical appearance. While there were exceptions, images of Jesus throughout the first five centuries A.D. usually portrayed him as young and clean-shaven with short hair. Figs. 137 through 141 below show how Jesus was typically depicted in the early centuries of Christianity.

(Fig. 137) *(Fig. 138)* *(Fig. 139)*

(Fig. 140) *(Fig. 141)*

(Fig. 137) Christ's face from a mosaic pavement of the fourth century found at Hinton St. Mary, Dorset, England. (Fig 138) Detail from Sarcophagus of Janius Bassus, A.D. 359. (Fig. 139) Portrait of Jesus, ca. 500. (Fig. 140) Christ healing A blind man. St. Appollinare Nuovo in Ravenna, ca. 510-520. (Fig. 141) Christ Enthroned. Apse mosaic, San Vitale, Ravenna, ca. 545.

The Traditional Image of Christ Becomes Established

After five centuries of variation, Christ's likeness abruptly changes to become more uniform. His hair, previously depicted as short, becomes long and parted in the middle, falling to his shoulders. Previously clean-shaven, his new image has a forked beard and a thin mustache that droops to join the beard. His face is longer and more refined with a nose that is also longer and more pronounced. His eyes become more deeply set and his whole countenance is set in a rigidly front-facing attitude.

Interestingly, starting in the sixth century, consistent patterns of anomalies begin to occur in the depictions of Christ. Some anomalies are added features; these anomalies include a three-sided square between the eyebrows, a V shape at the bridge of the nose, a second V within the three-sided square, an accentuated line between the nose and the upper lip, a heavy line under the lower lip, a hairless area between the lower lip and beard, and a traverse line across the throat. In other anomalies, features change: for instance, a raised right eyebrow, accentuated cheeks, an enlarged left nostril and heavily accentuated owlish eyes. These regularly appearing features of Christ can be seen on the visible negative image on the Shroud of Turin. Figure 142 and its corresponding numbered text shows Ian Wilson's full listing of these facial anomalies and their locations:

(Fig. 142)

(Fig. 142) Facial markings found on the Shroud

1. A transverse streak across the forehead.
2. The three-sided "square" on the forehead.
3. A V-shape at the bridge of the nose.
4. A second V-shape, inside the three-sided square.
5. A raised right eyebrow.
6 An accentuated left cheek.
7 An accentuated right cheek.
8. An enlarged left nostril.
9. An accentuated line between the nose and the upper lip.
10. A heavy line under the lower lip.
11. A hairless area between the lip and the beard.
12. The forked beard.

Facial markings found on the Shroud (con't)

13. A transverse line across the throat.
14. Heavily accentuated, owlish eyes.
15. Two loose strands of hair falling from the apex of the forehead.

Most intriguing is that many of these oddities have no apparent artistic purpose. Furthermore, many are irrelevant and detract from the naturalness of the face. Although each artistic representation does not contain every feature, the fact these consistent anomalous patterns appear indicates that artists through the centuries were studying and interpreting a similar source. Artists used the Mandylion as just such a primary source. Since all these features appear on the Shroud, a strong argument is made for declaring the Shroud and the Mandylion to be one and the same. This argument or premise is called the "iconographic theory." First developed by Vignon, he asserted in this theory that the Shroud had a definite existence and influence on artists well before the 1300s. He further postulated that the similarity of Jesus' features in various likenesses could not be explained in any other way.

The artists would have seen and modeled the negative image on the cloth. The well-focused and highly resolved positive image revealed by the photographic negative would not have been available to them, so the features on the cloth would be vague and somewhat indefinite. While the eyes of the Shroud cloth image appear to be open and staring, they are actually closed as is revealed by the positive (photographic negative). The early artists attempted to incorporate the Shroud's facial features as they appeared on the cloth, composing the best and most accurate representations they could. Their efforts made it appear as though the artists were diligently attempting to follow a definitive, superior representation of Christ.

Wilson believes it possible to statistically analyze the frequency of occurrence of these anomalous features in various works of art. In his studies, he compared portraits from the sixth, eighth, tenth, eleventh, and twelfth centuries. On these portraits he found between eight and fourteen of these odd features on each of them, yielding an impressive average of 80 percent incidence. Figs. 143 through 159 show many portraits still in existence today that bear these features.

(Fig. 143) (Fig. 144) (Fig. 145) (Fig. 146)

(Fig. 147) (Fig. 148) (Fig. 149) (Fig. 150)

(Fig. 151) (Fig. 152) (Fig. 153) (Fig. 154)

(Fig. 155) (Fig. 156) (Fig. 157) (Fig.158)

(Fig. 159)

Fig. 143. Christ Pantocrator mosaic, dome of church of Daphni, ca. 1050-1100.
Fig. 144. Christ Pantocrator mosaic, in the Apse of Cefalu Cathedral, Sicily, ca. 1148.
Fig. 145. The face of Jesus, Martorana, Palermo, Sicily, ca. 1148.
Fig. 146. Christ Enthroned, mosaic in the narthex of Hagai Sophia, Constantinople, late 800s.
Fig. 147. Christ the Merciful, icon in mosaic, Ehemals, Staatliche Museum, Berlin, 1000s.
Fig. 148. Icon at Dormitron Cathedral, Moscow, 1100s123
Fig. 149. Icon at St. Ambrose, Milan, 700s.
Fig. 150. Mandylion of the Commenus period, 1100s.
Fig. 151. Christ Enthroned, Monastery of Chilandari, Mt. Athos, Greece, 1200s.
Fig. 152. Mosaic of Jesus, National Museum, Florence Italy, 1100s.
Fig. 153. Pantocrator, mosaic, Holy Luke Monastery, 1100s.
Fig. 154. Christ Pantocrator, fresco, from the dome of the Karanlik Monastry Church, Cappadocia, early 1100s.
Fig. 155. Christ Enthroned, fresco, Church of St. Angelo in Formis, Capua, Italy, 900s.
Fig. 156. Icon at St. Bartholomew's, Genoa, 1200s.
Fig. 157. Bust of the Savior in the Niche of the Pallium, immediately above the Tomb of St. Peter, Vatican Grottoes, mosaic from the 700s.
Fig. 158. The face of Jesus, Martorana, Palermo, Sicily, 1100s.
Fig. 159. Icon at St. Catherine's Monastery, Mt. Sinai, ca. 500-700.

In his studies of Byzantine coins, Dr. Alan D. Whanger of Duke University also discovered similar features of the Shroud, appearing on coins, particularly on Justinian II coins minted between A.D. 692 and 695 and a gold solidus of Constantine VII struck in A.D. 945. In addition, Whanger has also worked with some very early portraits of the sixth century. In a unique process he developed called the "polarized image overlay technique," an image is superimposed over another to identify points of similarity. With this technique, Whanger identified from thirty-three to more than one hundred points of congruence when these various images are matched to the Shroud face.[6] For example, Fig. 160 reveals a sixth-century Byzantine icon and Fig.161 shows it overlaid on the Shroud image. So many points of congruence between the Byzantine icon, the Justinian II coin and the Shroud image exist, that Dr. Whanger concludes the artists of the Byzantine artifacts must have copied from the Shroud.[7]

(Fig. 160) Sixth century face of Christ.

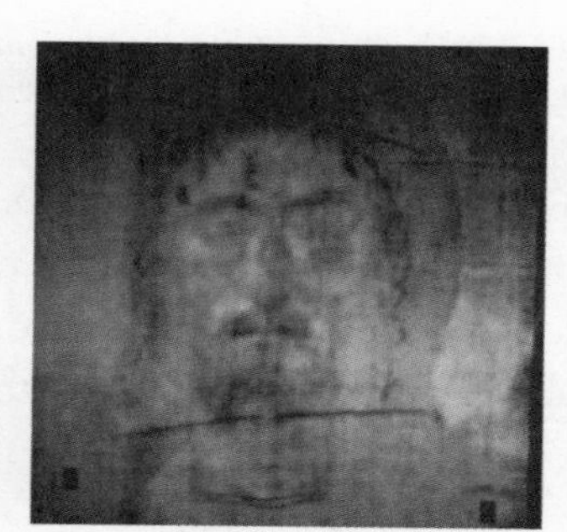

(Fig. 161) Overlay of sixth century icon on Shroud facial image.

The image overlay technique was also utilized on the Justinian II coin images. When they are superimposed over the Shroud face, all three images have a transverse line in the exact same spot. On the Justinian coin images, this is artistically represented as a wrinkle line on the figures' garments, which is a totally unnecessary feature. The Justinian II coins were minted in A.D. 695. If Whanger is correct, the image on the Shroud must date far earlier than the 1350s if it was used as a model for the coin.[8]

(Fig. 162) *(Fig. 163)* *(Fig. 164)*

(Figs. 162-164) Note the correlation when the image of the Justinian II solidus coin minted A.D. 692-695 is laid over the Shroud's facial image.

(Fig. 165) *(Fig. 166)*

Figs. 165 and 166 show two more Justinian II coins from c. A.D. 692 with similarities to the man in the Shroud.

While the argument can be made that the Shroud image was simply duplicating the coin features, and not vice versa, this would not be very plausible. If this theoretical Shroud artist, using a coin as his model, was trying to copy wrinkle lines from the coin images onto his model of the Shroud, the wrinkle lines would best be placed on a garment worn by the man on the Shroud. Why would this artist imitate the coins' wrinkled garment lines by instead portraying eleven feet of full-bodied, double-sided images of a crucified and naked man, with the frontal image, quite coincidently, having a transverse streak. This notion is ridiculous.

Dennis Mercieri, another Shroud researcher, also noticed similarities between the face on the Shroud image and a coin; this coin of gold, issued by Nicephorus II, was minted between A.D. 963 and 969. Mercieri wrote, "In linking the shroud with this Nicephorus coin, I assumed the shroud inspired the coin's design. If so, then the shroud dates to A.D. 969 or before."[9] Due to the prestige and power of their rank, both Justinian and Nicephorus would have been two of the very few in their times to have access to this sequestered, mysterious image of Christ on cloth. Based on this foundation established by Vignon and others, Ian Wilson first asserted in 1978 that the Shroud of Turin was actually the Mandylion or Image of Edessa.

The History of the Mandylion

The Mandylion, the most highly revered relic in the Byzantine Empire, disappeared from Constantinople the same time as the Shroud known to de Clari and Mesarites disappeared in 1204. During its tenure in Edessa, numerous miracles had been attributed to this cloth. The Mandylion's reputation as not made by the hand of man and bearing the divine imprint of Jesus was used to argue against the forces of iconoclasm, a force that would destroy countless images in the eighth and ninth centuries. This cloth was so highly revered that Romanus Lecapenus, Byzantine emperor in 943, was intent on bringing it to the capital of Constantinople where an already impressive collection of relics was assembled. For this mission, John Curcuas, the emperor's most able general, was dispatched to the city of Edessa, a city that had been taken from Byzantine control by the Muslims in 639.

Thus, the Byzantine Army besieged Edessa in what must have been one of the most unusual military missions in history, not to gain power or territory, but to obtain the cloth. Curcuas made several promises to Edessa's emir: the city would be spared, two hundred high-ranking Muslim prisoners would be released, twelve thousand silver crowns would be paid, and Edessa would be guaranteed perpetual immunity. All this was promised in exchange for only one thing-- the Image of Edessa. The Christian minority, which was tolerated in Edessa, was infuriated by this proposed exchange. Only after receiving permission from his superiors in Baghdad, did the emir finally agree. (However, before the Byzantines would leave Edessa, the emperor's designated representative, the bishop of nearby Samosata, would only accept the Mandylion if it were delivered with two known copies.)

Upon its arrival in Constantinople, the new emperor, Constantine Porphyrogenitus, commissioned a special feast-day sermon, "The Story of the Image of Edessa." This sermon recounted the history of the cloth to that time, and today it serves as one of our most important sources documenting its past. Upon its arrival in Constantinople, the Image of Edessa became known as the Mandylion. While in Constantinople, the image was kept in three different locations at various times: the Church of St. Mary of Blachernae, the Hagia Sophia, and the Chapel of Pharos in the Boucoleon Palace.

Wilson theorizes the Mandylion, or Image of Edessa, was actually the Shroud folded into an encasement or frame so that only the head of the image was visible, with its full-length feature not becoming more commonly known until sometime after the eleventh or twelfth century.

This is indicated by the illustration below from John Skylitzes, a Greek historian of the late eleventh century, who wrote about Byzantine emperors from the ninth to the eleventh centuries.

(Fig.167) Minature by John of Skylitzes of Mandylion's arrival in Constantinople in 944.

Figure 167 depicts the visible facial image of Jesus on the Image of Edessa or the Mandylion upon its arrival in 944. Yet, the picture clearly indicates this image was actually on a much larger and folded cloth. Wilson gives numerous plausible reasons for asserting the Shroud was the Mandylion so encased. First, in its sojourn the Image of Edessa/Mandylion would have traveled from the Jerusalem area to the eastern Anatolian steppeland of Edessa and back to Constantinople in western Turkey; all three of these locations directly correlate to the geographical origins of the pollen Dr. Max Frei found and identified on the Shroud.[10] This will be further discussed later in the chapter.

Moreover, the Mandylion and the Shroud have been described in the same terms. For instance, the Mandylion has been referred to as vague, watery or blurry — these same terms have been used to describe the Shroud image on cloth. When the Mandylion arrived in Constantinople in 944, two sons of the reigning emperor were recorded as being disappointed upon viewing it for the first time because it seemed extremely blurred and they were unable to distinguish its features.[11] The Mandylion, furthermore, was described in literary text as "an imprint," a description that also most accurately describes the Shroud. In the "Story of the Image of Edessa," the sermon's author described the image as being made by a "secretion without coloring or painter's art … it did not consist of earthly colors." The same can also be said of the Shroud.

Because its image was vague, the Mandylion was copied in various ways. These variances can be found in the length of the beard and the degree to which it forks. Other variations exist with the hair, falling vertically to the neck in some copies while splaying out to the sides in others. Likewise, variations exist with the eyes: sometimes they seem to be looking straight ahead, while at other times, they seem to be looking to the right or left.[12] It is impossible to know how many artists have copied the Mandylion directly. While some artists definitely had an opportunity to copy the image, other artists may have only copied the copies.

Furthermore, one of the most persistent and striking claims attributed to the Image of Edessa and the Mandylion throughout the centuries is that it was *acheiropoietos*, or "not made by hands."[13] Much later in the 1930s, Pope Pius XI, while speaking of the Shroud, stated that its image was "certainly not by the hand of man."[14] This is a very common conclusion of many who have visually studied the Shroud. This statement was made four decades before Wilson advocated his theory that the Image of Edessa, the Mandylion and the Shroud were the same cloth. For centuries, the Mandylion was described as *acheiropoietos*. In the 20th century after the Shroud was thoroughly examined, many scientists would arrive at the same conclusion. One STURP scientist, John Heller, who conducted extensive chemical analyses of the Shroud, spoke for many when he concluded the Shroud image "was not made by the hand of man."[15]

Another common feature found in portraits of Jesus dating from 540-940 further support Wilson's hypothesis. These portraits feature a trellis pattern surrounding the circular halo around Jesus' face.[16] The tenth century sermon mentioned previously, "The Story of the Image of Edessa," gives a plausible explanation for this feature. It describes Edessan King Abgar "fastening it to a board and embellishing it with the gold which is now to be seen."[17] The Mandylion apparently continued to be mounted and framed in this fashion while the cloth was in Constantinople. Byzantine art scholar, Professor Andre Grabar, has exhaustively studied copies of the Mandylion.[18] The copies made more than fifty years after the Mandylion's disappearance in 1204 show the Shroud suspended with the cloth limp and hanging free. However, copies of the Mandylion before its disappearance in 1204, show the cloth taut, with a fringe and frequently with a trellis pattern. In addition to adding decoration around the portrait, the fringe or trellis pattern would serve to keep the full-length Shroud hidden inside its frame.

The coloring of the Mandylion's copies is another indicator that the Shroud and Mandylion are the same object. The Shroud's original background is ivory white, the natural color of linen. While white is a common color with many shades, the consistent color of the background of many Mandylion copies is also ivory white. From a few feet away, the Shroud image looks sepia brown. However, close microscopic examination under white light reveals the true color of the Shroud body image fibers to be straw-yellow. The color images of the Mandylion copies also vary "from a sepia monochrome to a rust-brown monochrome, slightly deeper but otherwise virtually identical to the coloring of the image on the Shroud."[19] Another indicator the cloths are one in the same is copies of the Mandylion contain the same anomalous facial features, or Vignon markings, discussed earlier. Wilson asserts as many as thirteen of these anomalous features have appeared on copies of the Mandylion.

According to several references, the Image of Edessa/Mandylion was not necessarily a small cloth. In describing the Image of Edessa, one early author actually used the word *sindon*.[20] *Sindon*, meaning "shroud," is the word used in the gospels to describe Jesus' burial garment.[21] The original "Latin Abgar legend" also presumed the burial cloth was a cloth several yards long.[22] Writing circa 730, John of Damascus refers to the Image of Edessa as a likeness of Jesus upon a *himation*. This word means a full-length, oblong, outer garment that the Greeks wore over the shoulder and down to the feet.[23] In yet another reference to its size, this one from the late tenth century, Leo the Deacon refers to the Image of Edessa as a *peplos*, which means a full-size robe.[24]

In his theory, Wilson proposes that the full-length Shroud was doubled, then doubled twice again. This results in a cloth that is doubled in four sections. When folded in such a manner, the head appears disembodied in a wide landscape frame (3'7"), as opposed to a more typical portrait frame or format. Existing copies of the Mandylion prior to its disappearance in 1204 support this theory. There are only two existing copies that show the head displayed in a portrait, rather than a landscape orientation, as is seen in Fig.116.[25] This artistic portrayal of a head on a landscape-shaped background is unnatural and is contrary to the universal artistic conventions found in art history throughout the centuries. Not only is it visually unappealing, it is also a waste of artistic space, which in itself may be one of the more compelling reasons to support Wilson's hypothesis. For this unusual and contrary portrayal to persist through several centuries suggests that the artists copying the Mandylion must have had a common reason: they must have been trying to duplicate the appearance of the original Mandylion, which served as a model for all the others. All would not have arbitrarily decided to do this for no explicable reason.

Further support for Wilson's theory is found in the unique description of the Image of Edessa/Mandylion as τετραδιπλον. This word is first used in a sixth-century text[26] after the cloth is rediscovered after five centuries of absence. Later, in a tenth-century text written not long after the cloth's arrival in Constantinople, it is described similarly as ρακος τετραδιπλον.[27] This word is a compound of two very ordinary Greek words τετρα, meaning "four," and διπλον, meaning "doubled," hence "doubled in four." However, according to Cambridge University Professor G. W. H. Lampe, editor of the *Lexicon of Patristic Greek,* in all of literature the word τετραδιπλον is used *only* in connection with descriptions of the Image of Edessa/Mandylion. The rarity of this word's usage is not an accident. Rather its use *only* in descriptions of the Mandylion suggests that the author was trying to characterize what he may have been fortunate to observe — the manner in which a full-length cloth was folded within its frame in a doubled-in-four pattern. Otherwise, the use of this term in connection with the Mandylion has no apparent meaning.

This doubled-in-four configuration has been confirmed by scientific studies, specifically by Dr. Max Frei's pollen sampling of the Shroud, which determined that the facial area contained more pollen exposure than other parts of the cloth.[28] Furthermore, in another important study, Shroud scientist John Jackson examined the locations where Wilson's proposed folding would have left evidence on the cloth. He found that such a fold pattern probably remains on the Shroud today.[29] This doubled-in-four fold pattern would leave seven fold marks on the cloth. In his studies, Jackson utilized X-ray, reflectance and raking light (grazing angle illumination) photographs of the Shroud and identified four probable fold marks from Wilson's fold pattern. He found tentative fold marks at two other locations on the cloth. All were consistent with Wilson's doubling-in-four method. Jackson could not detect a seventh fold mark because the site where this final fold mark should be located was fire marked, water stained, and patched at this position. While additional scientific examination of the Shroud could reveal more

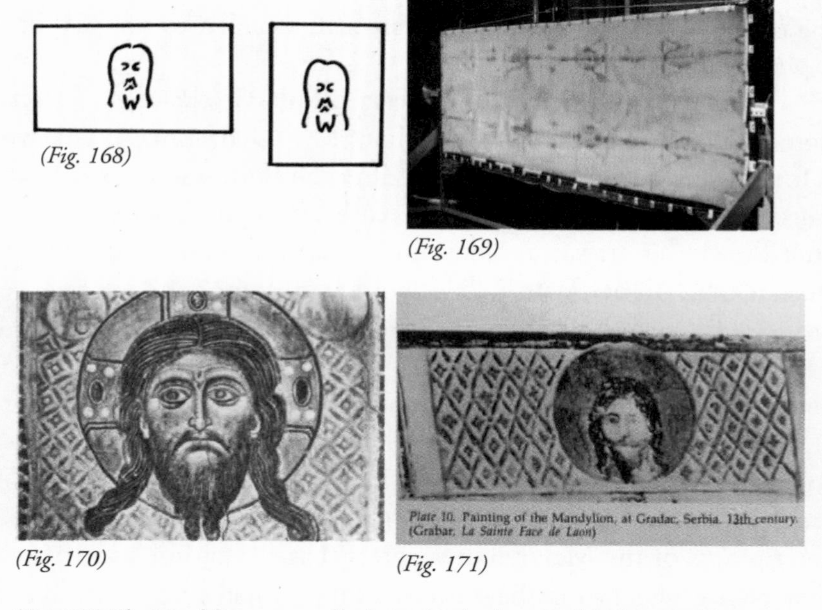

(Fig. 168) *(Fig. 169)* *(Fig. 170)* *(Fig. 171)*

(Fig. 168) The Mandylion was usually depicted in landscape rather than portrait aspect. (Fig. 169) Full-length Shroud as seen with the naked eye. (Figs. 170 and 171) Mandylion copies from 1199-1200s, enclosed in a trellis frame, which is how the Mandylion was kept for part of its history according to historians.

information on this matter, Wilson was convinced--"certainly there can no longer be claimed to be any absence of fold marks consistent with the Image of Edessa/Shroud identification hypothesis."[30]

Sadly, the Image of Edessa/Mandylion has been lost to history. However, if the Shroud and the Image of Edessa/Mandylion are, in fact, one and the same, the known history of the Image of Edessa/Mandylion, as explained in the next section, would complete almost the entire missing history of the Shroud of Turin.

All of this evidence is clearly consistent with Wilson's theory that the Shroud was the Mandylion. Over the centuries there were many more copies of the Mandylion and references to it that have not survived the record of history. (Perhaps, because of the Shroud's long-time fold configuration and the great number of copies as such, most references to the Mandylion are as a facial image.) The above evidence, of course, doesn't prove that the Shroud was the Mandylion, but it does indicate a likely presence and influence in history in this part of the world from the sixth to the 13th centuries.

Early History of the Mandylion or Image of Edessa

Numerous written accounts exist detailing the early years of the sacred cloth known as the Image of Edessa especially as it relates to the early, albeit brief Christianity of Edessa and the semi-legendary Abgar story.[31] Eusebius, author of the famous *History of the Church* (A.D. 325), also wrote the first account of the Abgar story. In it Eusebius states that his account is his own translation of the Edessan archives from the Syriac into Greek. The archives stated that King Abgar of Edessa, who ruled from A.D. 13 to 50, suffered greatly from an incurable disease. Abgar had heard of the many miracles that Jesus had performed, so he sent a messenger with a letter to Jesus, inviting him to come to Edessa to cure him. Unable to make the journey, Jesus sent a letter to Abgar, in which he promised to send one of his disciples to cure him after Jesus' mission on earth was complete.

When whole caravan loads of ancient Syriac manuscripts were recovered and retrieved from the desert of lower Egypt from the Nitrian Monastery in the 1840s, many more details of this legend emerged.

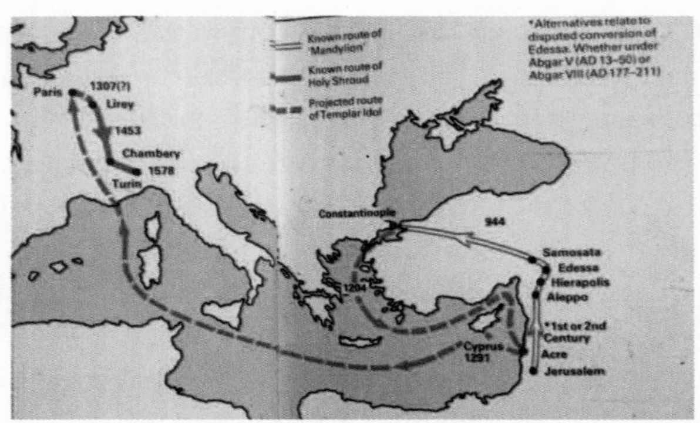

(Fig. 172) Map of the Shroud's various locations.

Many of these manuscripts contained accounts of the Abgar story written in the Syriac language,[32] the same language from which Eusebius' account was translated. Taken together, these manuscripts and the "Story of the Image of Edessa" provide an extensive account of the Abgar story, detailing the early evangelization of Edessa and the role the Image of Edessa played in its evangelization.

All of these accounts agree in this important aspect: following Jesus' departure from earth, the disciples sent the evangelist Thaddaeus (Addai, in the Syriac language) into Edessa. The more reliable versions refer to him as "one of the seventy" described in Luke 10:1. While in Edessa, Thaddaeus healed many people in the name of Christ and he lodged with a man named Tobias, whose father, also named Tobias, was a Jew originally from Palestine.[33] As is evidenced by the number of tombs, a substantial Jewish community resided in Edessa at that time. Abgar, hearing of the wonders that Thaddaeus had been performing, was reminded of Jesus' promise, so Thaddaeus was summoned to the King. As further chronicled in the "Story of the Image of Edessa," Thaddaeus brought with him the Mandylion and, before entering the throne room to see Abgar, "placed it on his forehead like a sign."

> Abgar saw him coming from a distance, and thought he saw a light shining from his face which no eye could stand, which the portrait Thaddaeus was wearing produced.
>
> Abgar was dumbfounded by the unbearable glow of the brightness, and, as though forgetting the ailments he had and the long paralysis of his legs, he at once got up

> from his bed and compelled himself to run. In making his paralyzed limbs go to meet Thaddaeus, he felt the same feeling, though in a different way, as those who saw that face flashing with lightening on Mount Tabor.
>
> And so, receiving the likeness from the apostle . . . immediately he felt . . . his leprosy cleansed and gone . . . Having been instructed then by the apostle more clearly of the doctrine of truth . . . he asked about the likeness portrayed on the linen cloth. For when he had carefully inspected it, he saw that it did not consist of earthly colors, and he was astounded at its power. . . .[34]

These accounts further state that, upon Abgar's request, Thaddaeus stayed and spoke of all the things that Jesus had said and done and of his mission on earth, in effect giving the first sermon at Edessa. Most accounts also state that Thaddaeus stayed after this first sermon, establishing the first church at Edessa and the birth of Christianity in this area. Aggai, who formerly made the silks and headdresses for Byzantine's pagan kings assisted Thaddaeus during this time. When Thaddaeus died, Aggai became the leader of the church in Edessa. Both Abgar V and his son, Ma'nu V, who ruled from A.D. 50 to 57, allowed Christianity the freedom to develop in a formerly pagan country and it thrived during their reigns. However, this freedom ceased when Ma'nu VI ascended to the throne when he chose to return to the pagan practices of old. Ma'nu VI ordered Aggai to "make me a headdress of gold, as you did for my fathers in former times." Refusing, Aggai stated, "I will not give up the ministry of Christ, which was committed to me by the disciple of Christ and make a headdress of wickedness.[35] This refusal would be his death sentence. Shortly thereafter, the King's men burst into the church as Aggai was preaching, breaking both of his legs and killing him.

The young and fragile Christian community disappeared from Edessa. At this same time, the Edessa Image also disappeared from recorded history. It would not reappear for hundreds of years. Between the death of Aggai and the sixth century, no one living or traveling in Edessa was aware that the Image of Edessa/Mandylion was in the city.

Several examples exist that show holy pilgrims traveling to Edessa who were unaware of the sacred image. One example is the pilgrim Egereia, who visited Edessa around the year 384 as part of her tour of Christian holy places.[36] Egereia describes her travels in the greatest of detail but makes no mention of an image-bearing cloth. Historian Sir Steven Runciman describes her as follows:

> She was a sightseer of a thoroughness unrivaled even by the modern American; and, had so interesting a relic then existed, she would certainly have referred to it.[37]

Another example is St. Ephraim, the so-called harp of the Syrian church. Living in Edessa in the late fourth century, he chronicled its history. However, in all the reams of ecclesiastical verse he wrote, not one mention is made of the Mandylion. Nor is the Mandylion mentioned by two other well-known literary sources from that time: the "Chronicle of Joshua the Stylite," written by the monk around 507, or any of numerous writings of Jacob of Serug, a most prolific writer in Edessa, who died about 521.

When the Image of Edessa was rediscovered in the sixth century, the reasons for the cloth's disappearance centuries earlier became more apparent. Found in a space above the city's western gate, the cloth, according to the "Story of the Image of Edessa," had been concealed and carefully bricked over. Also found hidden with the cloth was a brick-red tile imprinted with the face of Christ and a small lamp. Subsequently known as the Keramion, this tile was another object of historical significance. According to Wilson, the Keramion was one of many stone or clay heads of gods and gorgons commonly displayed over gateways in the Parthian Empire. (Example can still be seen today at Parthian Hatra.) The Keramion was kept in Hierapolis (the sister city of Edessa) for many years before also being transferred to Constantinople in 969.

Wilson also believes the Keramion was probably modeled upon Christ's likeness on the Mandylion and displayed above the city gate in Edessa until the reign of Ma'nu VI. With his return to paganism and subsequent persecution of Christians, to prevent its destruction the

Keramioin was removed from view and concealed in the brickwork. Since destruction of Christian relics was occurring, someone also hid the Mandylion there for safekeeping. This hiding place proved fortuitous for two reasons: not only did it save the Christian relics from destruction by the pagan ruler, but, for the next five centuries, it also provided a hermetically sealed environment.

This location turned out to be fortuitous for another reason. Severe floods in 201, 303, 413, and 525 caused extensive damage to Edessa's palaces and churches. Since the land rises steeply at what was Edessa's western gate and the Mandylion was sealed above that gate in one of the city's highest points, it remained unscathed. The flood of 525 was the most devastating flood that Edessa experienced: one third of the population died, and public buildings, palaces, churches and much of the city wall were destroyed. Due to this vast destruction, the wall and its outworks had to be rebuilt and it is during this rebuilding, Wilson postulates, that the Image of Edessa was rediscovered. As discussed earlier in this section, evidence of the cloth was not recorded by Jacob of Serug before his death in 521, nor was it mentioned by Joshua the Stylite in 507. However, the Mandylion was described by Evagrius and the role it played in the attack on the city by Chosroes the Persian in 544.

An important controversy between the Monophysites and the Orthodox was taking place at the time of the flood in 525, which may also help explain why the cloth remained unknown for several more years. The Monophysites, dominant in Edessa at the time, did not believe that Christ should be represented in icons or pictures because he was more spiritual than human. The Orthodox, on the other hand, argued that since Christ had both a human and divine nature, his human appearance could be represented in art. With the Monophysites in control, the rediscovery of the Image of Edessa may not have been made public for several years. However, after 544, the image was prominently mentioned after Evagrius described the important role of the Mandylion in the defense of the city.[38]

In his account, Evagrius relates how Chosroes built a massive mound of timber, higher than the city wall, which was to be moved next to the wall. This mound would serve as a platform from which his army could attack the city. Before it was moved to the city wall, however, the Edessans had tunneled under the wall below the mound with the intention of setting it on fire. Evagrius then goes on:

> The mine was completed; but they [the Edessans] failed in attempting to fire the wood, because the fire, having no exit whence it could obtain a supply of air, was unable to take hold of it. In this state of utter perplexity they brought out the divinely made image *not made by the hands of man*, which Christ our God sent to King Abgar when he desired to see him. Accordingly, having introduced this sacred likeness into the mine and washed it over with water, they sprinkled some upon the timber . . . the timber immediately caught the flame, and being in an instant reduced to cinders, communicated with that above, and the fire spread in all directions.[39] (italics added.)

This miraculous account recorded by Evagrius' marks the triumphant reappearance of the Mandylion in Edessa.

After its triumphant reappearance in the sixth century, the Edessan Image, once again, became an object to be studied and revered and those who saw it would attempt to capture its essence in words. From that point on, the Image of Edessa/Mandylion was described as archeiropoietas, "not made by hands" and imprinted on the cloth. As its reputation grew and increased, the Image of Edessa became the most valued possession of the city. Subsequently, artists throughout the Christian world sought to make copies of the Mandylion with its likeness of Christ. With this re-emergence of the Image of Edessa in the sixth century, the traditional likeness of Christ, as was discussed earlier in this chapter, began to appear in art.[40]

(Fig. 173) Tenth-century copy of the Edessa cloth.

Wilson made a brilliant, well-reasoned theory thirty-seven years ago to account for the Shroud's history from the first century to the 1350s. However, I disagree with one major aspect of it. Wilson speculates that the Shroud would have caused the disciples serious problems for two reasons. The first is, as a burial garment,

Jewish law would have deemed the cloth unclean. In addition, its image would have violated the Second Commandment. He believes the disciples sending the cloth to nearby Edessa would have solved their problem. Not only was Edessa outside the borders of the Roman Empire, its emperor Abgar had also shown an interest in Jesus, and images would not be forbidden there. However, he believes one final step was needed--the transformation of the burial garment into a portrait. The grave cloth of a convicted criminal executed under the most degrading circumstances would be repugnant to anyone, especially the king to whom the cloth was being sent. The prime candidate to perform this transformation, according to Wilson, would be Aggai, the maker of headdresses and silks. The earliest copies of the Mandylion contain a gold trellis work. This is typical of the trellis-style embellishments found on headdresses and costumes of Parthian monarchs. If Aggai indeed disguised the cloth as a portrait, Thaddaeus and Aggai would be the last to know that the image was a full burial shroud with Jesus' body imprinted on it. This fact would remain unknown for many centuries.

To me and many others, the weakest link to the Shroud's authenticity was the total absence of references to an image of Jesus on his burial garment following his resurrection. If this image existed, it should have been proclaimed both orally and in writing, not only in the Gospels and New Testament, but in many other sources. Even had an image been visible, the disciples would not have violated the Second Commandment: "You shall not make for yourself a graven image, or any likeness of anything that is in heaven above, or that is in the earth beneath ...; you shall not bow down to them or serve them." This is true for two important and obvious reasons. First, the disciples had not made any graven image, for Jesus, himself, would have left it during his resurrection. Second, there is no account of anyone worshipping the image itself. Therefore, the prohibition against worshipping a graven image or an idol had not been violated.

There is no record of any kind detailing an image of Jesus on a cloth at this time. The New Testament does not mention an image left behind by Jesus on his burial garment. While other sources, such as the apocryphal Gospel of the Hebrews written in the second century, do mention Jesus' burial garment as being given to the servant of the priest (Peter), they do not mention an image on this garment. In another tradition, the garment was acquired by Mary Magdalene. In a fourth century account by the apostle St. Nino, she reports the common belief during her youth in Jerusalem was that St. Luke had acquired the burial garment from Pilate's wife and then hid it.[41] While they certainly all three cannot be true, the historical accuracy of these sources is less important than the fact that nowhere in any of the three is mention made of an image being on the burial cloth.

No mention is made of an image of Jesus in the first account of the Abgar story, written by Eusebius c. A.D. 325, discussed previously in this chapter. However, the tradition of Christ's letter to Abgar clearly survived during this period. Reference to an "image not made by human hands" does not appear until the sixth century. From that point on, several accounts and versions appear of Christ's miraculous imprint on the cloth, the best-known being written after the cloth's tenth-century arrival in Constantinople. With the rediscovery of the image of Christ not made by human hands, but miraculously imprinted by Jesus, writers from the sixth century on give the cloth a first-century history. In fact, these are the only versions that can or do give the image a first-century origin.[42] Likewise, no paintings prior to the sixth century exist, which depict Abgar receiving the Image of Edessa. Only paintings completed after the sixth century depict this image. The absence of such a known image best explains the complete lack of a record or tradition of such an image at this time.

Remember, the yellow body image on the Shroud only consists of "chemically altered cellulose consisting of structures formed by dehydration, oxidation, and conjugation products of the linen itself."[43] The changes in cellulose that are known to be the result of aging are these same dehydrative and oxidative processes. The spectral reflectance curves for the body image and background can also be produced by laboratory simulations using controlled accelerated aging processes. The Shroud's chemistry is similar to the chemistry that causes yellowing of linen with age. In fact, the body image is visible only because it's caused by a more advanced decomposition process than the normal aging rate of the background linen itself."[44] In other words, the body image is a result of an aging process that develops over time. Since the image appears to have developed as the result of an aging process, the image that we see today would not have been present for many decades.

This lack of image on the Shroud in its early history actually gives it more authenticity. This offers a plausible, scientific explanation as to why the Gospels do not mention an image on Jesus' burial cloth. If all the disciples or apostles had was an unmarked burial garment, its significance would have been minimal. It could have easily sojourned to Edessa. As a burial cloth, to the Jews or anyone else, there would be little reason to keep it or subsequently, to mention it in the Gospels. However, due to the simple fact it had covered Jesus' body, they may have been reluctant to destroy it. It easily could have made its way to a near-by people who spoke virtually the same language, had an early interest in Christianity and lived astride a major east-west caravan route. While the second- to fourth-century accounts of Jesus' burial garment are somewhat contradictory, they are consistent with its continued existence after the time of the Gospels, albeit without an extant image.

As Wilson implied, the Shroud could very well have been hidden away in Edessa at the time for the very reasons he states. All Christian relics or symbols of any kind were at risk. Regardless, at over fourteen feet long, the Shroud would still have to be folded no matter where it was kept. When it was folded naturally (doubled in four), the face would have been the only part showing. Perhaps, once it was rediscovered at a time after an image of Christ had developed, a frame was then put on it, and it remained in the same fold pattern in which it was found. Alternately, it could even have been given its fold pattern at the time the newly formed image was discovered to camouflage the fact that it was a burial garment and would easily be recognized as such in that part of the world. This could have happened in the sixth century or any century previously.[45]

Since the most devastating of Edessa's floods occurred in 525, and accounts of the image of Christ in Edessa "not made by the hands of man" also start to appear in the sixth century, this appears to be the most logical time that the cloth and its image were rediscovered. Perhaps, the cloth was rediscovered earlier, but did not acquire its fame until it was publicly utilized to help save the town from attack.

One other less likely scenario should be considered. Possibly, the

Shroud had been exposed to the elements from the time it left Jerusalem until A.D. 57. Perhaps, by the end of Abgar's reign, or during the reigns of Ma'nu V and Ma'nu VI, due to this exposure, it was beginning to or had developed a more visible image than would have been observed in Jerusalem around A.D. 30. If this were the case, around A.D. 50 to 57 when the image first began to appear, Aggai could have naturally folded the cloth to display a visible facial area and conceal its nature as a burial cloth and subsequently framed it. Even if some kind of an image formed on the cloth while it was in Jerusalem, the disciples still may not have seen it. To actually see an image they would have needed to unfold and suspend the fourteen-foot cloth and stand six to ten feet away, something they may never have had the opportunity or inclination to do.

Another Shroud Route To Edessa

Alternative theories exist as to how the cloth could have reached Edessa. One such theory is proposed by historian and archaeologist William Meachem, who writes as follows: "By A.D. 66 the Judeo-Christians had migrated east of the Jordan, and thereafter little is known of them apart from their increasing isolation from the early church and their heretical tendencies. If the Shroud had been taken from Jerusalem by this group, its obscurity in the early centuries would be understandable."[46]

The exact provenance of the Image of Edessa prior to the sixth century cannot be known for certain at this juncture in history and may never be. However, we do know several facts. A very real Abgar V ruled in Edessa from A.D. 13 to 50. A very real Image of Edessa was documented and all accounts written described it as "not made by hands." What's more, the city of Edessa was located only 350 miles from Jerusalem and was even closer to one of the first Gentile Christian communities at Antioch. Edessa was situated at the intersection of the major trade routes of the day, and its people spoke a language almost identical to the Aramaic of the Palestinian Jews. All of these facts serve to confirm the accounts of Abgar's learning of Jesus and the early evangelization of Edessa, however brief, a very real possibility.

Proving the history of first century Edessa is problematic due to a lack of any other early references. According to Warren Carroll:

> Almost all the records of early Christianity (before 300) that have come down to us derive from the Graeco-Roman, not the Parthian world; and their total volume is small. Therefore, we are unlikely to find among them historical reports on a city in an alien realm, its people speaking an alien tongue, who had the faith only briefly, and then lost it for a long time. When the first Christian history properly speaking is written by Eusebius, the early conversion of Edessa is featured.[47]

Another reason for the lack of references in this early period could also be due to the burning of the great libraries. Crusaders destroyed the library of Constantinople in 1204 where information pertaining to the Mandylion quite possibly could have been contained. In addition, invaders inflicted heavy damage on the great libraries of the important early Christian center of Alexandria, once in 391 under Theodosius I and again in 642 during its fall to the Arabs. Perhaps in future years, more will be uncovered by historians and archaeologists, but much enlightening information has been destroyed forever.

In 944 the Christians in Edessa had been reluctant to relinquish the Image due to the powerful and protective reputation it had attained. However, had they not done so, the cloth almost certainly would not have survived. Two centuries after the Shroud was removed to Constantinople, Turkish Muslims seized and destroyed the city of Edessa in an attack unprecedented in its almost fifteen-hundred year existence. J.B. Chabot states, the city was so thoroughly looted that…

> . . . for a whole year they went about the town digging, searching secret places, foundations and roofs. They found many treasures hidden from the earliest times of the fathers and elders, and many [treasures] of which the citizens knew nothing. . . .[48]

The Christian population scattered as Christian civilization was obliterated. Had the cloth still been in Edessa, it, too, would probably have been one of the casualties.

The Full-Length Feature of the Shroud Becomes Known

The power of the Muslims continued to grow rapidly in the eleventh and twelfth centuries. In an effort to protect themselves from this growing power, Byzantine emperors actively sought allies in prominent visitors from the West. In the past, visitors had not been allowed to view precious relics of the empire. At this time, in a complete reversal of previous practices, guests were sometimes shown the empire's treasures. Also, during this same time, the first documentation of a full-length figure on the cloth is recorded. In one such document, an interpolation of an original eighth-century sermon recorded sometime before 1130, Pope Stephen III refers to the Mandylion:

> For the very same mediator between God and men [Christ], that he might in every way satisfy the king [Abgar], stretched his whole body on a cloth, white as snow, on which the glorious image of the Lord's face and the length of his whole body was so divinely transformed that it was sufficient for those who could not see the Lord bodily in the flesh, to see the transfiguration made on the cloth.[49]

Another twelfth-century example is found in a Vatican Library Codex; its reference to Christ's letter to Abgar reads in this way:

> If indeed you desire to look bodily upon my face, I send you a cloth on which know that the image not only of my face, but of my whole body had been divinely transformed.[50]

Yet another twelfth-century example can be found by the English Monk Ordericus Vitalis in his history of the church written in 1130:

> Abgar reigned as toparch of Edessa. To him the Lord Jesus . . . a most precious cloth with which he wiped the sweat from his face, and on which shone the Savior's features miraculously reproduced. This displayed to those who gazed on it the likeness and proportions of the body of the Lord.[51]

It is quite possible that the last reference may have derived from the first reference mentioned and, keep in mind, none of these references should be taken literally. Jesus impressing his naked, crucified body on the cloth while alive is absurd. In their attempts to explain the history of the Mandylion, these references are also highly implausible. However, and more importantly, they attest to the fact that, at this time, people had knowledge of a full-length figure of Jesus on the Mandylion cloth.

Another indication occurred simultaneously to these written references in which the Shroud also contained a full-length figure: the new Lamentation scenes in art.[52] In earlier depictions of Jesus' burial, he was invariably shown wound like a mummy in swathing bands, not in a large shroud-like cloth. In the twelfth century, with no other apparent explanation, artists began depicting scenes of Jesus' burial quite differently. His body is no longer wrapped in bands, but enveloped in a shroud. Several examples also show, for the first time ever, his hands crossed over his loins, consistently portraying the right hand crossed over the left.[53] This is exactly how the hands appear when seen on the Shroud image. An excellent illustration of these qualities is the famous Hungarian Pray manuscript, ca. 1192 to 1195 (below). Furthermore, as the Shroud does, the manuscript uncharacteristically depicts Jesus completely naked. There are two other unusual characteristics. First, Jesus' thumbs are clearly missing on both depictions. A second feature is a "hot poker hole" pattern in the shape of a "7," found on the herringbone type of pattern on the manuscript's sarcophagus. This pattern resembles similar holes on the Shroud, known to predate the fire marks of 1532.

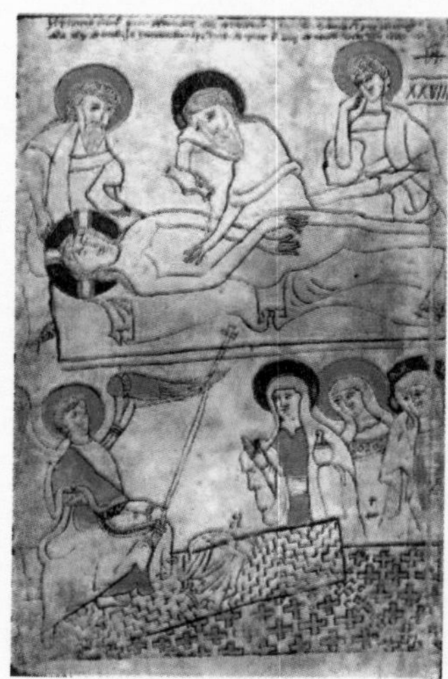

(Fig. 174) Hungarian Pray Manuscript, 1192-1195.

Other examples exist that also illustrate a full-length representation of Christ. Liturgical cloths, called *epitaphioi*, have this feature. While the earliest examples of *epitaphioi* survive from the thirteenth century, Wilson believes their similarity to the Lamentation scene and their consistent feature of Christ's body laid out in death with his hands crossed right over left at

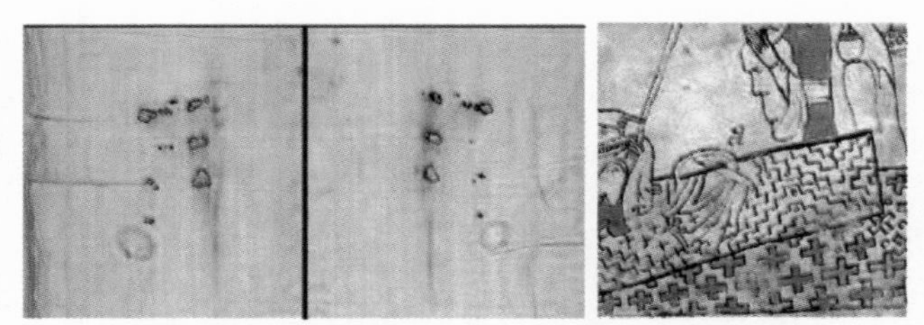

(Figs. 175 left and center and Fig. 176 right) Shroud holes and Manuscript holes.

the loins suggest a twelfth-century origin. A liturgical cloth in the Museum of the Serbian Orthodox Church, Belgrade (below) is an excellent example of just such a portrayal. It is important to note that, while the Shroud was first displayed in France in the mid-1350s, this artistry existed a full century before that first exhibit.

(Fig. 177) Liturgical cloth from the 1200s, whose scenes probably began in the 1100s. Both dates are well before the Shroud first appears in Europe in the 1350s.

Another example indicating that the full-length feature was known at this time can be found from an umbella, an ornamental tapestry canopy probably sent to Pope Celestine II from Byzantium in the twelfth century. The canopy was destroyed, but archivist Jacopo Gimaldo has preserved its appearance in his drawings. These reveal that the centerpiece of the umbella was a representation of Christ in exactly the manner of the Shroud.[54]

The fold marks, as well as a band of discoloration and small tack marks present on the Shroud, may also illustrate that the Shroud's full-length feature was known at this time. As discussed earlier, during the STURP examination in 1978, John Jackson first noticed these fold marks. In his overall Shroud study since then, Jackson has reconstructed the precise manner in which the cloth was folded. When the cloth is folded doubled-in-four, there also exist four closely spaced parallel lines beginning at, and immediately below, the crossed hands. Jackson found that these lines and the doubled-in-four fold marks all lined up if the Shroud was wound around a square block of wood, which ran the length of the cloth while it was folded inside its container as illustrated in Fig. 178.

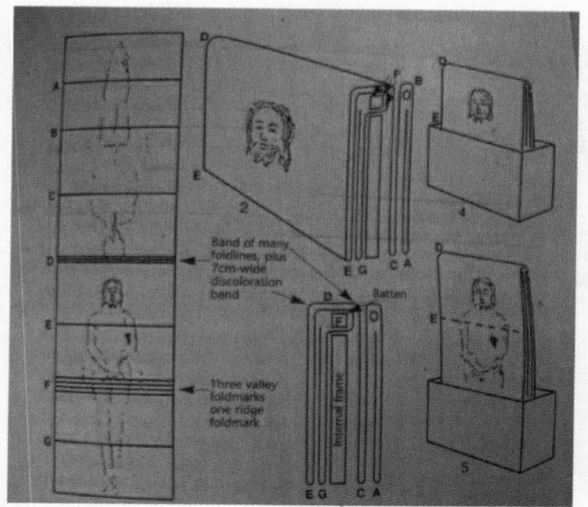

(Fig. 178) The fold marks on the Shroud suggest the manner in which the Mandylion was displayed.

For years, scholars could offer no explanation for keeping the Shroud folded in such an odd arrangement. However, in 1993, Shroud scholar Heinrich Pfeiffer, S.J., came up with a plausible theory. If the shroud was tacked to a frame above and behind these small blocks, the cloth could then be pulled straight up vertically and the image would "stand" or arise, but would stop at the level of the hands.[55] This theory would explain Robert de Clari's description of Jesus' shroud, which stood up straight or raised itself so that the figure of our Lord could be seen.

Jackson offers additional evidence for Pfeiffer's theory in a band of discoloration found on the Shroud of Turin that could have been wrapped around the block of wood under this fold arrangement. In addition, small round rust marks consistent with the locations where tacks would have been placed to secure the cloth to the backing boards are also present on the Shroud.[56]

Providing even more support for this theory is the Extreme Humility or Utmost Humiliation (Man of Pity or Man of Sorrows in the West) types of paintings, which began during the late eleventh and early twelfth centuries. A typical representation is illustrated in Fig. 179.[57] Some of these representations show a man rising out of a box, obviously too small to contain the rest of Jesus' body, but easily large enough to contain the rest of a folded cloth.[58] Because the Shroud is not a painting and contains both full-length frontal and dorsal images, Jackson thinks that the Shroud in the fold-and-display arrangement was the prototype for the Extreme Humility representations and not vice versa.

(Fig. 179) Man of Pity, ca. 1300, made in Constantinople. Presently at St. Coce in Gerusalem, Rome.

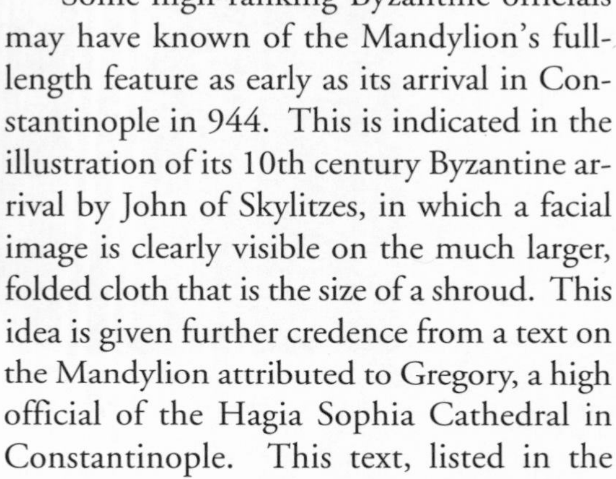

Some high-ranking Byzantine officials may have known of the Mandylion's full-length feature as early as its arrival in Constantinople in 944. This is indicated in the illustration of its 10th century Byzantine arrival by John of Skylitzes, in which a facial image is clearly visible on the much larger, folded cloth that is the size of a shroud. This idea is given further credence from a text on the Mandylion attributed to Gregory, a high official of the Hagia Sophia Cathedral in Constantinople. This text, listed in the Vatican Library as Cod. Vat. Graec. 511, pp. 143-50b, appears to date from the time the Mandylion was first received at Constantinople. The amount of detail describing the event, ceremony and those present, strongly suggests that the author was an eyewitness. While it was discovered by Gino Zaninotto, Werner Bulst actually presented the text of this document in his book.[59] The most interesting part is its reference to the image itself, in which its full-length feature must certainly have been observed. According to Bulst:

> After having related the legend of Abgar, Gregory describes and explains the image in a manner for which up to now we have found no parallel. . . . He speaks of the "side" [πλευρα] and of the "blood and water found there" ['αιμα και ’υδορ εκει], and the wound in the side. He must have seen the cloth up close and thus we may suppose that he belonged to the imperial delegation responsible [in 944] for the reception and verification of the Edessa image.[60]

In a paragraph on how the cloth was mounted in its frame, Bulst continues with this summation: "Gregory's attestation of the side wound confirms anew the identification of the Edessa image and the Turin Shroud. That the wound with its blood and water is not mentioned elsewhere is easily explained. The image was not shown in public."[61]

While in Constantinople, the cloth itself was never publicly exhibited. An inscription on a twelfth-century Byzantine communion veil offers a probable reason for this:

> If no Israelite might look directly on the countenance of Moses when he came down from the mountain where he had seen God, how shall I look upon Thy revered Body unveiled, how regard it?[62]

The Byzantines would have accepted not being allowed to view the cloth as perfectly normal. They were accustomed to the most sacred parts of the liturgy being performed out of the sight of the congregation. Furthermore, they were raised on classical tales of people being blinded after looking at gods with unveiled eyes. Only anonymous monk-artists were allowed private expositions in order to paint copies. On the two recorded occasions the Mandylion was involved in a public exposition, it was carried in a casket and was not removed for public viewing. It was treated similarly while the cloth was kept in Edessa; only the archbishop was allowed an occasional, momentary glimpse of the cloth.[63]

During this era of its history, the Image of Edessa and the Mandylion had been attributed as having great protective powers, both while in Edessa and Constantinople. Legend has it curing Abgar, and after being rediscovered five centuries later, saving Edessa and defeating Chosroes, the invader. Those who argued against iconoclasm, a force that destroyed countless images in the eighth and ninth centuries, cited it as the primary example.[64] Of course, the Byzantines wanted the cloth for its famous protective powers and its enormous reputation. In 1204 when faced with the threat of obliteration from the soldiers of the Fourth Crusade, who were inside and outside the walls of Constantinople, it is no wonder that the Shroud was publicly displayed for the first time at the Blachernae Palace, a rallying point for the citizens of Constantinople in times of distress.

Shroud Locations Between 1204 and the1350s

The Image of Edessa/Mandylion was a revered and documentable historical object until 1204, when it disappeared from Constantinople during the Fourth Crusade, which was primarily commanded by and composed of an army of the French. After this date, it was never mentioned or seen again. However, subsequently in the next century, a cloth appeared in France This cloth matched the previously mentioned written descriptions of a full-length figure of Christ and it also matched depictions of the Mandylion/Image of Edessa. What actually happened to the Shroud between 1204 and the mid-1350s, when it was displayed in Lirey, France, as the burial garment of Jesus? Several theories exist as to its provenance and these will be discussed in the following section.

One popular and tenable explanation is that the cloth fell into the hands of the Knights Templar.[65] Formed around 1118 for the purpose of protecting pilgrims traveling to and from the Holy Land, the Knights Templar was one of several great orders that arose out of the Crusades (The Knights Hospitalers and the Teutonic Knights were others). Men of the noblest and bravest blood desired to become a Knights Templar because they had fought with such valor in the Second and subsequent Crusades. They had earned the reputation of one of the bravest and fiercest of all groups fighting in the campaigns. The Knights Templar would go on to become one of the most powerful and intriguing groups in Europe. They used their power and influence to establish a series of virtually impregnable fortresses across Europe and the Near East. Their fortresses became useful storehouses for national treasurers and valuables of all kinds. The Templars were so influential that they appointed their own bishops and, as they held monastic privileges, were responsible only to the pope.[66]

In addition to their power and influence, the Templars also had enormous wealth and became the leading money lenders in Europe with bankers, kings and popes on their list of distinguished clients. They were also the principal financial source for the Fourth Crusade. From 1187 to 1291 the Templars' base of operation was a port called Acre in what is now northern Israel. Due to their wealth and the location from which they primarily operated, it is logical that the Templars came to possess the Shroud and other relics that had been obtained during the Fourth Crusade. Their heavily guarded fortresses would ensure secrecy regarding the whereabouts of such an object and in such a fortress the cloth could reside with the Templars indefinitely. Their almost infinite wealth and power would eliminate the need to find another home for the cloth as they would not need to use it for financial purposes so would have no need to borrow against it or sell it. Finally, there would be no problem of inheritance. Future, less wealthy, private owners of the Shroud would have all these issues.

The Knights Templar was a very select group. In addition to its members' impeccable reputations as business men, they also took vows of chastity. Templar knights were sworn upon pain of death never to reveal the details of their initiation ceremony. However, one aspect of this secret ceremony eventually leaked: during the ceremony, initiates were given a momentary glimpse of the supreme vision of God attainable on earth.[67] The Templars were so convinced that displaying this vision of Christ's image made Him present in a far more powerful way than it did in a normal Mass, they omitted the words of consecration during the Mass when the cloth was displayed.[68]

So just what was this image of "the supreme vision of God attainable on earth?" Normally viewed only by the Grand Master and the inner circle of the Templars, the vision was described as an idol or head. Ian Wilson believes and cites several examples that this vision was the Holy Image of Christ, before which the Templars prostrated themselves in adoration. Wilson, and others,[69] think that (like the Mandylion) copies of it were made for the various Templar chapters. Further support is given to this theory by several accounts. These speak of its being displayed at a ceremony taking place just after the Feast of Saints Peter and Paul on June 29.[70] The next feast that would have taken place in medieval times would have been the feast dedicated to the Holy Face, celebrated two days later on July 1.

In some contemporaneous accounts the Templar idol head is described as "terrifying." Wilson points out that during medieval times this reaction would have been commonplace with the most holy images of Christ often engendering fear. For example, in the twelfth century,

Pope Alexander III ordered that Gregory the Great's Acheropita image in the Sancta Sanctorum Chapel be veiled because it caused a dangerous, even life-threatening trembling. In a similar example from the Grail legend, the image of Christ caused the knight Galahad to tremble. It stands to reason that the badly bruised and beaten face of Christ on the cloth might also appear frightening.

Perhaps the strongest evidence supporting the theory that copies of the Mandylion had been distributed to Knights Templar chapters was discovered in 1951. A panel painting (Fig. 180) was found in a Templar ruin in Templecombe, England, matching many of the descriptions of the Templar idol: "bearded male head," "life-size," "painting on a plaque," "disembodied," "with a grizzled beard like a Templar."[71] The panel is distinctly medieval in style and is very similar in appearance to Byzantine copies of the Mandylion. This discovery of the "life-size" Templar painting negates any theory that the Templar idol may have been some form of bust. Multiple copies of the original exist as is evidenced by the variety of descriptions: some in gold cases, silver cases, wooden panels. An even more specific reference from a Minorite friar states that in England alone there were four: one in the sacristy of the Temple in London, one at "Bristleham," another at Temple Bruern in Lincolnshire, and another at a place beyond the Humber River.

As mentioned previously, from 1187-1291 the Templars operated from Acre, which also served as the site of their main treasury. When Acre fell, the treasury was first moved to Cyprus for a short stay; then in 1306, from Cyprus by sea to Marseilles; then to Paris, where it was settled at new Templar headquarters, at the Villenueve due Temple, a huge fortress opposite the Louvre. Throughout this period, the Templars continued to consolidate wealth and power until both rivaled that of a king. As is often the case when wealth and power flourish, soon accusations that the Templars were idolatrous and decadent began to flourish as well. Cropping up with these accusations, rumors began circulating in Europe that the Templars worshipped an idol at their headquarters, an idol like "... an old piece of skin, as though all embalmed and like polished cloth."[72] This description matches the depiction of the Shroud cloth. Further characterized as "pale and discolored," these words portraying the idol are also true of the Mandylion and the Shroud.

Whether because of their wealth, power, charges of idolatry or a combination of the three, the Knights Templars' days were numbered and in 1307 King Phillip IV of France made a surprise sweep against them. When initially the kings' men attacked the Paris headquarters, they met fierce resistance. Templars were arrested, tortured and forced into "confessions." However, only after the initial attack were the kings' men able to search for the Templar idol. Once they were able to do so later, the idol was not to be found. King Phillip encouraged the cataloguing of Templar goods that had been seized and taken to England. Yet again, the Templar idol could not be found among them.

So what had happened to the Templar idol? For some time the Templars were aware that charges against them existed and an attack was imminent. Therefore, it is easy to theorize that someone was appointed to guard the idol for safekeeping. Either during the attack on the main headquarters in Paris or sometime preceding it, the Templars or their designee removed the Shroud, cutting away its frame or the casket in which it was folded. In the attack on the Parisian headquarters, many prominent Templars were taken prisoner. Among them was Jacques de Molay, the Templars' Grand Master, who was arrested, confined and tortured. He and one of the other principal masters of the order were eventually burned at the stake. Who was this master who was also brutally killed? Hailing from Normandy, a town located only fifty miles away, his name was Geoffrey de Charnay. [73]

(Fig. 180) Painting of late thirteenth or early fourteenth century workmanship found at site of former Knights Templar in Templecombe, England.

(Fig. 181) Burning at the stake of Templar Order's Grand Masters Jacques de Molay and Geoffrey de Charny.

Was this elder Geoffrey de Charnay related to the younger Geoffrey de Charny that owned the Shroud when it was first exhibited? Even today the answer is unknown. While the spelling of the two names differs slightly, this is insignificant in that there was little standardization in spelling in medieval French. It would have been highly unusual for two quite prominent people in the same era to have nearly identical first and last names and not be related. Geoffrey de Charnay was considered noble and powerful due to his rank as Grand Master in the Knights Templar. Since he had taken a vow of chastity, he probably did not marry or produce any heirs. Even if he had heirs, all of his land and possessions would have been confiscated by Phillip IV prior to de Charnay's death in 1314. The younger Geoffrey made his first appearance in 1337 when his prominence as a French soldier is recorded in the military records. As with the older Geoffrey, this Geoffrey would also establish a noble reputation, but his was for being one of the bravest men of his era. Ironically, while the older Geoffrey died at the hands of a French King, the younger Geoffrey died at the side of French King John II (The Good), as he fought to save the king's life. The younger Geoffrey valiantly fought as though he sought to make a name for himself or as though he sought to regain some lost family honor. If for whatever reason, the younger, less prominent Geoffrey acquired the Shroud directly or from the older, more established Geoffrey, it might explain his reticence to reveal how he acquired it.

Another plausible connection exists between the Shroud and Geoffrey de Charnay. When Constantinople was sacked, the Shroud was kept at the palace of the Blachernae (Blakerna). This palace was taken by Henry of Hainault, who became Emperor of Constantinople a year later, when he succeeded his brother, Baudouin (Baldwin I). (Baldwin's reign was the beginning of the so-called "Latin Emperors" in Constantinople which continued until 1261.) Recall that after the devastating attack of 1204, the Byzantine Empire was in ruins. Lack of financial prosperity was common for rulers during the "Latin Empire," and regencies between rulers were also common. One source has the Shroud appearing on a list of relics in Constantinople in 1207.[74] While this reference is more vague and less certain than the earlier references in 1201 and 1204, if it is true, then it is possible that the Shroud stayed in Constantinople with the Latin emperors or regents. From 1216 to 1219, there was a regency between emperorships with Nargeaud de Toucey as regent; from 1228 to 1237 in another regency, Nargeaud's son, Phillipe de Toucey, served in this capacity. Geoffrey de Charny's first wife, Jeanne de Toucey, shared a name with these regents. Her uncle, Guillaume de Toucey, was appointed Canon of the church at Lirey at the request of Geoffrey de Charny. The advocates of this theory believe that Geoffrey's connection with the Shroud came about through his wife's family.[75]

A third plausible way in which the Shroud could have come to France from Constantinople is mentioned by Hungarian Oxford scholar Dr. Eugen Csocsan de Varallja.[76] The Byzantine Emperor Isaac II Angelus' wife was the much younger, Hungarian-born Empress Mary-Margaret. Isaac II Angelus died in the crusaders' attack upon Constantinople. A few short weeks after the emperor's death, Mary-Margaret married Boniface de Montferrat, who led the Fourth Crusaders. They later moved to Thessalonica where Mary-Margaret founded the Church of the Archeiropoietas (Church of the Image of Edessa), which is known today as the Ancient Friday Church.[77] It is possible this church may have been founded to house the Image of Edessa/Shroud, which they could have taken with them from Constantinople. Yet another connection exists here. Three years after Boniface died, Mary-Margaret wed Nicholas de Saint-Omer, a son of the titular prince of Galilee and a kinsman of Geoffrey de Saint-Omer. This Geoffrey was one of the two founders of the Order of the Knights Templar.[78] Later, Mary-Margaret and Nicholas' son William would also become involved with the Order of Knights Templar.

Regardless of how it got there, the fact remains that the Shroud had traveled to France from Constantinople. Whether it traveled by way of Geoffrey de Charnay of Normandy, the de Toucey family, the Empress Mary-Margaret, or some other route (three other theories have been expressed by various authors[79]), we may never know and furthermore, it is not really that important. Any of these theories could rationally explain how the Shroud traveled from Constantinople to France. Geographically, the distances are not far, and countless contacts between the two regions could account for the cloth's sojourns. If

Geoffrey de Charny's ancestors received the Shroud through the Templar connection or through the efforts of a regency, some dishonesty or trickery may have come into play. Consequently, Geoffrey may never have been given an explanation, a sort of "don't ask, don't tell." Even if an explanation was given, he may have been reluctant to divulge it. In any event its provenance was never explained or recorded. All chances of explanation expired with Geoffrey de Charny at the Battle of Poitiers in 1356 when the numerically-superior French forces were unexpectedly defeated, and Geoffrey de Charny lost his life in that battle.

While it is sometimes thought that the first exposition of the Shroud was not held until after Geoffrey's death, evidence exists that contradicts this theory. A badge exists that contains the full-length frontal and dorsal figures of the Shroud (See Fig. 182). This badge also bears his shield, along with his wife's coat of arms. The presence of his shield would only be allowed if he was alive as was dictated in the rules of heraldry.

(Fig. 182) Pilgrim's medallion with full-length frontal and dorsal figures of the Shroud represented. The shields contain the arms of the family of Geoffrey de Charny and his wife, Jeanne de Vergy. The roundel in the center represented the empty tomb. From a damaged amulet found in the Seine in 1855 and presently in the Musée de Cluny.

(Fig. 183) Geoffrey II de Charny.

In 1389, thirty-five years after Geoffrey de Charny's death, his wife and son wanted to exhibit the Shroud in Lirey once again. Coincident to the exhibition, a document appeared known as the "d'Arcis Memorandum." The effect of this memorandum continues even to this day. From the time of its first appearance, all subsequent documents and opinions asserting the Shroud is a painting revolve around this memorandum.[80] Said to be written by Bishop Pierre d'Arcis of Troyes to Pope Clement VII, it is dated to late 1389 by the learned French scholar Ulysse Chevalier. (Troyes is a town in France, located twelve miles from Lirey.) According to this memorandum, at the time of the previous exhibit, thirty-four years earlier, then Bishop Henri de Poitiers had discovered "after diligent inquiry and examination how the said cloth had been cunningly painted, the truth being attested by the artist who had painted it…."[81] The supposed recipient of this memorandum, the Avignon Pope Clement VII was the first of the "antipopes" of the Great Western Schism. As such, he was recognized as pope in France but not in Rome.[82]

The Memorandum continues with d'Arcis expounding at length about the avariciousness of the dean at the modest church at Lirey. D'Arcis had a reason for this attack. Apparently, the exhibitors had not asked him, the local bishop, for permission to display the Shroud. Furthermore, they later ignored his command to discontinue the Shroud exhibit. This, coupled with the fact that the Shroud exhibit was attracting pilgrims not only from Troyes, but also from throughout the entire area, had d'Arcis quite upset.

D'Arcis could produce no actual documents from Henri de Poitiers, nor could he provide an actual name for his supposed artist. Historian Daniel Scavone believes that d'Arcis may have had a hidden agenda.[83] D'Arcis had been accused of wanting the cloth for his own benefit, and while he denied in the memo that he wanted the cloth for his own gain, he may have had a reason for needing it. According to Scavone, in 1389 the nave of the unfinished Cathedral in Troyes collapsed, and records show that was the only year that expenses were greater than income. The collapse of the nave was devastating, so devastating in fact that no major effort would be made for sixty years to complete the cathedral. As Scavone says, "Recall, too, that the Memorandum alludes again and again to the avarice of the Lirey canons: Bishop d'Arcis could be telling more about himself than about the canons."

Most Shroud scholars are unaware that a renowned Shroud critic, Ulysse Chevalier, manipulated this memorandum when he presented it in the early part of the 20th century.[84] Two handwritten copies of the memorandum attributed to d'Arcis exist, "Folio 137" and "Folio 138." Folio 138 is a heavily edited and marked first draft with underlinings, some parts crossed out and some violent expressions cancelled. It is unsigned, undated and even unaddressed to its intended recipient. Chevalier carefully transcribed this d'Arcis' version and published it in his very influential book. However, Chevalier affixed the heading from Folio 137 titled, "The Truth about the Cloth of Lirey, which was and now is being exhibited and about which I intend to write to our Lord the Pope in the following manner and as briefly as possible," and placed it onto Folio 138 (italics added). Thus, the document seen in Chevalier's book does not even exist, for it is actually a combination of two documents. Folio 137, the second draft, is neater than the first, and with its proper heading removed and affixed to the earlier messy draft (Folio 138), Chevalier gives the impression that Folio 137 was sent to Clement VII. Chevalier declares, "This Memorandum must have reached the pope at the end of 1389."[85] However, Folio 137 was neither dated nor signed.[86] Moreover, it was merely addressed to a scribe for editing, but there is no evidence that it was ever sent to the scribe.[87]

(Fig. 184) Copy of the second draft of D'Acris Memorandum, Folio 137.

While two copies of the rough draft exist, there is no proof that the letter was ever sent to the pope. No copy of such a letter exists in the Vatican Archives; the Troyes diocesan records; the works of Nicholas Camuzat (the historian for the diocese); the Bibliotheque Nationale of Paris, (where the copies of the rough draft were found); or anywhere else.[88] Moreover, nowhere in any of the documents of Pope Clement VII, concerning the Shroud and its exhibition, is there a word about such a previous communication from the bishop.

Even operating under the assumption that the rough draft was changed into a letter and sent to the Pope, there are still numerous other problems with its contents. The first issue is with its reference to the earlier exhibition and inquiry at the church in Lirey. In his letter, supposedly written in 1389, d'Arcis' alludes to the previous exhibition and inquiry taking place about thirty-four years earlier, which would put the year at 1355. However, on May 28, 1356, documents confirm that Bishop Henri de Poitiers, the man who is supposed to have discovered the great Shroud fraud, presided at the consecration of the Lirey church. At this consecration, Bishop de Poitiers gave his unqualified and lavish blessings to the church, praising its canons and founder. Had a forgery episode occurred at this church the year before, would the Bishop have been singing its praises? This seems highly unlikely.[89]

Furthermore, no records exist of an inquiry ever having taken place. Since he was the successor to the same office held by Henri de Poitiers, had records existed, d'Arcis would easily have had access to them. If he had knowledge of any supportive evidence, surely he would have put that information in his letter. Likewise, no record of an investigation exists in any of the work of the diocesan historian, Nicholas Camuzat.[90] Should the exhibition and inquiry have occurred in a year other than 1355 is not the issue. Having no documentation or record of such an investigation, regardless of the year it took place, remains the fundamental problem.

In addition to the problems with his memorandum detailed above, the veracity of d'Arcis' claims is also questionable because they are based on several layers of hearsay. In the first layer, D'Arcis was never able to produce an actual statement of confession by the unnamed artist that had supposedly painted the image. The second layer is tied to the first. D'Arcis also never revealed the source of this statement or provided corroboration for the truthfulness of over thirty year old information. It is quite possible he may have been taking mere rumor and stating it as fact or he may even have been fabricating evidence to further his own agenda.

As is detailed throughout this book, medical, archeological, photographic and scientific evidence points to the images on the Shroud as those of an actual human body. Even with the technological advances at our disposal today, these images could not have been painted.

Numerous reasons, as delineated in the first few chapters alone, would have made it impossible for a medieval forger to have painted mutually exclusive features on the Shroud. The images appear to have developed over time and seem to have been caused by radiation, which are just two among scores of reasons making it *impossible* for an artist to have painted the Shroud's images or blood marks. D'Arcis' claims not only lack any evidence to support them, but they are completely refuted by extensive, unfakable evidence. Unfortunately, Bishop D'Arcis' claims were believed by many people for centuries until the Shroud's first comprehensive scientific examination.

Pollen Analysis Confirms Shroud's Middle Eastern History

As was alluded to earlier in the chapter, Dr. Max Frei's pollen studies of the Shroud provide further scientific evidence of the provenance of the Shroud. His studies strongly support the theory that the Shroud/Mandylion was in Palestine and Turkey prior to its first documented presence in France in the 1350s. Dr. Frei found fifty-eight pollen grains from plant species on the Shroud; however, only seventeen of these — less than one third — can be found in France or Italy.[91] The presence of seventeen pollen grains from these countries is to be expected since the Shroud has been in either France or Italy for the last 650 years. Surprisingly, only a minority of pollens are native to Western Europe. The majority are native to the Middle East, including the countries of Turkey and Israel. The distribution of pollen grains from plants in Edessa (eighteen) and Constantinople (thirteen) is very similar to that of western European plants (seventeen). Such a proportional distribution of the pollens is to be expected if the Shroud was indeed the folded and framed Mandylion/Image of Edessa, as theorized by Wilson and many others.

Although the Shroud has been around for hundreds of years, its exposure to the elements has been limited. Although displayed scores of times in Europe, it resided most of the time inside a closed container. The same is true of the Mandylion. While it spent hundreds of years in both Edessa and Constantinople, it, too, was rarely exposed to these climates. The cloth would have received four principal exposures to the Turkish climate. The first would have been when it was initially brought to Edessa; the second, when it was transported west to Constantinople; and third, when it was displayed during the attack on Edessa by Chosroes in 544 and, fourth and finally, the sacking of Constantinople in 1204.

Pollens can be blown for vast distances. However, this does not appear to be the case with the Shroud and the large number of pollen grains from plants grown in Jerusalem and the Middle East. Foreign contamination while the Shroud was in Europe is unlikely for several reasons. First, the vast majority of the Middle Eastern species identified by Dr. Frei are from low-lying herbs and shrubs.[92] Archaeologist Paul Maloney explains, "The pollens which do travel far are more likely those from trees which raise their pollen sources to the wind rather than low-lying shrubs which lie relatively protected from such winds."[93] Secondly, many of these Middle Eastern species found by Dr. Frei are insect pollinated. They would not be expected in a wind-distributed assemblage.[94] Even if these two reasons were completely disregarded, foreign contamination could never account for the pollen found on the Shroud. Between The Middle East and Western Europe are several Mediterranean basins and high mountain ranges, which create a countervailing wind system between these two areas. As a result, to reach France or Italy, a pollen grain from Jerusalem would have to travel through more than 2,500 kilometers of these complex and countervailing winds.

In the unlikely event a pollen grain had survived the wind system between the two regions, keep in mind the Shroud was rarely exposed to the outdoors. This unlikely pollen grain would have had to land on the Shroud on one of the few days the cloth was exposed to the outdoors during the many centuries it was in Europe. Furthermore, a wind carrying an errant Middle Eastern pollen grain would also contain many more pollen grains from France and Italy. Making this scenario even more improbable is that the pollens on the Shroud are from plants that bloom in different seasons of the year. So this improbable pollen scenario would have had to happen repeatedly, providing yet more evidence as to why a foreign contamination theory could never account for the vast majority of pollens on the Shroud being native to the Middle East.

The Shroud has likely existed for almost two millennia. In those two millennia much has been lost regarding the history of the Shroud and we will probably never know its exact location for every year of its existence. However, in this chapter, we have detailed an historically documented, plausible provenance of the Shroud of Turin from first-century Jerusalem to present-day Turin. This provenance is based on all currently known and acknowledged evidence available from such widely varying fields as art, history, geography, science, textile studies, literature, even literature of a semi-legendary nature. We should continue to investigate all such relevant areas; however, if the scientific tests recommended in this book indicate the Shroud dates to the first century, then we would have answered the overriding question. We should also not lose sight of other more important questions: who was this victim and what happened to him? Who did this? When and where did these events occur? Finally, how was the scientific and medical evidence encoded and captured in the cloth and what caused its images?

Was a miraculous radiating event the cause, an event consistent with every aspect of the resurrection of the historical Jesus Christ? If it was, then how did this event manage to encode the extensive, mutually exclusive features of the body images and blood marks of this variously wounded and crucified man after he suffered, died and was buried in a Shroud?

11

THE HISTORICALLY CONSISTENT HYPOTHESIS

—~—

How the Body Images Were Encoded

We have discussed extensively the underlying scientific, medical, archaeological and historical evidence relating to the Shroud of Turin, most of which was discovered or became visible only by modern technology. Based on this evidence, one exceptional hypothesis has been developed and refined over two decades that can account for all of the Shroud's primary and secondary body image features, its approximately 130 unparalleled, still-red blood marks, its reverse or outer side imaging at the hands and face, the skeletal features, the possible coin and flower images, the excellent condition of the cloth and its aberrant medieval C-14 dating result.

This hypothesis is known as the Historically Consistent Hypothesis, which is the only one that can account for these features on the Shroud. The ability of this hypothesis to account for all of these unique features will be the primary focus of this chapter. Although this hypothesis was derived strictly from the evidence contained on the Shroud, its physical processes are consistent with key historical events reported to have happened to the historical Jesus Christ. Only after physicist Arthur Lind and I formulated the processes and results within this chapter did we realize that several related historical events that have also been attributed to Jesus can also be explained by related processes within this hypothesis. No other hypothesis, let alone any scientists, philosophers, theologians or others have ever attempted to account for these events before.

This new hypothesis is not only consistent with several historical events, but it makes certain predictions with which to judge the validity of this or related models. Moreover, these predictions can be evaluated in a future round of scientific examination of the Shroud. According to our hypothesis, the Shroud and its physical environment behave according to standard physics principles. The method itself, and what occurs to the body, will be presented in terms of currently accepted scientific principles — some of which are just being discovered. Yet, as pointed out by STURP physicist, John Jackson, who has tested and evaluated more methods of image formation of the Shroud than any other person, "The real test of any hypothesis is . . . its ability to explain observations, make predictions, and provide insight into how reality is constructed."[1]

The Historically Consistent Hypothesis observes that only radiation can account for all the Shroud's body image features, its off-image features and the still-red color of all its blood marks. It further observes that the source of this radiation could only be the dead body wrapped within this burial cloth and that the radiation primarily consisted of protons and neutrons. This hypothesis asserts that if the body suddenly disappeared while leaving behind a very small amount of the basic particles of matter, that all of the Shroud's primary and secondary body image features would be encoded.

The body obviously left the cloth in which it was wrapped, yet no decomposition stains have been detected anywhere on it. In addition, since the body was in rigor mortis (which would postpone decomposition staining) when the frontal and dorsal images were encoded, the body clearly appears to have left within two to three days after having been wrapped within this cloth. If the cloth had been removed from the body by any human or mechanical means — some, most or all of its intimately encoded blood marks, especially its numerous scourge marks, would have broken or smeared at their edges. Modern scientists and physicians have known for over a century that the body left the cloth within two to three days and that it did so in a very mysterious manner.

An unprecedented event appears to have occurred to this body immediately before or during its disappearance that caused the man's unique, full-length body images to be encoded on his burial shroud. Two hypotheses clearly stand apart from all others by accounting for all the Shroud's primary *and* secondary body image features. Both of these hypotheses require radiation to be emitted from an instantaneously disappearing body — a very unconventional event. Yet, this is the primary reason these are the only hypotheses that can account for the Shroud's body images. Like so many other unfakable features and circumstances involving the man in the Shroud, his unconventional disappearance is also quite consistent with several events that reportedly occurred to the historical Jesus Christ.

The first of these hypotheses, the Cloth Collapse Hypothesis was developed by Dr. John Jackson only after he observed that all naturalistic and artificial image forming hypotheses had been tested and failed.[2] Dr. Jackson asserts that if the body of the man in the Shroud became insubstantial and emitted ultraviolet light, the Shroud's primary and secondary body image features could have been encoded on the cloth. While ultraviolet light is a good candidate, it has some shortcomings; however, the cloth's momentary collapse into an area of radiation once occupied by the body is a critically important concept.

The Historically Consistent Hypothesis proposes that the body of the man in the Shroud disappeared, leaving behind a very small amount of some basic particles of matter such as protons, neutrons and, perhaps, electrons and gamma rays.* Nuclear analysis indicates that only about 3.0×10^{18} (neutrons or protons) would have been released from the body. This is an insignificant fraction of the weight of a human body. This number is approximately 0.000000015% of the total number of neutrons (and protons) that were in the body.**

Primary Body Image Features

As the body disappeared, it would have lost its mass, allowing the cloth to collapse or fall momentarily into an area of radiation once

---

***And possibly other electromagnetic radiation such as X-rays, ultraviolet, visible and infrared radiation.**

****As the main building blocks of matter, comprising 99.9% of the mass or weight of all atomic nuclei, protons and neutrons are found in immeasurable abundance in all bodies. Protons and neutrons comprise the nuclei of the more than 55 elements in the human body and jointly comprise all but the lightest one of these.[3] If the basic building blocks of matter were given off (or if only neutrons were released) from the body, regardless how infinitesimally small the fraction, both neutrons and protons would necessarily be given off with protons comprising about 55% and neutrons about 45% of their total.**

occupied by the body. As the cloth fell through the radiant body region, the very penetrating neutrons, electrons and gamma rays would have passed completely through the cloth without encoding body images. However, the easily-absorbed protons would have evenly deposited their energy to produce a uniform straw-yellow coloration, but only on the topmost two or three fibers.

(Fig. 185) Superficial, straw-yellow coloration at nose.

The parts of the draped cloth that were originally closest to the body would have laid over the highest parts of the supine body (i.e., the tip of the nose and lips). Those parts of the cloth would have fallen through the radiating body region longer than the other parts of the cloth. The next closest parts of the cloth (e.g., chin, forehead and cheeks) would have received not quite so much radiation and so on. However, even parts of the cloth that were not originally touching the body would have fallen into the radiating region and received some radiation. Each encoded point on the cloth would receive a radiation dose in proportion to the time it was within the region. Thus, a precise, three-dimensional frontal body image would have been encoded with every degree of brightness/darkness at every point on the cloth's frontal body image being directly correlated with the original distances that the draped cloth was from the underlying body.

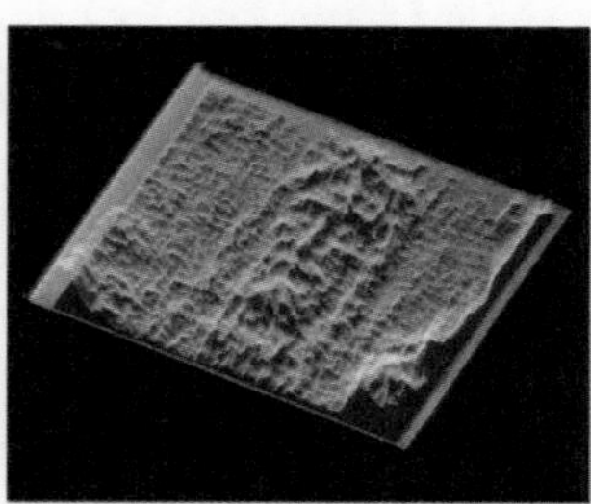

(Figs. 186, left and 187, right) The direct correlation with brightness and three-dimensional relief is seen best on the facial image

Since all parts of the draped cloth would fall by gravity, the frontal body image would also have been formed along vertical paths through the space between the cloth and the body. As we saw earlier, when the two men in the figure below let go of each end of the cloth, the top part conformed roughly to the contours of the underlying body. Yet, regardless whether the contoured cloth was sloping downward, upward, or was relatively flat, all parts of the frontal body image were encoded in a vertical straight-line direction to their underlying points on the body.

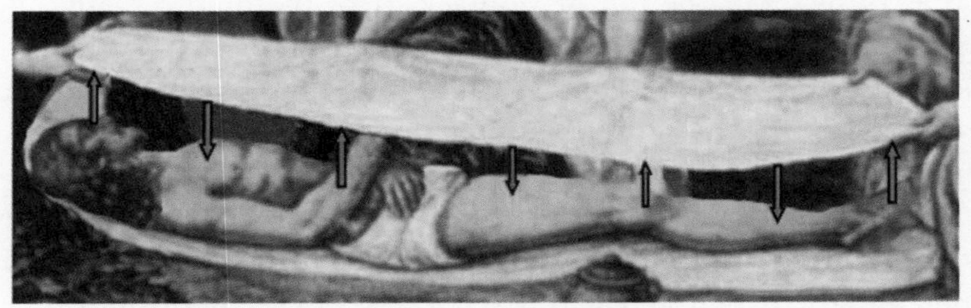

(Fig. 188) All parts of the frontal body image were encoded with vertical directionality.

Indeed, scientists found that only the vertical, straight-line direction from the draped cloth to the underlying body would consistently correspond with the observable points on the Shroud. Any other paths resulted in blurred or distorted images that lacked the high resolution seen on the Shroud image. In addition, only a vertical encoding direction from the body to the cloth resulted in an anatomically reasonable, three-dimensional body image such as that on the Shroud.[4] The Shroud's collapse by gravity would explain its consistent vertical directionality throughout the length and width of its frontal body image.

The vertical collapse of the draped Shroud would also explain the absence of two-dimensional (latitudinal and longitudinal) directionality on its body image, and how its directionality was also encoded through the space between the originally draped cloth and the underlying body.

As we just saw, STURP scientists observed three decades ago that if the Shroud's body image was not encoded in a vertical, straight-line direction, it would be blurred and its high resolution would be absent. Since the source of the radiation is the disappearing body, only the easily-absorbed protons directly beneath each part of the collapsing cloth would be encoded. This would result in a highly-detailed, negative image encoded on the cloth.

When the body suddenly disappeared, a brief vacuum would have been created [5] that would draw the dorsal cloth up a short distance into the radiant body region encoding body image (and blood marks) on the dorsal side.[6] Moreover, if the frontal and dorsal parts of the body disappeared vertically from their outer surfaces inwardly toward the middle parts of the supine body, this would have created downward forces throughout the top or frontal part of the draped cloth, and upward forces throughout the bottom or dorsal parts of the cloth. Because the body disappeared in a fraction of a second, these vertical forces would have been much stronger than those of gravity and the small upward vacuum that was just mentioned.[7]

Radiation emanating from the length and width of the body wrapped within this burial garment would leave negative images containing left/right and light/dark reversal throughout both inner portions of the cloth. The right side of the man's reclined body would appear on the left side of his negative images on the cloth. The darkest or most yellowed parts on the cloth's negative image (those closest to the body) would appear lightest when seen on the positive image.

Since the radiation was emitted from the entire body, the amount of time that the frontal part of the image was exposed to the radiation would be approximately equal to the dorsal part. This would have encoded body images that were independent of weight or pressure, yet had approximately equal intensities.

Radiation, of course, can operate effectively over various surfaces. While proton or other forms of radiation from the body could encode the man's skin, hair and skeletal features, only neutrons released from within particle radiation could explain subtle features from coins or flowers that other forms of radiation are not capable of encoding.

Since the entire image-encoding event was completed within a fraction of a second, while the cloth was only partially through the body region, only surface features such as the skin and hair, and internal or skeletal features closest to the surfaces of the reclined body (e.g., the hands, face, teeth and spine) would become encoded. This explains why the man's internal organs were not encoded, as well as the absence of body features more than four centimeters from the draped cloth, such as the insides of his legs below his hands. Some of the skeletal images may be somewhat convoluted, such as the thumb, because the surface skin and tissue of the hand above it would be encoded first. After the brief image-encoding event was complete, the part of the cloth draped over the front of the man continued to fall to the ground by gravity.

Since the cloth traveled straight down or rose directly up, neither the crown of the man's head nor the sides of his body would be encoded on either the frontal or dorsal images. However, where either part of the cloth briefly intersected at the above edges of the radiant body region, sharp boundaries would be encoded. In fact, as STURP scientists noted 33 years ago, only mapping or plotting the Shroud's body image in a vertical direction from the body can explain the lack of side images and several other subtle body image features.[8]

Secondary Body Image Features

The man's folded hands and his still-upraised left leg would have created a small tent-effect at these parts of the draped cloth. This could explain why the inner side of the man's left leg below his hands is barely encoded and appears much thinner than the upper part of the right leg. While the inner part of the right leg is only partially encoded below the hands, it appears much wider and even fatter than the upper part of the left leg (see Figs. 189 and 190). After 33 years of studying the Shroud,

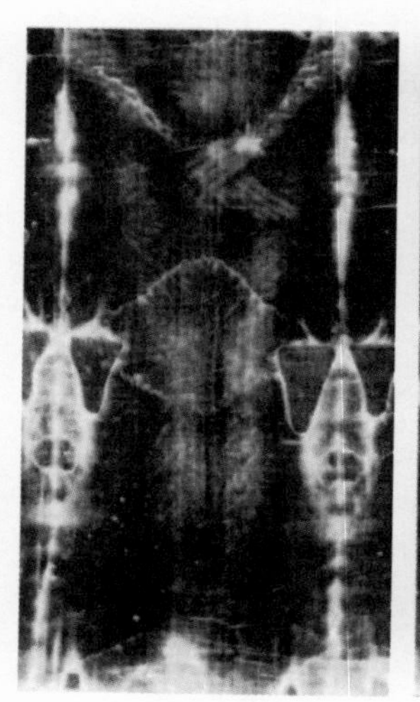

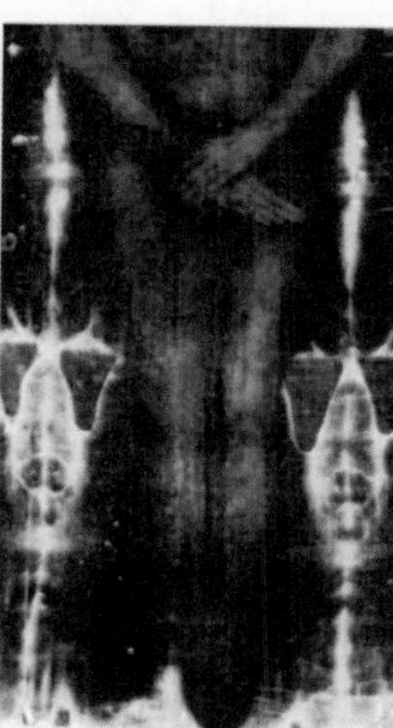

(Figs. 189, left and 190, right) A slightly reconstructed image of the man in the Shroud without the water stain appears at right, next to the Shroud image, better clarifying the appearance of the man's upper legs.

I noticed the difference in size on the positive image — found only above the knees and only on the frontal image — only after I saw Professor Giulio Fanti's reconstructed image side by side with the Shroud's detailed positive image. These distortions were never reported before from observations with the naked eye on the negative images.[9]

The distorted appearances of both legs on the frontal images of the man in the Shroud could be due in part to the original drape of the cloth where it would be flatter over the upper part of the right leg and more upright at the upper part of the left leg. However, the collapse of the draped cloth into the radiant regions of the upper legs would further enhance these original draped positions on the cloth's frontal image. After the negative image developed over time, and the cloth was stretched out and photographed in the 20th century, the upper part of the front right leg appears more than twice as thick as the upper part of the front left leg on the detailed positive image.

This area covers a fairly considerable distance from just below the man's hands to just above his knees. If you look at the man's legs above his knees on the dorsal image, they appear to be the same size. When you look at the man's legs below his knees on the frontal and dorsal images, they also appear to be about the same size. The clear encoding distortions on the upper part of the front of the man's legs are just recently being observed on the positive images and understood by scientists and sindonologists in the late 20th and early 21st centuries. Only a draped cloth collapsing into a radiant body region can account for the odd image distortions at this one particular location. Since Christ's legs have never been portrayed in a lopsided, distorted way, why would a medieval forger encode the legs like this, especially only above the knees on the frontal side?

Like the upper leg distortions on the frontal body image, there are a number of other odd features encoded on the Shroud's body images, which I refer to as secondary body image features. Since these strange features are not found on other images, and have no apparent purpose, it makes no sense for a forger to encode them. Nor can these unusual features be explained under any naturalistic method of encoding. These peculiar features actually appear to be nothing more than image distortions or motion blurs; however, they can be easily explained by the Historically Consistent Hypothesis.

As seen on the negative, positive and three-dimensional images of the face, part of the beard is upturned. The most likely explanation for this and the gaps along the sides of the face is that a small chin band held the mouth closed. Notice also on the positive image that vertical lines run down from the chin, especially below the right side of the man's beard, which is not so upturned. As part of the contoured cloth lying over the chin fell, it would flatten and acquire such lines or motion blurs at the edge of the chin. Before the radiation ceased, more lines could have been left on the area of the cloth that did not have to fall through as much beard before flattening. This would be the area on the man's right side, immediately next to where the beard was turned up. Notice also the wide, rectangular area of body image near the man's throat below his chin and beard. This could have been caused by the cloth falling or moving at this location during the encoding process, leaving this appearance after the cloth is straightened or flattened.

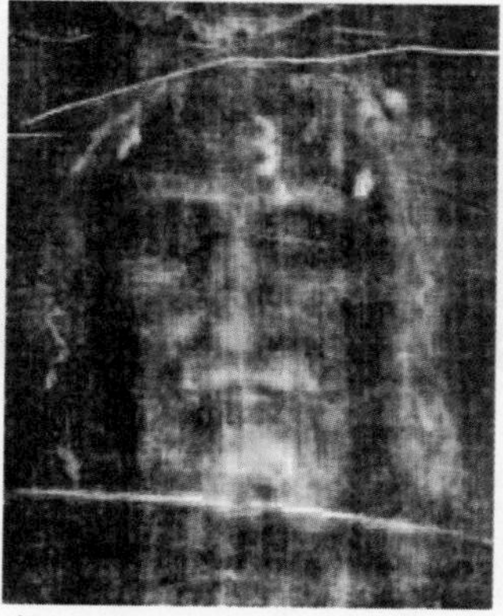

(Fig. 191)

Notice also the odd-shaped feature encoded as part of the body image next to the neck area and below the end of the length of hair on the man's left side. This could be a displaced hair image caused by the cloth moving under this method. There are also two faint body images in the blank space off the left side of the man's face, next to the eyebrow and cheek bone. They, too, could be from motion blur by the cloth in this region, but their faintness and the lack of any such image on the right side, could be due to the chin band slowing the complete collapse of the overlaying Shroud cloth.

Other subtle forms of distortion exist on the Shroud image that are accounted for by the Historically Consistent Hypothesis, such as the length of the man's fingers. Under this hypothesis, the protons emanating from these surface bones became encoded as the cloth passed through this highly-elevated portion of the supine body. Dr. Giles Carter was the first to observe that the man's fingers remained bent from the crucifixion. After a two-dimensional cloth falls through and encodes curved fingers, when the cloth is then flattened, it results in a longer area of the cloth having been used to encode the fingers than if the fingers had been straight. A simple experiment with a cloth tape measure bears this out. If you measure from the top of your wrist to the end of your bent fingers, and then measure again with your straight fingers, the first measurement will be longer. Thus, the encoded fingers look somewhat longer when encoded under this hypothesis.

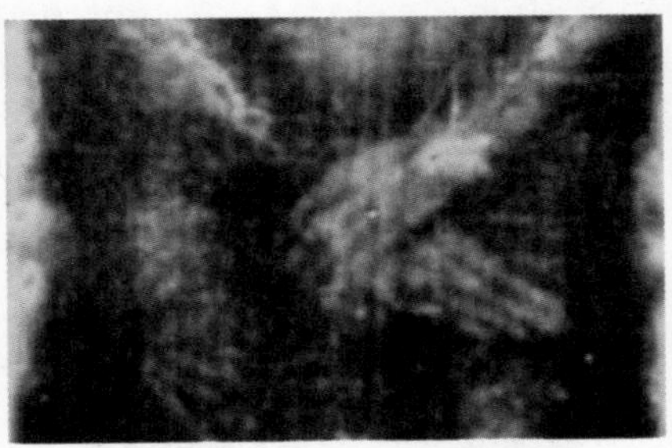

(Fig. 192)

Why would a medieval forger encode unusually long fingers and make the right leg twice the width of the left leg above the knee on the frontal image, if he wanted his representations to appear realistic? Why would he encode vertical lines running down from the chin, where they are not typically found? Why would he encode as body image the rectangular shaped feature at the neck and the odd-shaped feature next to the neck area and below the left length of hair? There are no logical answers to these questions (that are in addition to all the questions we asked regarding the primary image features earlier). Nor are there any explanations how these various secondary body image features could result under any naturalistic method. In fact, like so many other features found on the Shroud, advocates of artistic and naturalistic methods do not even attempt to explain these secondary features. However, the Historically Consistent Hypothesis readily explains all of the Shroud's primary and secondary body image features as a natural consequence of its collapsing into a momentary field of radiation created by the disappearing body wrapped within it.

We saw that as the draped cloth collapsed into the radiant body region, the parts that covered the highest points on the reclined body would fall the farthest distance and for the longest time within the field of radiation. These would be the parts of the cloth that draped over the man's head and his hands. As these two parts of the Shroud linen fell through the field of radiation, their outer sides would also have become exposed to this radiation.

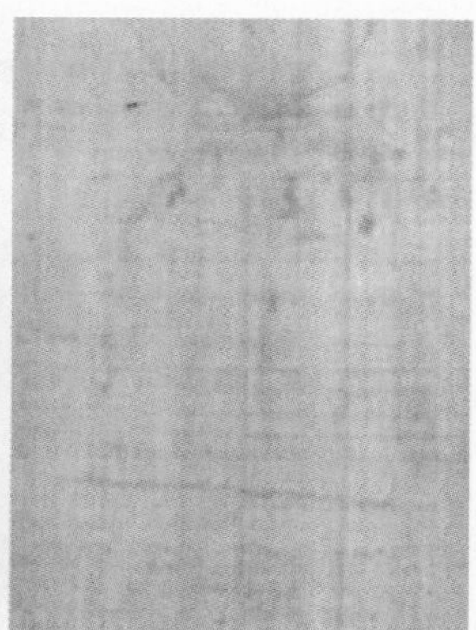

(Fig. 193, left) Arrow indicates outer side of Shroud behind its backing cloth. (Fig. 194, right) Outer side imaging of the man's hair above and around his face, his eyebrows and nose, and possibly other facial features may be visible above.

While such imaging would not be three-dimensional, vertically directional or have focused resolution, there would be some discoloration on the outer parts of the cloth at these locations. After photographing the reverse side of the Shroud without its backing cloth in 2002, Professor Fanti identified imaging at these areas, one of which is seen in Fig.194.[10]

The Shroud's collapse into a radiant region of protons (alpha particles, deuterium or other highly charged particles) or UV radiation (as in Dr. Jackson's hypothesis) from the disappearing body wrapped within it would also best explain its outer side imaging at the hands and face.[11]

Non-Body Image Features

We also saw that as the Shroud collapsed into the field of particle radiation, the penetrating neutrons (gamma rays and electrons) would strengthen the linen cloth in its non-crystalline regions, helping in several ways to explain its excellent condition. As we also saw, the neutrons could ricochet within the tomb and penetrate the Shroud more than once. This would cause a limited number of molecular bonds within this region to break and reform, thus cross-linking these molecules and giving the Shroud greater resistance to the various effects of aging. Only the Historically Consistent Hypothesis explains this important effect on the Shroud.

Even more remarkable qualities that are found throughout the Shroud are best explained by particle radiation. While the Shroud's non-image or background has yellowed and darkened with age, its body images are caused by a more accelerated aging process.[12] Scientists have also established that the application of radiation (light or heat) to cellulose will artificially darken it in what amounts to a rapid simulation of the aging process.[13] Dr. Rinaudo was able to reproduce the Shroud's background color on experimental linen by artificially aging or heating it at 150° C for ten hours. He was also able to irradiate cloth with protons that left the irradiated linen white without any discoloration. However, after artificial aging or heating, the proton irradiated linen produced a superficial, straw-yellow color like that on the Shroud's body image.[14] Rinaudo noted, "these experiments suggested that aging might be an *essential process* to unveil an initially concealed image into a visible one." (emphasis added) [15]

While the Gospels clearly discuss Jesus' burial shroud, neither they nor any other books in the New Testament mention an image of Jesus on it. Similarly, several apocryphal and other sources that mention Jesus' burial shroud do not mention any kind of an image of Jesus on it. Actually, if an image-forming process encodes an initial image, as opposed to one that develops over time, then this process would appear to be *inconsistent* with the process that encoded the images on the Shroud. Furthermore, an initial image would be inconsistent with the Gospels. (Ultraviolet light initially produces a white image that is whiter than the natural color of new linen.) This is one more reason why only particle radiation emanating from the body of the man in the Shroud can account for all of its unique features and is consistent with all the events and circumstances in the historical accounts of Jesus' resurrection.

Interestingly, the Shroud's uniform straw-yellow coloration could not be removed when STURP scientists applied an entire range of acids, bases, organic solvents, oxidants and reductants to it. Finally, when a very potent reductant, diimide, was applied, its straw-yellow coloration instantly went away. In order to duplicate just this one aspect of the Shroud's body image features, a medieval forger would have to naturally or artistically encode body image coloration, whose chemical composition was strong enough to survive an entire range of 20th century acids, bases, solvents, oxidants and reductants, yet disappear immediately with diimide. Rinaudo's proton irradiated linen sample reacted in the same way.[16] This is the only known encoding agent that I'm aware of for which this claim can be made.

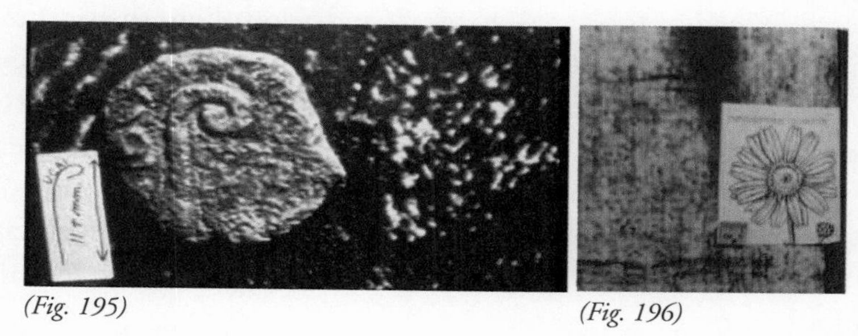

(Fig. 195) *(Fig. 196)*

Particle radiation emanating from the disappearing body could also explain how the faint coin or flower images could have been encoded on the Shroud (Figs. 195 and 196). When some of the many neutrons that flew out of the body region hit the coin or flowers, they could have caused these objects to leave faint images on the cloth in several ways. For example, when a neutron hits the nucleus of copper, the primary component of ancient bronze coins, the nucleus can absorb the neutron and give off either a proton, alpha particle, deuterium or a low-energy gamma ray. Each of these first three particles encodes superficial images on the cloth and, if given off the coin's surface, could possibly encode its faint features. Similarly, flowers contain trace amounts of heavier elements such as iron, calcium and potassium. When any of the countless neutrons hit these three heavier elements, each could absorb the neutrons and give off protons and alpha particles. Any protons or alpha particles given off the flowers' surfaces could possibly encode superficial images on the Shroud.

Low-energy gamma rays are also given off by the heavier elements in flowers, as well as in coins that, in turn, cause ultraviolet rays or long-wave X-rays to be given off that could encode superficial images. Gamma rays could also cause short-wave X-rays or visible light to be given off that could encode coin or flower features more than two or three fibers deep.[17] (Many of the non-penetrating X-rays and ultraviolet rays would not escape from the coin, but would not have nearly as much difficulty with the flowers.) It may be easier to leave flower images than coin images under all the above processes.

Flowers have a tendency to leave images on cellulose and some have left images on paper. (See also Chapter Five of *The Resurrection of the Shroud.*) Throughout the image-encoding process, the flowers would have maintained or increased their original closeness to the frontal and dorsal sides of the Shroud, where their images have been observed. The coin over the eye could have made only a faint image on the cloth from its original position. (There could have been another coin over the left eye, but it did not leave an impression that is clearly detectable at present.) This coin would have fallen through the body region faster than the cloth after briefly and faintly encoding the letters, staff or lituus and of a Pontius Pilate lepton.

Only neutrons can possibly initiate coin and flower images that may be encoded on the Shroud. However, something else quite interesting is also revealed by studying the area of the man's eyes. Notice on the positive images of the face in Fig. 197 that the man's closed eyes appear

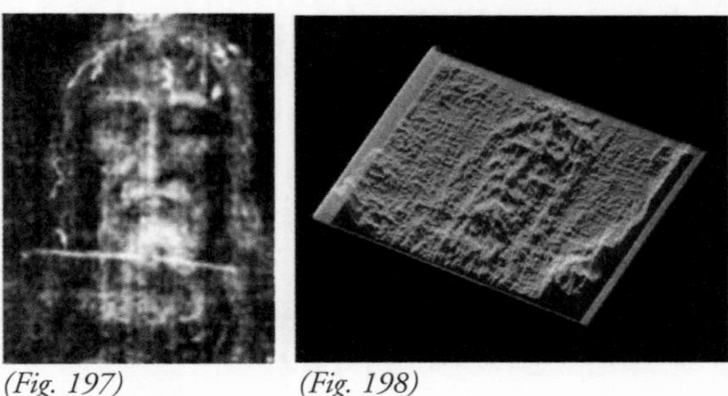

(Fig. 197) *(Fig. 198)*

normal — that is eyeballs or round curved objects appear under the eyelids. In fact, these round curved eyeballs appear at the same locations on the three-dimensional image of the face seen in Fig. 198. (The coin itself would be too small and would lack sufficient relief to encode this type of three-dimensional image.)

The overall area of the eyes is three-dimensional, vertically directional and possesses the photographic resolution of the body images. These facts indicate that the coin features were encoded in one manner and the eyeballs in another. Under the Historically Consistent Hypothesis, the coin's features are faintly encoded at the beginning of the event. After the coin fell through the body, the cloth collapsed through the same area encoding the eyelids and eyeballs (over the faint letters, lituus and motif) in the same manner as it did the rest of the body image.

Only a hypothesis in which the Shroud collapsed into a radiant region of protons and neutrons given off from a disappearing body, can explain both of the remarkable sets of features observed over the eyes. Neither feature could have possibly been observed, let alone forged, by a medieval forger since modern technology revealed them only in the 20th century. Nor could they have occurred naturally.

We have seen how the protons in this hypothesis will produce the Shroud's uniform, superficial coloration on cloth whose fibers and threads lacked any cementation, added pigments or material of any kind. Where body image fibers crossed, underlying fibers were protected and remained white as did the inner part of the proton-encoded fibers. Although the colored fibers are uniformly encoded, they are weaker and more corroded than the non-imaged fibers. Proton irradiated samples also absorb the cloth's natural fluorescence and duplicate the Shroud's microchemistry results of dehydrated, oxidized cellulose.[18] Proton radiation broke the natural single bonds of carbon, oxygen and hydrogen within linen, allowing the first two atoms to double-bind with each other or themselves. These double-bonded or conjugated carbonyls absorb and reflect light as the straw-yellow color that is seen on the Shroud.[19] Proton radiation noted in the Historically Consistent Hypothesis can account for all 32 Shroud body image features that were listed in Chapter Five.[20]

The Historically Consistent Hypothesis can not only explain all 32 Shroud body image features, but its physical processes embody every aspect of the very intricate encoding relationship that necessarily existed within the very short spaces between this burial cloth and the body wrapped within it. The intricate aspects of this encoding relationship were emphasized in Chapter Three. Like the body image features in this chapter, the next chapter will show how each of the remarkable blood mark features could be physically encoded in the Shroud. We will see a very similar process utilized that also embodies the very intricate relationship that necessarily existed within the very short spaces between the draped Shroud cloth and the body wrapped within it.

As we saw, the neutrons that are released within this process will increase the C-14 content within each location on the Shroud linen, in a manner that correlates with their distances from and their positions upon the body wrapped within this burial garment. The new C-14 atoms created during this intimate process will remain within the molecular structures of the Shroud and will not be removed by stringent pretreatment cleaning, heat or other conditions incurred by this burial cloth. Only neutron radiation, contained in the Historically Consistent Hypothesis, has scientifically demonstrated an explanation for the Shroud's aberrant medieval C-14 or radiocarbon dating.

As we saw, the neutrons released from within the particle radiation will also explain the still-reddish color of the Shroud's centuries old blood. This reddish color is not only apparent to anyone who has ever seen the Shroud, but when it is exposed to sunlight, its blood marks appear even redder.[21] The late Dr. Carlo Goldoni concluded after a series of experiments that when blood marks are first exposed to neutron radiation and then to ultraviolet light (such as the Shroud would naturally receive from sunlight during exhibitions), it resulted in the blood marks having a bright red coloration. Further experiments should be undertaken to build upon these initial results.[22]

Only the Historically Consistent Hypothesis can explain the Shroud's full-length body images, including *all* of their primary and secondary body image features; its skeletal features; its outer side discoloration at the hands and face; its possible coin and flower images; its excellent condition; its aberrant medieval C-14 dating; and the still-red color of its blood marks.

Needless to say, no artistic or naturalistic method comes close to duplicating all of these features. Only hypotheses that involve radiation emanating from the dead human body wrapped within the Shroud can account for all of its primary body image features. Only hypotheses involving a burial cloth *collapsing* into a radiant region once occupied by a disappearing body can account for all the Shroud's primary and secondary body image features, its skeletal features and its outer side imaging. However, only the Historically Consistent Hypothesis, involving a burial cloth collapsing into a field of *particle* radiation that consists of protons and neutrons emanating from a disappearing human body, can also account for the Shroud's still-red color of its blood marks; its possible coin and flower images; its excellent condition; and its aberrant medieval radiocarbon dating.

We will see in the next chapter how only the Historically Consistent Hypothesis could also explain all of the pristine blood marks in the Shroud, Jesus' appearance on Easter Sunday, and possibly provide insight into Jesus' resurrection, his sudden disappearances and reappearances reported after his resurrection, which no hypothesis has even attempted to explain before.

12

SECOND PHASE OF THE HISTORICALLY CONSISTENT HYPOTHESIS

Introduction

The first phase of the Historically Consistent Hypothesis discussed in Chapter Eleven states that if the body of the man in the Shroud disappeared while giving off less than one millionth of one percent (0.000000015%) of the protons and neutrons in his body, it would have had astounding results on this cloth. This event would have encoded the full-length body image with all of their primary and secondary image features; their skeletal features; the outer side imaging at the hands and face; and the possible coin and flower images. This event would also have explained the cloth's aberrant C-14 dating; the cloth's excellent condition; and the still-red color of its centuries-old bloodstains located throughout both images.

As we have seen, scientific testing can be applied to the entire Shroud, and to its cloth, blood and charred samples at the atomic and molecular levels that could actually prove or confirm that such an unprecedented, miraculous event occurred to the dead *body* of the man wrapped in the Shroud of Turin. These tests combined with such testing on any limestone samples found on the outer or inner side of the cloth, and from Jesus' reputed burial tombs could also prove or confirm when this event occurred and where it happened. The entire sequence of events that happened to the multi-wounded, crucified dead man, under all the same circumstances of time, location, people and instruments, along with a miraculous radiating event could confirm that the individual wrapped in the Shroud could only have been the historical Jesus Christ.

This chapter adds a second phase to the Historically Consistent Hypothesis that is completely independent of the first phase of the hypothesis. If part or all of the second phase is shown to be incorrect or not to have occurred, it would have no bearing on the independent validity or the occurrence of the above miraculous event in the first phase of the hypothesis. I would have given the second phase a different name; however, the proposed events and the resulting consequences of the second phase are much like those of the first phase of the Historically Consistent Hypothesis. They, too, are consistent with critical historical events that, if they occurred, are more relevant and significant to humanity than any other events in history.

The second phase of the Historically Consistent Hypothesis states that the blood that bled onto the skin or scalp of the man in the Shroud also disappeared with his body. (If the blood marks had not disappeared along with the body, they would have fallen by gravity to the dorsal side of the cloth. Thus, none of these blood marks would have been encoded on either the frontal or dorsal sides of the cloth.) This phase also asserts that both the man's body and his shed blood reappeared. While the body reappeared outside of the burial shroud or the tomb, the shed blood marks reappeared in their original positions once occupied on the body, and thus, transferred onto the enveloping Shroud. We will see that this type of transfer can encode and embed all of the remarkable and unique features found on the Shroud's 130 blood marks that have never been duplicated or explained before. This blood transfer will also be seen to be consistent with the unprecedented reappearance of Jesus on Easter Sunday.

The disappearance and reappearance of the dead man in the Shroud is not only consistent with the disappearance and reappearance of Jesus Christ during his resurrection on Easter morning, but with Jesus' other sudden disappearances and reappearances after his resurrection. This phase of the Historically Consistent Hypothesis also offers an explanation of how Jesus accomplished these feats without violating currently understood concepts of science.

Like the first phase of this hypothesis, the second phase is also

(Fig. 199) Jesus appears before doubting Thomas.

derived from observations on the body images and blood marks on the Shroud of Turin. Most of the observations in this chapter concern the wide variety of pristine blood marks that are seen on the Shroud itself and its very revealing positive images. These observations are also seen with photomicroscopy, photographic enlargers and under UV fluorescent and other forms of lighting.

Blood Mark Characteristics

The nearly unanimous conclusion of pathologists, physicians and anatomists who studied the Shroud since the beginning of the 20th century is that this cloth wrapped a dead human body. The following are just some of the signs that a dead human male was wrapped in this burial cloth.

- the arterial and venous blood flows that correspond to arteries and veins in the head;
- the different types of bruises and swelling identified on the face;
- the flow of watery fluid from the pleural cavity and of blood from the right auricle, which fills with blood on death;
- the photographically revealed abrasions on the knee, leg and across the shoulder blades;
- the abnormally expanded rib cage indicating asphyxia;
- the enlarged pectoral or chest muscles drawn in toward the collarbone and arms;
- the contraction of the thumbs from an injury to the median nerve;
- numerous skeletal features;
- the many signs of rigor mortis;

- the postmortem bleeding;
- the microscopically precise reactions around the approximately 100 scourge marks throughout the body;
- the variety of coagulated bloodstains with serum surrounding borders and clot retraction rings that occur with actual wounds and blood flows, that are found throughout both the front and back of the body, and revealed only by modern scientific technology;
- the precise alignment of all the blood marks on the positive images; and
- the identification of human hemoglobin, human albumin, human whole blood serum, human immunoglobulins, and human DNA from the man's blood marks.*

If a medieval forger were to encode and align all of these features throughout the full-length, front and back images of a dead human body as precisely as those seen on the Shroud's positive images, he could only have done it with a human body. If a forger was going to encode the Shroud's blood marks with human hemoglobin, human albumin, human DNA, and human immunoglobulins, along with human whole blood serum at their various borders, then he'd have to also use human blood on this human body.

The conclusion that these are real wounds and real blood flows from a real human being can be made just by looking at the man's side wound. Recall that for centuries people would have only seen the Shroud's largest blood flow on the left side of the man. (The Bible does not say on which side Jesus was pierced.) Only after 1898 was it revealed that this large stain was actually on the man's right side. Computer imaging technology applied by STURP would even show the line at the top of an elliptical-shaped lesion that matched the Roman lancea. Modern medicine has been able to reveal that if a lancea pierced a dead man at this particular location on his right side, that clear, postmortem watery fluid would escape from his pleural cavity, and that postmortem blood would also escape from the right auricle (which fills with

---

*See Chapters Two and Three for more discussion.

blood on death). Since a medieval forger could not have known that the blood stain and watery fluid were actually on the man's right side, or have encoded features that he could not even see, or have encoded all of the above features precisely on the left side, so that centuries later they would line up perfectly on the man's right — we can conclude that an artist could not have encoded all of these features. Of course, such intricate encoding and alignment could only have resulted from this burial *cloth* originally laying over the right side of a *man* who incurred such a post mortem wound.

Almost all of the bulleted items and even more evidence in Chapters Two and Three show the piercings, blood flows, reactions and positions of a real human body who incurred numerous wounds and bled profusely before and after he died. The man's wounds and abrasions were inflicted at various times with various instruments resulting in approximately 130 blood marks scattered throughout the body. Like the above postmortem flows from the side wound; the arterial and venous bleeding from arteries and veins in the forehead; the wrist wound in the space of Destot that apparently damaged the median nerve; and the different angles of flows on the forearms, the blood marks are almost perfectly aligned on his dead human body. This alignment becomes even more apparent on the positive images of the man.

Although these blood marks vary in intensity, they are congealed and pristine while showing no signs of breaking or smearing at their edges. They are not just encoded, but are embedded in the *cloth* with the same shape and configuration as when they formed and coagulated on the *body*. Observe the statement of Dr. Pierre Barbet, whose pioneering medical studies on the Shroud began in the early 1930s. As a battlefield surgeon in World War I, Barbet saw many bloodstains on cloth, but noted that the Shroud's distinctive blood was comprised of "stains with clearly marked edges which with such outstanding truthfulness reproduce the shape of the clots *as they were formed naturally on the skin."* (emphasis added)[1]

Blood marks with serum-surrounding borders and clot retraction rings are revealed by photographic enlargers and ultraviolet lighting on blood marks throughout the entire body. The same technologies, along with chemical tests and microscopes, reveal upraised edges with indented centers and serum-surrounding borders around each of the scourge marks throughout the man's body.[2] None of these features are visible with the naked eye. A medieval artist would not have had access to photographic equipment, a microscope, or an ultraviolet light source because none of these tools would be invented for several more centuries. There are approximately 100 scourge marks on the man in the Shroud. The inaccurate representation of just *one* of them would reveal an unnatural physiological reaction and expose the work as an artistic creation. Only *intimate* contact at some point in time between the blood and the Shroud throughout the length and width of both sides of the body could have yielded such intricate, detailed scourge and blood marks on both sides of the cloth.

The reader should understand that only the blood marks that were visible with the naked eye would have initially appeared on the cloth, and they would have appeared by themselves. As such, they would have been unimpressive bloodstains without any context of location or form on either the frontal or dorsal body images. As we saw, the body images consist of oxidized, dehydrated cellulose with conjugated carbonyls, whose straw-yellow coloration developed over time. The apostles or disciples who saw the Shroud in Jesus' tomb or possibly afterward would not have been impressed by any abstract bloodstains or with how they corresponded or aligned with any negative or positive images of a body. The faint negative body images would not have been visible on the cloth in their day.* What impressed these people was Jesus' sudden disappearance from his burial cloth and his reappearance on Easter Sunday.

The sudden disappearance of the body wrapped within the Shroud was a critical component in encoding all of the man's primary and secondary body image features in the Historically Consistent Hypothesis. As we saw in the last chapter, only cloth collapse hypotheses can account for all of these numerous body image features. The body's disappearance is not only critical for image formation on the Shroud of Turin, but it is also consistent with Jesus' unexplained disappearance from his burial shroud in the Gospel accounts of his resurrection. It is important to note that the evidence acquired from more than a century of

---

***As a burial shroud, it would also have been considered unclean.**

scientific, medical, archaeological and historical examination of the Shroud's body images and their blood marks is not merely consistent with its authenticity as the burial shroud of Jesus Christ. This evidence is also consistent with his passion, crucifixion, death, burial and resurrection as described in these historical sources.

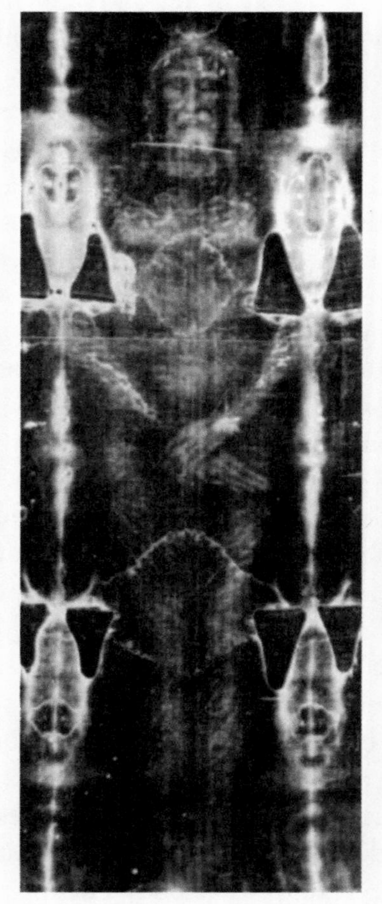

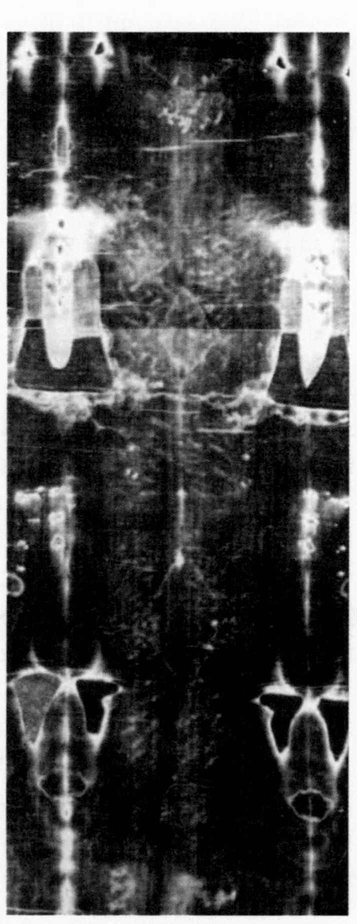

(Figs. 200, left and 201, right) The numerous blood marks throughout the Shroud linen cloth correspond and align perfectly on the man's positive images, revealing many intricate details.

This hypothesis also proposes that the man in the Shroud was Jesus Christ who reappeared outside of the tomb shortly after he disappeared from inside of it. While the disappearance or reappearance of the man in the Shroud are critical assumptions for all the features hypothetically accounted for under both phases of the Historically Consistent Hypothesis, neither part of the hypothesis is required to explain what actually happened to the man during or between these events, or the resurrection. However, this hypothesis will volunteer a possible explanation. While one part of the hypothesis assumes that the body disappeared, the other part assumes that it reappeared.

How Blood Marks Transfer From the Body to the Cloth

Like the body wrapped within the Shroud, its various blood marks also disappear and reappear according to the second or concluding phase of this hypothesis, only they do so within the enveloped burial cloth. If the blood marks did not reappear in this way, they would not become encoded in the cloth. Moreover, they would not become embedded with all of their unique features. We will see that this process even accounts for some odd, yet distinctive blood marks seen on the Shroud's negative and positive images. This aspect of the hypothesis was also devised only after observing that naturalistic and artistic methods have been tested since the beginning of the 20th century, but all have failed to duplicate the Shroud's blood marks, body images and its other remarkable features. It was also devised only after studying the unparalleled properties of the cloth's blood marks and their alignment on both of the man's positive images.

Arthur Lind, retired Fellow at McDonnell Douglas Aerospace and the Boeing Company, spent most of his career as an experimental scientist working on one top secret defense project after another. Recently, he conducted the most thorough and innovative series of experiments to date to duplicate the Shroud's remarkable blood marks in cloth by naturalistic and artistic means.[3] Despite his best efforts, he was unable to duplicate all of these remarkable features on even one blood mark by naturalistic or artistic methods, let alone encode all of the features found on the Shroud's approximately 130 blood marks.

Countless natural experiments in natural laboratories have actually been conducted long before modern times. Millions of people have been bloodied and/or buried under a variety of circumstances and covered with shrouds, blankets, sheets, shirts, jackets, soldiers' uniforms, bandages, etc.; yet none have left any images approaching the full-length frontal and dorsal images on the Shroud or their 130 corresponding blood marks. Dr. Lind and I spent hours in the famous Museum of Egyptology in Turin looking at countless ancient burial garments, but their bloodstains didn't begin to have all the unique features found on the Shroud's bloodstains. Many other ancient cloths are also in existence that do not begin to compare in so many respects with the Shroud's unparalleled blood marks.

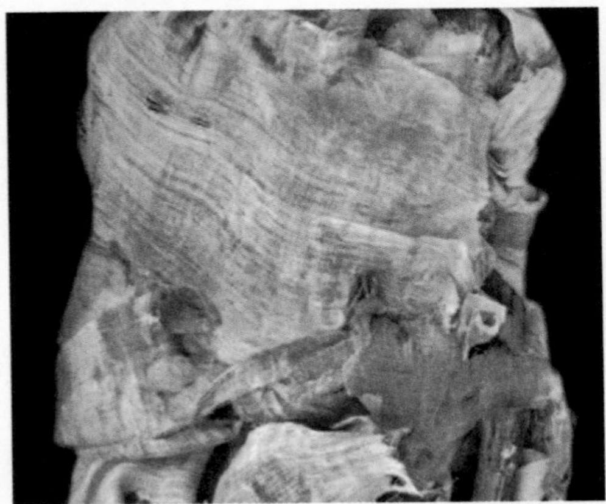

(Fig. 202) Ancient mummy cloth from c. 1200 B.C.

The sudden disappearance of the body and its blood marks, followed by the sudden reappearance of the blood marks within the enveloping, collapsing and rising Shroud, could explain a number of their unique features. This sequence could explain how they appear in this burial *cloth* in the same shape and configuration as when they formed and coagulated on the *body.* The reformation or reappearance of the blood marks in the linen cloth could also explain the microscopically precise, invisible to the naked eye reactions such as upraised edges, indented centers and serum surrounding borders seen around each of the scourge marks with photographic enlargers and ultraviolet lighting. The reappearance or reformation could also explain the coagulated bloodstains with serum and clot retraction rings on blood marks distributed throughout the Shroud — even where the cloth would not have originally been in *contact* with the body. Direct contact alone cannot embed all the Shroud's various blood marks with their complete and intimate features. (Even if such intimate contact were somehow acquired between the cloth and all of its blood and scourge marks, the cloth could not have been removed from the body by any human or mechanical means without smearing or breaking some, most or all of these scourge and other blood marks at their edges.)

Process Encodes Blood Marks Not Originally Touching the Cloth

Many of the Shroud's blood marks would not even have been in contact with the body. For example, twenty different blood marks have been identified across the back of the man's head. A round surface will not make contact at all points on a cloth spread upon a hard limestone floor. Most of the contact on the dorsal side of a reclined body will be at the shoulder blades, buttocks and upper legs. The top parts of the shoulders and much of the lower back will not make contact with the cloth or floor, especially if the man's arms have been folded over his groin and his legs are pointing down with the left one upraised. However, blood, scourge marks and/or shoulder abrasions are quite noticeable on the tops of the shoulders and the lower back of the man in the Shroud. Much of his upraised left leg would not have been in natural contact with the dorsal side of the cloth, but scourge marks are also clearly evident at these locations.[4]

The blood marks on the Shroud are so embedded, they can be seen on the opposite (or outer) sides of the burial linen that draped over and laid under the bloodied corpse. In 2002, the Shroud's patches and backing cloth from 1534 were removed. This revealed the full outer (or opposite) side of the cloth for the first time in five centuries. This allowed all fourteen feet of the inner and outer sides of the Shroud to be photographed for the first time. Most of the man's blood marks are clearly visible on the outer side of this linen cloth in approximately the same shape and form as those on the inner side. The numerous blood flows from the front of the man's head, his side wound, his wrist, both of his arms and the front of both of his feet are clearly visible on the outer side of the cloth that draped over his body. The blood marks and/or blood flows from the back of the man's head, the small of his back and the back of both of his feet are also easily visible on the outer dorsal side of the burial cloth. Many of the scourge marks on the back of the man's calves and legs and on his back are also visible on the outer side of the cloth that laid under the man.

Many of these same blood marks would not even have been in contact with the *inner* side of the cloth, yet are visible on the outer side of this burial garment! Even where there was contact, the jelly-like, coagulated blood marks could not have been embedded *into* the cloth and onto its outer side by pressing against all the blood marks on both sides of the linen. This, too, would have broken, smeared and altered the blood marks and their edges.

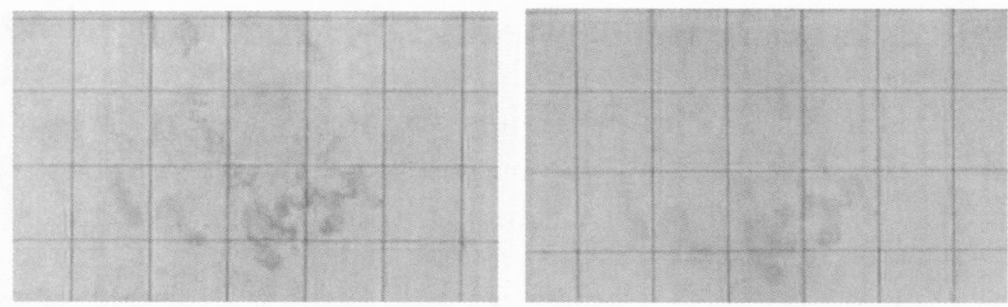

(Figs. 203, left and 204, right) Blood marks from the inner and outer sides of the Shroud at the back of the head.

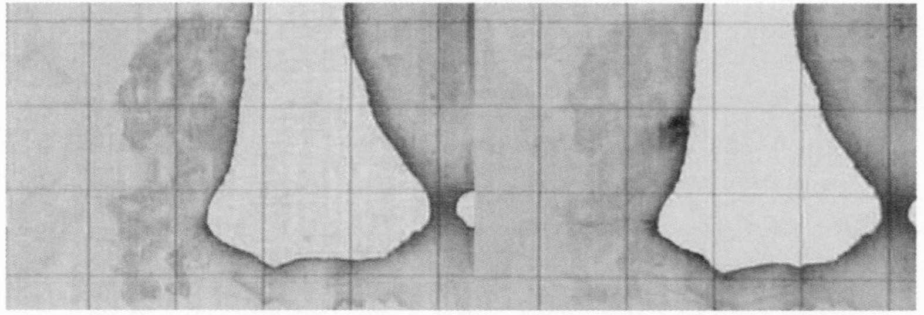

(Fig. 205) Blood marks from the inner and outer sides of the Shroud at the side wound.

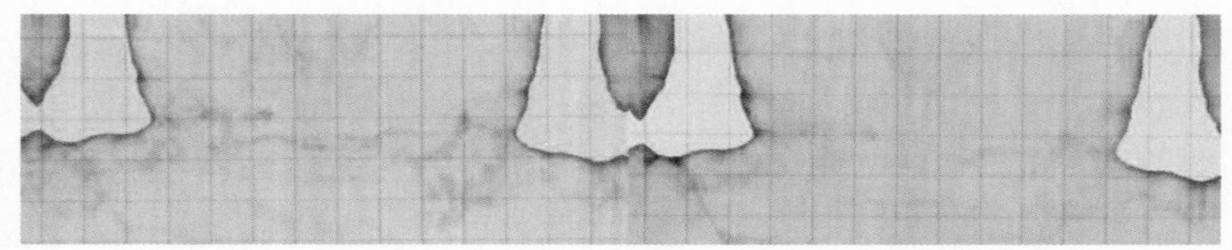

(Fig. 206) Blood flows are even visible on both the inner and outer sides of the Shroud at the small of the back.

Under the Historically Consistent Hypothesis, when the frontal and dorsal surfaces of the body suddenly disappeared inwardly toward each other, a strong vacuum was created that caused the frontal side of the cloth to collapse even faster into the radiant region once occupied by the body (while the dorsal side was drawn up into this region). Because the blood marks consist of the same DNA, atoms, molecules, etc. as the body, they can also disappear. As we discussed earlier, the entire process occurs very quickly. When the blood marks reappear within a fraction of a second, they become trapped or embedded in the falling (or rising) cloth in the same shape and configuration as when they formed and coagulated on the body and are not smeared or broken at their edges. Because the cloth collapsed on its frontal side and rose on its dorsal side, even the blood marks that were not originally in contact with the Shroud become embedded within it. If the coagulated blood marks located on the outside of the man's body reappeared at their original locations within the enveloped burial cloth (while the body itself reappeared outside the cloth), it would account for the numerous pristine human blood marks that align expertly with both body images and are found throughout both inner and outer sides of the linen Shroud. [5]

This disappearance and reappearance sequence also superbly explains the incorrect location of the blood marks that are clearly encoded *on or over* the man's hair on both the frontal and dorsal images. Let us first look at the approximately twenty blood marks that can be seen encoded over or on the hair at the back of the man's head. If they were caused by sharp pointed objects, they would have bled at the *scalp*. They would not have bled from the scalp into and onto the *outer* region of the *hair*. The second phase of this hypothesis can explain how these blood marks became incorrectly embedded onto the back of the man's hair. When the body suddenly disappeared, the dorsal side of the cloth started rising at the back of the head. As it passed through the radiant region once occupied by the body, the hair would have become encoded on the inner side of the cloth. When the rising cloth reached the scalp, it would have then acquired the reappearing, pristine blood marks, which became embedded over the already encoded hair as seen in Fig. 207.

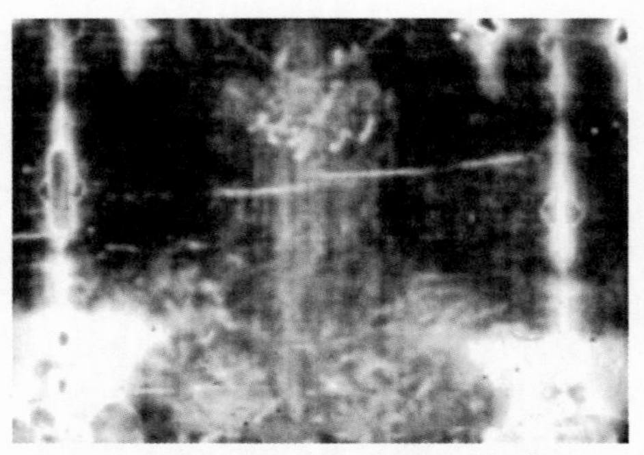

(Fig. 207) Approximately 20 human blood marks appear misplaced on or over the back of the man's hair.

The blood that is encoded on the man's hair above his forehead on the frontal image is also incorrectly located on the outer *surface* of the hair rather than on the scalp. This is especially apparent in the flow or spurt of arterial bleeding[6] seen on the man's hair above his right eye on the positive image (Fig. 208). Blood, of course, would only spurt from an artery located on the scalp. In addition, this blood would not have bled from the scalp into and onto the outer surface of the hair. The negative image to the right (Fig. 209) shows the original location of these wounds under the hair but on the scalp, according to Dr. Sebastiano Rodante, who studied the Shroud and its blood marks for five decades.

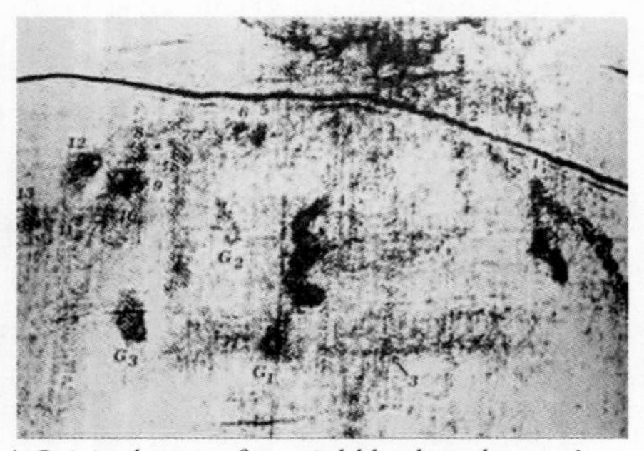

(Figs. 208, left and 209, right) Original spurt of arterial blood on the man's scalp at right is misplaced on or over his hair on the positive image on left. Venous blood marks on the opposite side of his scalp are also seen misplaced into the hair on the positive image.

The second phase of the Historically Consistent Hypothesis explains the transfer of the blood marks seen on the man's hair at left on the positive image above both sides of his forehead. When the hair, forehead and blood marks at the top of the man's scalp disappeared, the inner side of the collapsing cloth would have fallen into this radiant region thereby first encoding the man's hair. When the underlying blood immediately reappeared at the same location they once occupied on the scalp, the collapsing cloth would then acquire them in the same shape and form as they appeared on the scalp, but they would now be embedded over the already-encoded hair.

As with many odd, secondary body image features discussed in Chapter Eleven, these misplaced blood marks would not have naturally appeared on or over the front or back of the man's hair. Nor would a forger encode them in such an unnatural manner, or in a way that's never been portrayed on Jesus or anyone else.

Process Could Encode Blood Clots as Body Image

Like the body's skin and hair, when the blood marks disappear under this hypothesis, they give off a very tiny fraction of their protons and neutrons. If these very attenuating protons were in contact or within a few millimeters of the cloth, they would have deposited their energy uniformly and superficially onto the cloth. However, when the blood marks immediately reappeared, they would have covered the superficial areas of the cloth that received some of their protons. These blood-covered areas could not oxidize and dehydrate, and thus would not leave straw yellow coloring over time of the radiating blood marks on the cloth. However, if one of these blood marks subsequently fell completely off the linen, allowing the underlying cloth to be exposed to air, this part of the cloth could then oxidize and dehydrate (degrade). Straw-yellowed *images* of these blood marks could become visible over time at these particular locations on the cloth.

Some blood marks could become encoded in the cloth as degraded cellulose in another manner without the blood being physically present or embedded in the Shroud. If blood marks originally resided on the edge or side of a body part, say the inner part of the forearm, the cloth may have originally draped a few millimeters away from, and may not have laid flat over all of these blood marks. Under the Historically Consistent Hypothesis, when these blood marks disappeared and gave off small amounts of proton and neutron radiation, these protons would have reached the linen cloth. However, when the blood marks reappeared, they may not have become physically embedded into the falling, nonparallel cloth at this particular location. Over time, *images* of blood marks could appear at these particular locations on the cloth.

Although the present data does not allow certainty, physicist John Jackson has identified three or more locations on the inner arms of the man in the Shroud toward the elbow joints where "it seems that blood clots generated images of themselves, such as the skin and hair did where cloth/clot contact was very doubtful."[7] One of these blood marks, consisting of degraded cellulose, may also have had a very small amount of blood within it (suggesting it may have fallen off the cloth after it reappeared on it). According to Dr. Jackson, blood marks encoded as degraded cellulose could possibly be found at other locations on the Shroud. He calls for careful examination and confirmation of these blood mark images in the next scientific examination of the Shroud.

As we saw earlier, these particular blood marks were not even visible until the late 20th century. They could not have been encoded by a forger nor could they have occurred naturally. The Historically Consistent Hypothesis seems to be the only hypothesis that can readily account for the Shroud's blood mark images appearing as degraded cellulose. This hypothesis, in which the blood marks disappear and then immediately reappear in their original locations, can account for the Shroud's various types of blood marks with all of their unique properties. This hypothesis not only explains how most of the blood marks are embedded in the cloth, but why some are encoded as degraded cellulose. This hypothesis even explains the somewhat odd appearances of the blood marks located on or over the man's hair on both sides of his head.

More Consistencies Between Hypothesis and Historical Accounts

The reappearance of the shed blood within this burial Shroud and the body's physical appearance outside of it are *both* consistent with how Jesus looked when he reappeared on Easter morning. All four Gospel accounts indicate that Jesus completely left his burial tomb at the time of his resurrection. It is also compelling to note that none of the various Gospel accounts of the first sightings of Jesus (by several different people) on Easter Sunday describe him as having any of the bloodstains on his person. Jesus necessarily would have had bloodstains from his scourging, crowning with thorns, nailing in both of his wrists and feet, and stabbing in his side. The man in the Shroud, who suffered the same wounds as Jesus, is literally bloodied from head to feet on both sides of his body. Most, if not all, of the blood marks would have remained on Jesus' body in whole or in part.[8] Their presence in his burial shroud, but absence on him on Easter morning, is consistent with their transfer at the time of his disappearance or resurrection. The blood marks on the Shroud, the Gospels and the Historically Consistent Hypothesis are all consistent with the Shroud's authenticity as Jesus' burial cloth and with his reappearance outside of his burial shroud without his blood marks.

For *all* the reasons and consistencies discussed in Chapters Eleven and Twelve, it is proposed that the Historically Consistent Hypothesis, or one similar to it, could reflect several aspects of the resurrection of Jesus Christ reported in the Gospels to have occurred after his passion, crucifixion, death and burial. The unprecedented information derived from the Shroud of Turin allows this hypothesis to attempt to provide some of the consequences, physical processes and principles possibly involved in the Resurrection that were previously unknown to and absent from the Gospel accounts. Elements of the Historically Consistent Hypothesis or the Resurrection, such as the body's disappearance, which play critical roles in explaining aspects of the Shroud's body images and blood marks, are also utilized in explaining a possible small surface earthquake on Easter morning that did not harm the burial Shroud or tomb. These elements could also possibly explain Jesus' disappearances and reappearances after his resurrection.

As we saw in Chapters Five and Eleven, the instantaneous disappearance of the body would not have necessarily caused an explosion. This was true even where electromagnetic radiation or particle radiation was also emitted from the body, as we saw in physicist John Jackson's and biophysicist Jean Rinaudo's image forming models. The Historically Consistent Hypothesis asserts that less than one *millionth* (1/1,000,000) of *one* percent of the total neutrons in the man's body were released when it disappeared. While most of these remaining neutrons would have penetrated within the interior structure of the tomb as much as a meter, their energy would have easily been absorbed within the limestone burial tomb.[9]

Nuclear analysis indicates that a release of only $3.0 \times 10_{18}$ neutrons from the body at thermal energy would be needed to shift the C-14 dating of the Shroud by more than 1200 years.[10] This is the approximate amount of neutrons (and protons) that physicist Arthur Lind originally calculated would be released by the body in the Historically Consistent Hypothesis.[11] This number of neutrons released from the body is only 0.000000015% of the total number of neutrons that were in the body. According to Rucker, this and a somewhat larger amount of energy would be easily absorbed into the limestone walls of the tomb without creating any noticeable change.[12] Dr. Kitty Little agrees that such a small amount of energy would not have harmed the cloth or its burial tomb. However, she first proposed that such a localized event could have caused a small *surface* earthquake similar to a second one that seems to have occurred at the time of Jesus' resurrection as described in Matthew 28:2, which did not harm Jesus' burial cloth or his tomb.[13]

(Fig. 210) Drawing of typical newly hewn tomb of Second Temple period.

The sudden disappearance and reappearance of the dead man wrapped in the Shroud as proposed in the Historically Consistent Hypothesis would be consistent with the Gospel accounts of the resurrection of Jesus Christ. Although numerous miracles and deeds were attributed to Jesus in the Gospels, his body was not reported to suddenly disappear and/or reappear until the accounts of his resurrection and afterward. In Luke 24:31, Jesus' entire body was said to literally vanish after he broke bread with two of the disciples on the road to Emmaus; however he soon reappeared at another location among the apostles, thereby startling and frightening them (Luke 24:36-43). In two separate instances in John 20, Jesus also physically appeared in person fter the apostles had gathered within closed rooms.

Although the resurrection is mentioned repeatedly in the Gospels and the New Testament, this does not mean that it actually occurred. However, manuscript, archaeological and many other historical studies and comparisons have independently established these sources as or among the most reliable and accurate sources in all of ancient history. Since many unfakable body image and blood mark features on this burial cloth are uniquely accounted for by the Historically Consistent Hypothesis; the many unforgeable consistencies between the events that happened to the man in the Shroud and to Jesus as described in the Gospels; and the Shroud's non-image features that are only accounted for by this hypothesis, including its aberrant radiocarbon dating — a hypothesis that the resurrection of the historical Jesus Christ caused the unprecedented body images, blood marks and other extensive evidence throughout this burial cloth is only logical and *should* be given consideration. This consideration is further warranted when all naturalistic and artistic methods have failed and no other images or blood marks exist that are remotely comparable to those on the Shroud of Turin.

As stated earlier, this hypothesis is not required to explain what actually happened to the man in the Shroud between his disappearance and reappearance or to Jesus during his resurrection. One possible scientific explanation, however, was that he transitioned from our four dimensional reality to an alternate dimensionality.[14] Modern physics related to super symmetry and string theory generally postulates there are from 10 to 26 dimensions. In theory, a small amount of radiation could have been released from the body as it transitioned into the alternate dimensionality, according to Rucker. The body continues to exist in the alternate dimensionality and can return to our dimensionality. Because the atoms in the body do not go through a nuclear disintegration when they disappear there is not an explosion. As discussed earlier, while a small amount of energy would have been released when

99.999999985% of the body transitioned into an alternate dimensionality, this would not have harmed the Shroud or the burial tomb. No such explosion was reported at the time of Jesus' resurrection nor was his burial shroud or tomb damaged. Neither were any explosions reported in the several Gospel examples where Jesus suddenly disappeared and reappeared after his resurrection.

Rucker speculates that this transition into an alternate dimensionality could have "caused oscillations or ripples in the space-time field or continuum that defines our dimensionality, like ripples on the surface of a lake that result when a stone is thrown into the water."[15] Such an oscillating warp "could have caused a general shaking of the ground around the tomb, as waves on the water could cause a toy boat to move up and down."[16] Because the source of the waves was from the surface, and not deep underground, they would have diminished as they moved away from the tomb. In this way, others in Jerusalem may not have even felt any effects from such a localized earthquake.

Only a power like that of God could ostensibly cause Jesus' dead body to resurrect from the dead or transition to and from an alternate dimensionality. The Gospels record in several places that, while alive, Jesus predicted God would resurrect him after he endured the same kinds of suffering and death from which he and the man in the Shroud died.[17] Like Jesus, the man in the Shroud was dead when an unprecedented event occurred to him. If this event involved the sudden disappearance of his corpse, the opening of a window or portal to an alternate dimensionality, the resurrection or any other miraculous event to the dead body, then God would be the apparent, as well as the predicted cause of these events.[18] If the man in the Shroud was Jesus Christ, the Son of God, with all the immortal powers ascribed to him in the Gospels and New Testament, he would be the apparent cause of his own post-resurrection disappearances and reappearances. The New Testament records ten or more reappearances by Jesus.[19]

Rucker compares transitioning from one dimensionality to another to a person going from one room to another and shutting the door behind him. The atoms in the body continue to exist the entire time. There is no disintegration of the body, no explosion and nothing is damaged. It occurs as the result of His will.[20]

The disappearance and reappearance of the blood marks within the burial cloth could have involved a similar transition. Alternatively, Rucker asserts the man's blood marks could have been thrust vertically onto the frontal and dorsal sides of the cloth by a burst of collimated radiation from the body. This same burst would re-liquefy the congealed blood marks by heating and particle interaction, so that when they reached the Shroud they soaked into the cloth.[21]

Traveling to or through an alternate dimension is not necessary to explain the Shroud's primary and secondary body images, its 1988 carbon dating, its excellent condition, its outer side imaging, its skeletal features or its possible coin and flower images. Under the Historically Consistent Hypothesis, once the body instantaneously disappeared and the very small amount of particle radiation was released, all of the above consequences would occur. The second part of the overall hypothesis observes that the body and blood marks return or reappear, but the hypothesis is not required to explain how this occurs. Traveling to or from an alternate dimensionality is only a possible explanation for the disappearance and reappearance of the body and blood marks of the man in the Shroud, of Jesus at his resurrection, or his sudden disappearances and/or reappearances described in the Gospels. (God or the Son of God could also have possibly directed some type of analogous, limited, non-explosive space-time travel that we have not realized or discovered from a physics point of view.)[22]

The important advantage in transitioning to an alternate dimensionality is that the body does not have to dematerialize (or rematerialize) as it does in the analogous, limited space-time travel discussed in endnote 22. It is easier for the reader to grasp this critical point with the alternate dimensionality explanation, even though no explosion occurs under either possible explanation.

How Much Radiation Did the Blood Marks Receive

As we saw, if the shed blood marks on the skin and scalp of the man in the Shroud did not disappear from and reappear in their original positions, they would not have embedded in the moving cloth and aligned so superbly on both of its body images. If the embedding

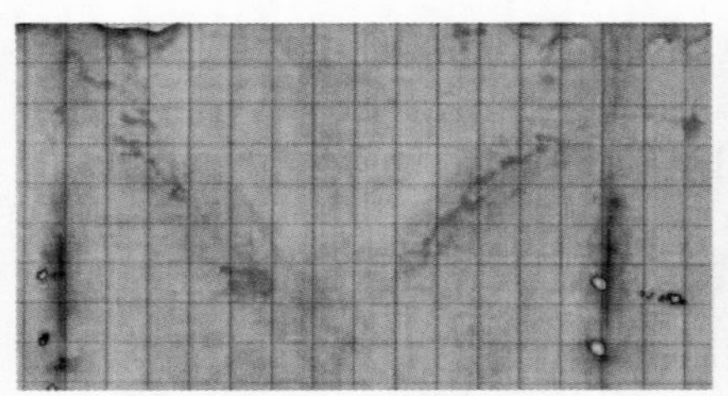
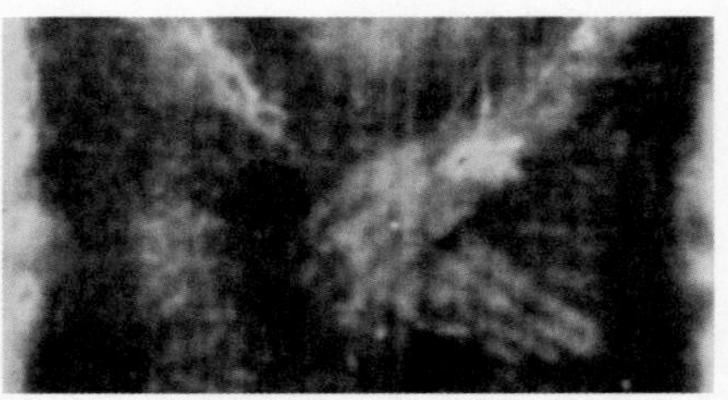

(Figs. 211, left and 212, right) Nail wounds align over wrists on body image at location where nail could enter the Space of Destot and strike the median nerve.

process had not acted this way, the blood marks would not be seen on or over the front and back of the man's hair. The alignments on the body (and misalignments on or over the hair), as well as everything else about the body images become more apparent on the positive images. The positive images show us how the dead man appeared as he laid within the Shroud when the unprecedented image-encoding event occurred.

If these blood marks disappeared *precisely* simultaneously with the body and reappeared immediately after the radiation ceased (and the neutrons stopped ricocheting), then the blood marks would radiocarbon date to their actual age, which is most likely the first century. (Only the cloth was carbon dated in 1988.) The Cl-36 to Cl-35 or Ca-41 to Ca-40 and the C-14 to C-12 ratios within these blood marks would indicate whether they had disappeared during this momentary radiating event. The same ratios from cloth or charred material at contiguous and other locations on the cloth would confirm whether such a neutron radiating event occurred from the body. However, since scientific testing and human history tell us the Shroud's blood marks could not be encoded naturally or artistically, and since they still have a reddish color after all of these centuries, I suspect they *did* receive some neutron radiation. This *partial* amount of neutron radiation could have been acquired by the blood marks if they disappeared slightly after the body began to disappear, or they reappeared just before the radiation ceased (or the neutrons stopped ricocheting). Because the blood marks are so much smaller than the much longer, wider, thicker and more complicated body, perhaps they could have disappeared and reappeared slightly quicker than the body, and thus, acquired some neutron radiation before they disappeared and/or after they reappeared.

One possibility is that the body began disappearing and giving off uniform radiation from its frontal and dorsal surfaces, which is where almost all the blood marks were located. At this point, the blood marks on the outer surfaces of the body would receive neutron radiation for the tiniest fraction of a second before they, too, disappeared. As the top and bottom of the supine body continued disappearing inwardly toward each other, the cloth would continue to follow by collapsing on its frontal side and rising on its dorsal side. When the blood marks reappeared within the frontal and dorsal sides of the moving, enveloping cloth, they could also possibly have received some radiation from the inner part of the disappearing body, or from some neutrons still ricocheting within the tomb.

Extensive Monte Carlo Neutron Particle Modeling of Cl-36 to Cl-35, Ca-41 to Ca-40 and C-14 to C-12 ratios in linen, charred material and blood throughout the entire Shroud of Turin, under a variety of possible physical conditions for the enveloped cloth and the blood marks, could aid in determining if and when the blood marks disappeared from the body and reappeared on the Shroud linen in all their precise locations.[23]

The amount of neutrons that irradiated the blood samples from various locations on the Shroud would reveal critical information. If the earlier-described testing on the cloth, charred materials, blood and limestone described in Chapters Six, Seven and Eight clearly indicate that both the Shroud and its blood marks were present when a miraculous neutron radiating event emanated from the body, it would be a resounding discovery. Scientists may even be able to calculate that the Shroud linen cloth and its blood marks originate from the first century and that the miraculous radiating event also occurred in the first century. If the results indicate that the blood marks received the same corresponding amounts of neutron radiation as the burial cloth, then it means the blood marks remained within the enveloped Shroud during the entire radiating event.

This would indicate they did not disappear and reappear as part of the Historically Consistent Hypothesis asserts. This would indicate that the blood marks were encoded and embedded in some other unique manner, perhaps as indicated earlier by Robert Rucker. Regardless of the specific manner, they would still have been transferred from a body and embedded into the Shroud during a miraculous event in which particle radiation was emitted from a disappearing body. If the test results show that the blood marks were irradiated by neutron radiation from the body, but the amounts they received were correspondingly less than what the Shroud linen received and less than Rucker's calculation, then these results would be consistent with the blood disappearing slightly after the body and/or reappearing slightly before its radiation ceased or its neutrons stopped ricocheting within the tomb.[24]

The concluding phase of the Historically Consistent Hypothesis can explain how the various blood marks on the Shroud of Turin appear on this cloth in the same shape and form as they first appeared on the body. These coagulated, human blood marks have never been reproduced by scientists or artists by any means and have never occurred naturally. The man's wounds were inflicted with different instruments and at different times, and thus vary in size, shape and density. The largest blood flow is post mortem; however, clear watery fluid also flowed from it. Some faint blood marks appear as lines and even as degraded cellulose on the body image.

After all these centuries, the Shroud's blood marks still have a reddish color. These various blood marks are not only encoded with unbroken edges with serum at their borders, but are embedded in almost the same shape and form on the length and width of the outer side of the cloth as they are on the length and width of the inner side of the cloth. Many of the man's blood marks would not even have been in contact with the cloth, yet are encoded with the same intricate precision as the rest of the blood marks on the cloth. Photography, photographic enlargers, microscopes and ultraviolet fluorescent lighting reveals the delicate upraised edges, indented centers and serum surrounding borders on all the scourge marks throughout the frontal and dorsal sides of the cloth, which are not visible with the human eye.

All of these features were intricately encoded and embedded into the cloth through the very short distances that existed between the body and the cloth. Only the Historically Consistent Hypothesis has accounted for all of these unique features found throughout the variety of the Shroud's 130 blood marks. This hypothesis even explains how many of the man's scalp wounds become misaligned over and onto both sides of the man's hair, and how some blood marks could have been encoded as degraded cellulose, like the body images.

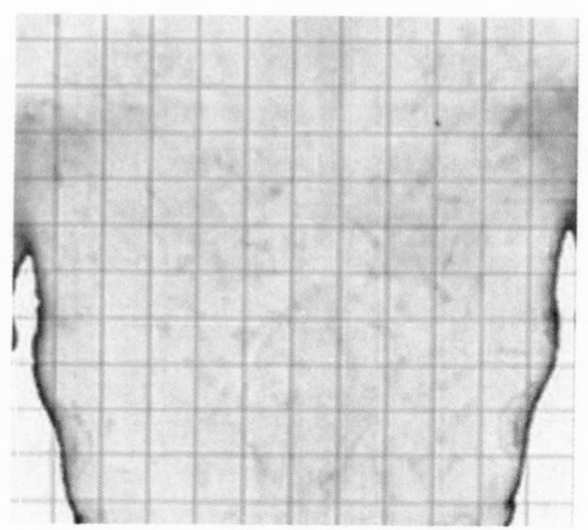

(Fig. 213) Scourge marks as seen with the visible eye on the inner part of the Shroud.

The reader should understand, however, that encoding and embedding the blood marks in the Shroud is just one aspect of the concluding phase of the Historically Consistent Hypothesis. If the blood marks were not encoded and embedded in the Shroud by their disappearance and reappearance, this would not affect the validity of all other aspects of both phases of the Historically Consistent Hypothesis.

Summary of Leading Hypotheses

By far and away, hypotheses involving the elements of radiation, particle radiation or a cloth that collapses into a field of radiation account for far more of the Shroud's many unique features than any other kinds of models or hypotheses. The Protonic Model and the Cloth Collapse Hypothesis are two such hypotheses; however, even these hypotheses have several shortcomings.

Dr. Rinaudo's Protonic Model (discussed more fully in Chapter 10 of *The Resurrection of the Shroud*) asserts that protons and neutrons at the surface of the body were given off in vertically collimated or straight line directions from both the frontal and dorsal surfaces of the man in the Shroud.[25] Rinaudo hypothesizes that vertically polarized gamma rays were given off inside the body with energies of 4.5 MeV, which initiated this reaction. When the gamma rays struck the nuclei of the deuterium atoms at the body's surfaces, their protons and neutrons would have been released at 1.135 MeV, in a known scientific reaction.

Dr. Rinaudo's original and innovative experiments with and analysis of protons and neutrons were ground breaking. I only have a few comments as to the shortcomings of his image forming hypothesis. His hypothesis does not account for any of the Shroud's pristine 130 blood marks scattered throughout the body images. In addition, because the cloth does not collapse into the radiation, it cannot account for any of the numerous secondary body image features on the frontal body images. For the same reason, this model cannot explain the discoloration on the outer side of the cloth at the hands and the face. Without the cloth's collapse and the possible release of gamma rays in this model, the Shroud's possible coin and flower images would not be encoded, nor would the cloth be strengthened to the extent it is in the Historically Consistent model. Without the Shroud's collapse or rise into the radiant body region, its skeletal features would not be encoded, or if encoded, not as well as in a cloth collapse model. The image's high resolution and its three-dimensional information would not be as distinct under this model as in a collapsing model.

Dr. Jackson's Cloth Collapse Model was the most astute physical model ever devised in sindonology.[26] The only physical shortcoming to this physical model is that it does not account for the Shroud's 130 pristine blood marks.

The other shortcomings of the Cloth Collapse Hypothesis derive from its use of ultraviolet radiation instead of particle radiation. Ultraviolet radiation cannot explain the Shroud's radiocarbon dating unlike the neutrons within particle radiation. Neither can ultraviolet light strengthen the cloth nor encode the Shroud's possible coin and flower images, unlike the neutrons (and gamma rays) within particle radiation. Unlike protons, ultraviolet radiation cannot irradiate linen without leaving white discoloration on the cloth. Rinaudo's proton radiation did not leave any initial discoloration on linen. However, after baking at 150° for ten hours (or artificially aging), his proton-irradiated samples, consisting of conjugated carbonyls, also darkened to the straw-yellow coloration observed on the Shroud. In addition, this coloration withstood acids, bases, oxidants and reductants, and disappeared immediately with diimide — like the Shroud's body image coloration.[27] Only proton irradiated cloth samples have been demonstrated to have this quality, as well.

No other hypothesis compares to the Historically Consistent Hypothesis. Naturalistic and artistic hypotheses do not remotely compare. Only hypotheses that assume that the body gave off a particular form of radiation or disappeared, can even compare to it. Even with these assumptions, all other physical objects and their physical environment (i.e., cloth, air, gravity, radiation, chemical modification of cellulose, radioactive conversions) behave according to scientific laws.

The Historically Consistent Hypothesis can explain all of the Shroud's unique primary and secondary body image features. It can also account for other image features such as the discoloration on the outer side of the cloth at the man's hands and face, his skeletal features, or the possible coin and flower features. It can also account for the still-red coloration of the centuries-old blood marks and the excellent condition of the centuries-old burial cloth. Moreover, it can easily explain the Shroud's aberrant medieval C-14 dating. All of these features are elegantly accounted for with the simple assumption(s) that the body disappeared as it gave off a very small amount of the basic particles of matter.

As we saw in Chapters Six, Seven and Eight, an entire series of sophisticated scientific tests can be conducted at the molecular and atomic levels on the Shroud of Turin and its samples that could *prove* such a miraculous event actually occurred. Moreover, these scientific test results, along with the unprecedented findings and evidence from over a century of investigation, could confirm that this miraculous event happened to the historical Jesus Christ. The new and comprehensive evidence could also easily corroborate that all of the events comprising the passion, crucifixion, death, burial and resurrection of Jesus Christ occurred precisely as they are described in the most attested and reliable sources of ancient history.

The first phase of the Historically Consistent Hypothesis is not dependent upon the second or concluding phase of the hypothesis. This concluding phase assumes that the body and blood of the man in the Shroud or Jesus Christ reappeared, with the blood reappearing within his burial Shroud and the body reappearing outside of his burial tomb. This would explain all the unique features on the Shroud's blood marks that no hypothesis has ever explained or accounted for previously. It would also explain and be consistent with the absence of blood on Jesus in several historical accounts from the different people who saw him on Easter morning. This hypothesis would also be consistent with and add possible insight into the resurrection, the earthquake on Easter morning and several subsequent disappearances and reappearances described in the Gospels that no hypothesis, let alone scientist, sindonologist or historian has ever attempted to explain before.

13

INCOMPARABLE EVIDENCE

As we have seen throughout this book, there is an extensive amount of corroborating evidence from the fields of science, medicine, archaeology and history that the Shroud is the burial cloth of Jesus Christ. All of the unprecedented scientific and medical evidence derived from thousands of tests on the Shroud and its samples is consistent with this conclusion, except for its radiocarbon dating results. This consistency is found in the same critical premortem and postmortem events happening to the man in the Shroud and to Jesus in their same vicinity, time and place, and with the same participants, instruments and other circumstances. Significantly, something unique happened to the man in the Shroud that distributed *all* the unfakable evidence of the entire series of events throughout this cloth. This unprecedented event even happened following the same sequence of events that occurred before the miraculous resurrection of Jesus Christ. The extensive evidence on this cloth is not only consistent with, but also comprises *unforgeable* evidence of every aspect of the passion, crucifixion, death, burial and resurrecrtion of the historical Jesus Christ.

This evidence will especially stand out at the atomic and molecular levels, as it already does at the fiber and thread levels. It also stands out visually when one looks at the entire cloth, particularly when viewing the full-length, frontal and dorsal body images and their 130 corresponding blood marks with various forms of modern technology. Most of the evidence, whether at the atomic, molecular, fiber, thread, blood mark or body image levels, was not even known or discovered until modern technology was applied to the Shroud.

Before we examine the textual authenticity and reliability of the historical sources describing the events that happened to Jesus Christ, let us look at a summary of the objective and independent evidence that has never been duplicated naturally or artistically in any age before or after its discovery by modern science. I don't mean to belabor, yet I cannot eliminate any of the evidence or its relevance to the various issues and events from the overall analysis. Please also keep in mind that the evidence from the Shroud is not only unfakable, but the Gospel and New Testament writers would not have had any *idea* of the extensive evidence on this burial cloth.

Summary of the Evidence from the Shroud

From the arterial and venous blood flows that correspond to arteries and veins in the head; the different types of bruises and swelling identified on the face; the flow on the right side of the chest of watery fluid from the pleural cavity and of blood from the right auricle (which fills with blood on death); the photographically revealed abrasions on the knee, leg and across the shoulder blades; the abnormally expanded rib cage indicating asphyxia; the enlarged pectoral or chest muscles drawn in toward the collarbone and arms; the contraction of the thumbs from an injury to the median nerve; numerous skeletal features; the many signs of rigor mortis; the postmortem bleeding; the microscopically precise reactions around the approximately 100 scourge marks on the body; the variety of coagulated bloodstains with serum surrounding borders and clot retraction rings (that occur with actual wounds and blood flows) that are found throughout both the front and back of the body; the precise alignment of the blood marks on the positive images; and the identification of human hemoglobin, human albumin, human whole blood serum, human immunoglobulins, and human DNA from the man's blood marks: all clearly show that the images and blood marks are those of a human male.

The swelling on both cheeks; the triangular wound on the right cheek; and the bruised, swollen and deviated nose show the victim was beaten about the face. The more than thirty wounds on the top and middle of the back of the head and on the top, middle and sides of the forehead indicate that his head was pierced with many sharp objects. These wounds are what one would expect if a crown of thorns had been placed over the victim. Numerous images and pollens from thorns have also been identified on the Shroud by Israeli experts.

There are approximately 100 wounds, in dumbbell-shaped patterns running parallel and diagonal across the body in groups of two or three. Photographic enlargement, microscopic examination and ultraviolet lighting show that the skin was torn open and that clot formation and retraction occurred at these sites. These wounds are signs of a savage scourging, probably by two separate individuals.

Two broad excoriated areas present across the victim's shoulder blades are consistent with a large, rough object being placed upon the shoulders. Many crucifixion victims were forced to carry their own crossbeams to the execution site. Scratches, lesions, and abrasions found on the front of the man's knees, along with dirt on the knee regions and nose, suggest that the victim endured several falls.

Pathologists and physicians are convinced the victim was crucified. This is apparent from the nail wounds in the wrists; the blood flows toward the elbows at angles formed by a body moving repeatedly in a seesaw motion; the nail wounds in the feet, which both point down; the abnormally expanded rib cage; the enlarged pectoral muscles; the upraised left leg; the beating and scourging (which often preceded crucifixions); and the vertical position of the victim at the time of death.

Virtually all medical authorities who have studied the Shroud agree that the victim died. This is most apparent from the findings of rigor mortis present on numerous places throughout the body, such as the left leg, thighs, buttocks, torso, thumb, head, expanded rib cage and the pectoral muscles. His death is further indicated by the postmortem blood flow on the foot and the postmortem side wound. Neither the unusual accumulation or mixture of blood and watery fluid that escaped from the unswollen right side wound, nor the gravitational manner of their flow, would have been present if the victim had not already been dead. Unlike most crucifixion victims, it was not necessary to break this victim's legs to kill him.

The man's wounds were inflicted at different times and with different instruments. Although the blood marks vary in size, shape and intensity, they also bear the markings of having flowed or been inflicted while the body was in the vertical position. Like the body images, the blood marks are unique. Despite the passage of many centuries, they are still red and seem to become redder with exposure to sunlight. The blood marks are pristine and coagulated with unbroken edges. They appear on the *cloth* in the same shape and form as when they flowed and formed on the *body.* Like the body images, many intricate details on the blood marks are seen only on the photographs or by other forms of modern technology. The blood marks had intimate contact with the cloth *even* where the body was not touching the cloth. The coagulated blood marks also appear on the outer side of the cloth in almost the same size and shape as they appear on the inner side. When the blood marks are left/right reversed with photography, they align perfectly on the man's positive images, which reveal how he appeared when the image forming process occurred. The blood marks reveal they had a unique intricate relationship with both the cloth *and* the body wrapped within it after the man's death.

The instruments used on this crucifixion victim were those commonly used by the Roman military guard, which performed such executions for the government. The scourge marks on the man match in detail the size and shape of marks from the Roman *flagrum*. The side wound matches the head of a Roman *lancea*, which is the very instrument described in the Gospels as having been used by the Roman guard on Jesus. Mocking and tormenting of crucifixion victims by the Roman military guard was common and this victim appears to have endured this treatment in the form of a crown of thorns placed over his head.

A further indication that the execution was performed by the Romans was the individual burial that was allowed for this victim, similar to that of another Roman crucifixion victim in Jerusalem, Yehohanan. The evidence that the Shroud victim's crucifixion took place in Jerusalem in the first century is another indication that it was performed by the Romans, since they alone, as military rulers in the area at this time, had the power of execution. The victim was not a Roman citizen, however, for scourging with a *flagrum* was forbidden to be performed on Roman citizens.

There are many indications that the man was Jewish. His physiognomy is Jewish. He had a beard and his hair was shoulder length, parted in the middle and caught at the back of his head, all traits found in Jewish men of antiquity. The man in the Shroud appears to have been buried with a chin band around his jaw, a proper Jewish burial custom.

In addition, his burial posture matches that of skeletons found at the first-century Jewish community at Qumran. The use of a single linen shroud is also consistent with ancient Jewish burial practices, as is the custom of not washing the body of a victim of violent death in which blood that flowed during life and after death is present.

The events that the Shroud victim endured appeared to have taken place in the Middle East. The crown of thorns that was placed on the man's head appears to have been a full crown as was worn in the East, unlike the wreathlet type found in the Western world. One of these thorny plants, which grows only in Israel and parts of the Middle East, left many pollens, and possibly its image, on the Shroud. The numerous indications that the victim was a Jew denote a Middle Eastern location since that is where most Jews resided in the first century. Limestone found in burial tombs in Jerusalem, including the same rock shelf in which Jesus was reputed to have been buried, matches the limestone found on the Shroud. The cloth appears to have been made in Jerusalem, since the vast majority of pollens found on the Shroud grow there. Using coins at Jewish burials was a common practice in Second Temple Jerusalem, as was the occasional custom of placing them over the victim's eyes. The flowers in bloom around the victim collectively grow only in Jerusalem and bloom only in the spring.

From the transmission photos taken of the entire Shroud, STURP scientists thought the cloth was much older than the patches that were sewn onto it in 1534. These scientists thought the Shroud appeared much older than the 1350s, when it first surfaced in Europe. After studying X-radiographs of the Shroud and the crudeness of its weave, and comparing it to the more modern backing cloth and patches sewn onto it, John Tyrer, head of textile investigations at the Manchester Testing House for more than twenty-five years, also concluded the Shroud was much older than the Middle Ages. From the comparison of the features found on the Shroud image with the features found on many religious icons, it appears that the Shroud existed prior to the thirteenth century and as early as the sixth century. By comparing the Shroud's image features with those found on three Justinian II coins, it appears the Shroud existed prior to A.D. 695.

The evidence that the crucifixion was likely performed by Roman military guards indicates that the crucifixion occurred before A.D. 315, since Emperor Constantine banned such crucifixions throughout the Roman Empire at that time. After examining the entire cloth and its component parts extensively, Dr. Flury-Lemberg concludes the Shroud appears to have been a high quality product of the first century with particular features that resemble those found in textile fragments of Masada, whose destruction occurred in 73 A.D. The evidence that the Roman executioners allowed an individual burial of a Jewish crucifixion victim in Palestine until A.D. 66 is an indication that these events took place before then. The possible appearance of a Pontius Pilate lepton minted between A.D. 29 and 32 over the right eye of the man in the Shroud relates quite specifically to the first century A.D. The unprecedented image-causing event that accounts for the unfakable body images, blood marks and many other features is consistent with the first century historical accounts of the astounding resurrection of Jesus Christ in 30-33 A.D.

Dozens of remarkable, unprecedented features are found on two full-length images, which centuries later, revealed high resolution and three-dimensional information encoded in a vertically directional manner through the short spaces between the body and the cloth. The encrypting agent acted over skin, hair, bones and other material, which all appear to be imaged on the cloth without any enlargement. Yet, the images do not consist of any material. They consist of uniformly encoded individual fibers, 360° in circumference, but only on the topmost fibers of the cloth's threads. The encrypting agent broke the individually-bonded atoms within the superficial fibers, which allowed them to double-bind, and then reflect the straw-yellow coloration that developed over time. Not only are all of these features imprinted, but approximately 32 extraordinary body image features are superficially encoded throughout tens of thousands of fibers on threads across both human body images. The features were delicately impressed by an agent that acted in the third dimension of depth over and across the reclined or supine body. As we saw, *only* radiation can duplicate or explain all of these mutually exclusive image features.

Because both body images lack any enlargement or magnification despite the short distances between the body and the cloth, the source of the radiation was the body wrapped in the cloth. Because only the body images were uniquely encoded on the surrounding cloth or film, the body alone was the source of the radiation. Since the three-dimensional and vertically-directional features throughout the full-length frontal body image are directly correlated with the underlying body, then *only* the body could have been the source of the radiation.

Since the 32 extraordinary features are found *exclusively* on the length and width of the frontal and dorsal human body images; both body images are encoded *only* on the inner sides of the cloth that wrapped the dead body; and that *only* radiation can duplicate or explain all of the exceptional body image features, we can be certain that the body was the source of the radiation that imprinted these body images.

Radiation emanating from the dead body of the man wrapped within this burial shroud would not only have been an unprecedented event, it would have been a miraculous event. This event would not only have encoded more than 30 image features from both frontal and dorsal body surfaces, but also internal skeletal features near the body surfaces. This event would also have physically captured all the unique blood marks from the body onto the cloth. This miraculous occurrence would even have captured the various events that happened to the man and their circumstances such as his scourging, crowning with thorns, crucifixion, and postmortem wounds and flows, along with his shoulder abrasions, and scratches, cuts and dirt on his legs, feet, knees and nose, probably due to falls that he could not brake with his hands. It even photographically captured the battered expression and appearance of the corpse as it laid in this burial shroud at the moment of this miraculous image-encoding event.

The body obviously left the Shroud and did so within two to three days for there are no decomposition stains anywhere on the cloth. The miraculous event that encoded the man's images had to have occurred prior to or during its disappearance. Since the man's body was also still in rigor mortis when its images were encoded, this miraculous event could have occurred in less than 48 hours. We saw many indications from the body images in Chapter Eleven that the body's disappearance happened suddenly during the image-encoding event. The Shroud's unique blood marks also confirm this, for if the cloth had been removed from the corpse by any human or mechanical means, it would have broken or smeared some, most or all of the edges of its blood marks.

Only image forming hypotheses involving radiation from a disappearing body can explain all of the Shroud's primary body image features or all the quirky, odd secondary image features that are associated with the cloth's momentary collapse into a field of radiation once occupied by the body. These hypotheses can also explain the still-red color of the Shroud's blood marks even after the passage of many centuries. They can also explain the possible discoloration on the outer side of the Shroud at the man's hands and feet. Only a hypothesis involving particle (proton and neutron) radiation from a disappearing body can also explain the excellent condition of this long, linen cloth, its possible coin and flower images, and, more importantly, its aberrant radiocarbon dating results.

The invalidity of the Shroud's 1988 radiocarbon dating could easily be proven in any number of ways using AMS, or the same technology used in carbon dating. This technology could also confirm whether the Shroud was irradiated with neutrons, as well as the actual age of this cloth and its blood marks. Moreover, a miraculous event in which particle radiation emanated from the length, width and depth of the dead body wrapped within the Shroud of Turin could also be proven by the same technology. This technology could also *confirm* when this event occurred, where it occurred and to whom it occurred.

This and other new technology that could test the Shroud at the molecular level could also confirm that such a miraculous event did take place and that it could have caused or explained the Shroud's body images, its blood marks and many outstanding questions relating to this cloth, as well as the most important sequence of events in all of human history.[1]

Historical Evidence

All of these events and their surrounding circumstances are described in the most attested sources of ancient history. These sources are comprised of numerous books written by various authors, who not only describe the events and their circumstances, but supply the names of the witnesses and the participants. These accounts were not only compiled during the lifetimes of the witnesses and participants, but these witnesses wrote several of the most important books. These events and circumstances were also spoken about publicly on countless occasions by these same witnesses and participants throughout much of the known world.

None of these important events were ever challenged or denied by these witnesses or by hostile witnesses. No individuals or groups of individuals who witnessed the events in the Gospels and New Testament arose to contest the description of the events, the participants, witnesses, times, places or other important factual circumstances. The central message to all of these presentations and writings was that Jesus Christ was beaten, scourged, crowned with thorns, crucified, killed and buried in a nearby tomb, and then resurrected from the dead. Furthermore, all of the participants, witnesses and authors consistently agreed that Jesus himself repeatedly said that if, among other things, we believed in these events that we, too, would have life after death.

One way of judging whether the Gospels and New Testament accurately reflect the contents of these original documents and presentations is to compare them to the other great works of antiquity in terms of the numbers of surviving manuscripts and their closeness in time to the originals. Manuscripts are handwritten documents or handwritten copies of these original documents. In ancient times and for many centuries until the advent of printing presses, all copies or publications had to be written by hand. In the following table by Professor Carl Conrad, Department of Classics, Washington University, we compare the handwritten copies of the books comprising the Gospels and New Testament to the greatest works of the greatest writers of the contemporary Greek and Roman eras of the classical age of history spanning from 800 B.C. to 476 A.D. These various writings or manuscripts are the primary sources for our knowledge of these significant eras of history. This bibliographic comparison will list the ancient author, the date their greatest writings were composed, the number of such manuscripts in existence today and the lapse in time between the earliest copy and the original composition of the work.

As clearly seen, the earliest copies of all the treasured works of Greek and Roman antiquity do not compare to the Gospels and the New Testament. In every instance, the lapse of time between the composition of the original Greek or Roman works and the earliest copies that we have are many centuries. Frequently, there is more than a thousand year time lapse. Furthermore, the number of these centuries-old copies are very few in comparison. As seen, the Gospel and New Testament manuscripts that still exist greatly outnumber these contemporary, classical sources. In addition, the New Testament was written far closer to the time of the original works, as early as twenty years from the original document.

Textual Attestation of Ancient Authors

Author	Work Composed	Manuscripts/ Papyri	Earliest	Time Lapse
Homer, poet (Greek) lived: ca B.C. 700				
Iliad	ca. B.C. 800?	188 manuscripts	ca. A.D. 950	1750 years
		372 papyri	ca. B.C. 50	750
Odyssey	ca. B.C. 700?	80 manuscripts	ca. A.D. 950	1650 years
		30 papyri	ca. B.C. 50	650 years
Hesiod, poet (Greek) lived: ca. B.C. 700				
Theogony	ca. B.C. 700?	79 manuscripts	ca. A.D. 950	1650 years
		32 papyri	ca. A.D. 50	750 years
Works & Days	ca. B.C. 700?	260 manuscripts	ca. A.D. 950	1650 years
		22 papyri	ca. A.D. 50	750 years
Aeschylus, tragic poet (Greek)lived: B.C. 525-456				
7 Tragedies	B.C. 484-456	30 manuscripts	ca. A.D. 1050	1500 years
Herodotus, historian (Greek) lived: B.C. 480-425				
Histories	ca. B.C. 450-430	9 manuscripts	ca. A.D. 950	1380 years
		18 papyri	ca. A.D. 50	480 years
Sophocles, tragic poet (Greek) lived: B.C. 496-406				
7 Tragedies	ca. B.C. 441-406	200+ manuscripts	ca. A.D. 950	1350 years
Thucydides, historian (Greek) lived: ca. B.C. 460-400				
Peloponnesian War	ca. B.C. 425-400	70+ manuscripts	ca. A.D. 950	1350 years

Author	Work Composed	Manuscripts/ Papyri	Earliest	Time Lapse
Euripides, tragic poet (Greek) lived: ca. B.C. 485-406 21 Tragedies	ca. B.C. 438-406	43 manuscripts	ca. A.D. 1150	1550 years
Aristophanes, comic poet (Greek) lived: ca. B.C. 450-385 11 Comedies	ca. B.C. 427-388	6 major manuscripts	ca. A.D. 1000	1400 years
		numerous papyri	ca. A.D. 250	750 years
Lysias, Orator (Greek) lived: ca. B.C. 459-380 Orations	B.C. 407-380	10+ manuscripts	ca. A.D. 1150	1500 years
Xenophon, historian (Greek) lived: ca. B.C. 430-354 Hellenica	to B.C. 362	12 major manuscripts	ca. A.D. 1350	1700 years
		3 papyri	ca. A.D. 50	500 years
Plato, philosopher (Greek) lived: ca. B.C. 429-347 Dialogues	B.C. 393-350	7 manuscripts	ca. A.D. 900	1250 years
Demosthenes, orator (Greek) lived: B.C. 384-322 Orations	B.C. 363-340	200+ manuscripts	ca. A.D. 900	1250 years
Aristotle, philosopher (Greek) lived: B.C. 384-322 Metaphysics	B.C. 348-320	3 major manuscripts	ca. A.D. 950	1280 years

Author	Work Composed	Manuscripts/ Papyri	Earliest	Time Lapse
Euclid, mathematician (Greek) Elements	B.C. 300	6 major manuscripts	ca. A.D. 888	1175 years
Epicurus, philosopher (Greek) lived: B.C. 342-271 Letters	B.C. 300	6 manuscripts	ca. A.D. 1150	1450 years
Polybius, historian (Greek) lived: B.C. 203-120 Universal History	to B.C. 145	22+ manuscripts	ca. A.D. 950	1100 years
Cicero, orator (Latin) lived: B.C. 106-43 Orations in Catilinam	B.C. 63	13+ manuscripts	ca. A.D. 850	900 years
De Oratore	B.C. 55	6+ manuscripts	ca. A.D. 850	900 years
De Republica	B.C. 51	1 manuscript	ca. A.D. 350	400 years
De Officiis	B.C. 44	7+ manuscripts	ca A.D. 800	850 years
Catullus, poet (Latin) lived: ca. B.C. 84-54 Poems	B.C. 60-54	4 manuscripts	ca. A.D. 1300	1350 years
Lucretius, poet (Latin) lived: ca. B.C. 94-55 De Rerum Natura	ca. B.C. 60	10 manuscripts	ca. A.D. 800	860 years
Sallust, historian (Latin) lived: B.C. 86-34 Bellum Catilinae	B.C. 43	20 manuscripts	ca. A.D. 850	900 years

Author	Work Composed	Manuscripts/ Papyri	Earliest	Time Lapse
Vergil, poet (Latin) lived: B.C. 70-19 Aeneid	B.C. 42-19	19 major manuscripts	ca. A.D. 150	170 years
Horace, poet (Latin) lived: B.C. 65-8 Poems	B.C. 35-13	250 manuscripts	ca. A.D. 850	875 years
Livy, historian (Latin) lived: B.C. 59-A.D. 17 Histories	B.C. 29-A.D. 11	15 manuscripts	ca. A.D. 350	350 years
Propertius, poet (Latin) lived: ca. B.C. 50-16 Elegies	B.C. 33-A.D. 16	18 manuscripts	ca. A.D. 1200	1215 years
Tibullus, poet (Latin) lived: ca. B.C. 48-19 Elegies	ca. B.C. 28	9 major manuscripts	ca. A.D. 1050	1080 years
Ovid, poet (Latin) lived: B.C. 46-A.D. 17 Metamorphoses	A.D. 8	17+ manuscripts	ca. A.D. 860	850 years
Lucan, poet (Latin) lived: A.D. 39-65 Pharsalia	A.D. 62	17+ manuscripts	ca. A.D. 850	800 years
Seneca the Younger, philosopher, poet, administrator (Latin) lived: ca. B.C. 5-A.D. 65 Moral Letters	A.D. 64-65	14 manuscripts	9th century	800 years

Author	Work Composed	Manuscripts/ Papyri	Earliest	Time Lapse
Pliny the Elder, natural historian (Latin) lived: A.D. 23-79 Natural History	A.D. 77	200 manuscripts	5th century	400 years
Martial, composer of epigrams (Latin) lived: A.D. 40-104 Epigrams	A.D. 80-102	14 manuscripts	9th century	800 years
Josephus, historian (Greek) lived: ca. A.D. 37-100 Jewish Antiquities	A.D. 93-94	8 manuscripts	ca. A.D. 1050	950 years
Pliny the Younger, civil servant and imperial administrator (Latin) lived: A.D. 61-113 Letters	A.D. 96-109	14 major manuscripts	6th century	450 years
Tacitus, historian (Latin) lived: ca. A.D. 55-117 Annals	ca. A.D. 115-117	36 manuscripts	ca. A.D. 850	730 years
Plutarch, historian-biographer (Greek) lived: ca. A.D. 46-125 Parallel Lives	A.D. 105-115	11+ manuscripts	ca. A.D. 1100	1000 years
Suetonius, historian-biographer (Latin) lived: ca. A.D. 69-140 Lives of the 12 Caesars	ca. A.D. 121	16+ manuscripts	ca. A.D. 850	730 years

Author	Work Composed	Manuscripts/ Papyri	Earliest	Time Lapse
Florus, historian (Roman) lived: A.D. 120 Epitome of Rome	ca. A.D. 120	27+ manuscripts	ca. A.D. 850	730 years
Juvenal, satiric poet (Latin) lived: ca. A.D. 60-130 Satires	A.D. 110-130	17 manuscripts	A.D. 875-900	750 years
Claudius Ptolemy, mathematician, astronomer, geographer (Latin) Mathematike Syntaxis	A.D. 121-151	6 manuscripts	9th century	700 years
Appian, historian (Greek) lived: ca. A.D. 160 Roman History	ca. A.D. 160	8+ manuscripts	ca. A.D. 1100	840 years
Galen, philosophical and medical writer (Latin) lived: A.D. 129-199 De usu partium	A.D. 129-199	9 manuscripts	ca. A.D. 150	750 years
New Testament	ca. A.D. 50-140	85 papyri	ca. A.D. 110	20 years
(Totals as of 1972)[2, 3]		268 majuscule mss. 2792 minuscule mss.	ca. A.D. 250-350 A.D. 835	100+ years 800 years

New Testament scholar F. F. Bruce of the University of Edinburgh notes the obvious when he states, "There is no body of ancient literature in the world which enjoys such a wealth of good textual attestation as does the Gospels and the New Testament."[4] We can be more confident that the New Testament accurately reflects the contents of the original documents than we can of any other works of ancient literature. As noted by John Warwick Montgomery, the former Director of Studies for the International Institute of Human Rights, "To be skeptical of the resultant text of the New Testament books is to allow all of classical antiquity to slip into obscurity, for no documents of the ancient period are as well attested bibliographically as the New Testament."[5]

The chart does not include Gospel and New Testament quotations that still exist from various commentaries and sermons during the first three centuries. Practically the entire New Testament could be reconstructed just from these surviving quotations. The chart also does not include the congruent contemporary translations of the Gospels and New Testament into Latin, Coptic, Syriac and all the major languages of its time. In every sense, the Gospels and New Testament were the most widely circulated documents of their day. We can be more confident that they accurately reflect the contents of the original documents than we can for any other sources of ancient history.

The earliest known written accounts of Jesus' resurrection appeared within about twenty years of its occurrence. This account was well within the lifetimes of the numerous witnesses to this transforming event.[6] Jesus appeared many times in the flesh after his very public crucifixion and death. He appeared to Mary Magdalene; to the women returning from the tomb; to Peter; to all the apostles; to the Emmaus disciples; to James; to Paul and to more than 500 other people.[7] In these and other appearances before the apostles and disciples, Jesus could suddenly arrive and disappear, yet was in the flesh. He ate with them and allowed them to see and touch the wounds on his body.

Jesus' appearances after his publicly witnessed death completely transformed his apostles. From clearly being afraid to be associated with him after his capture, and despondent upon his death — they became emboldened, preaching and testifying throughout the land of Jesus' resurrection. This witnessing and preaching was at great risk for these men were not only arrested and persecuted, but killed for their testimony. Yet, none ever recanted. Among the worst of those who persecuted Jesus' disciples was Saul of Tarsus. Ironically, this man, who would become known as Paul, would leave us the first written accounts of Jesus' resurrection, which included several of Jesus' appearances on earth that he learned of from Jesus' apostles and disciples.

Almost all of the more than 500 witnesses to Jesus' resurrection were still alive when his apostles and disciples boldly testified to this transforming event and to his other words and deeds. Most of these more than 500 people were still alive when Paul wrote of Jesus' numerous appearances following his death. If these accounts were inaccurate or incorrect, we could expect the participants and witnesses, including any hostile witnesses, to refute the accounts of these words, deeds or appearances. While some did not believe these miraculous accounts, none of the witnesses to these deeds refuted them. The transformations of Jesus' apostles and disciples could not have taken place as a result of a lie or a fabrication. Jesus' numerous appearances after his very public death transformed them into some of the most dedicated and well-known witnesses in history, as it did to Paul, to whom Jesus also appeared.

Christian churches also clearly became established throughout the land during the disciples' lifetimes. Jesus' resurrection from the dead was at the heart and center of preaching by the early Christian church. This was not a later legend that was added by future generations, but was at the core of the original teachings by the original witnesses. Sunday, the day of Jesus' resurrection became the Sabbath day for the earliest Christians, replacing the centuries-long Sabbath that began at sundown Friday evening and ended on Saturday evening.

Archaeology has also authenticated the Gospels and New Testament by verifying countless details of various circumstances and statements contained within them. While the profession of archaeology did not develop until the 19th century, it has confirmed the existence of countless public officials, titles, dates of service, markets, temples, customs, physical locations, etc. exactly as they are mentioned in the New Testament. Frequently, when the New Testament has been accused of inaccuracy, archaeology has subsequently confirmed the preciseness and accuracy of its contents and the inaccuracy of the conflicting sources.

Archaeology has consistently confirmed the accuracy of the Gospels and the New Testament to the smallest of details on countless items of the type that eyewitnesses' contemporary accounts would consistently contain.

Evidence much older than the Gospels and the New Testament also supports the divine cause and nature of all the evidence, the body images, the blood marks, the suffering and all the events that occurred to Jesus and the man in the Shroud. All of these events appear to have been prophesized centuries before their occurrence in the Old Testament. (See Appendix I)

Another New Testament event involving Jesus and three of the apostles is also worth mentioning. At the Transfiguration of Jesus on a mountain, his face shone like the sun and his clothes became dazzling white. This radiation was clearly attributed to God the Father, but it did not harm Jesus or immediately affect his clothing.[8]

In the following discussions we will be referring to archaeological and historical evidence, along with scientific and medical evidence. While it is easy to understand that the latter two fields relate to the evidence derived from the Shroud, there is both archaeological and historical evidence that we discussed earlier centering on and relating to the New Testament, and there is also a great deal of archaeological and historical evidence in Chapters Four and Ten of this book. While the historical evidence in Chapter Ten relates to the Shroud's provenance, the rest of the archaeological and historical evidence relates to the unique portrayal of events that happened to the man buried in the Shroud, his identity as Jesus and the identical critical events that occurred to Christ.

Pivotal Moment in History

Never before has a vast amount of objective and independent evidence existed to corroborate the occurrence of critical ancient events of human history. Archaeology and history have combined before to support the authenticity of historical items and to shed some light on past events. However, extensive, sophisticated evidence from the fields of science and medicine have never combined with archaeology and history in such a comprehensive manner to corroborate as unprecedented or controversial events as these. Nor has a comprehensive amount of unforgeable scientific evidence ever combined with the most attested and authentic sources of ancient history.

The incomparable evidence discussed throughout this book and summarized in this chapter clearly and consistently supports the authenticity of the Shroud as Jesus' burial garment. It further indicates that all of the wounds and suffering, as well as the crucifixion, death, burial and resurrection of the historical Jesus Christ occurred, just as these events are recorded in the most accurate and reliable sources of antiquity. The even more sophisticated tests that have been proposed in this book would completely refute the only scientific test result among thousands that is inconsistent with this extensive evidence.

Never has such lopsided evidence existed on a more important issue or matter in human history. Worse yet, most of the world is unaware of the overwhelming evidence in favor of the Shroud's authenticity and the critical sequence of events that happened to this man wrapped within it. Rarely, if ever, has one aberrant, controversial test result been considered correct and all other unfakable scientific evidence incorrect. The sophisticated atomic and molecular testing proposed in this book can completely refute the cloth's radiocarbon dating and dramatically prove the occurrence of these unique events in history. These new techniques and technologies are developing and can be perfected. When they are, they would not only refute the Shroud's medieval dating, but they could scientifically confirm the actual age of the cloth and its blood. Furthermore, they could confirm *when* this unprecedented event occurred and *where* it happened. In addition, the proposed testing could prove that this unprecedented event was a *miraculous* event in which particle radiation emanated from the length, width and depth of the dead, crucified body of the historical Jesus Christ. Everyone would acknowledge that this was a miraculous event. (These test results would even indicate that the corpse of Jesus Christ disappeared from his burial shroud during the miraculous event that caused his images to develop over time.)

The proposed testing could answer all of the outstanding issues about the Shroud. All of the unanimous, unforgeable scientific, medical and archaeological evidence derived from the Shroud could also confirm beyond all reasonable doubt that every aspect of the passion, crucifixion, death, burial and resurrection occurred precisely as these events are described in the most authenticated sources of history. It would only be right that the most extensive, unfakable and corroborated evidence in history existed for the most important events in all of history.

Some people have wanted the Shroud to be carbon dated in several locations before atomic, molecular or other testing occurs. While these different results would invalidate the Shroud's 1988 date, no one would know which date was correct. In fact, if the Shroud had been irradiated with particle radiation, *none* of the C-14 or radiocarbon dates would be correct. Carbon dating by itself would not prove that the Shroud had been irradiated with neutron or particle radiation, or that particle radiation miraculously emanated from the length, width and depth of the crucified body in the Shroud. Neither would it indicate that the body disappeared at the time of this miraculous event. Carbon dating the Shroud by itself would not tell us when this miraculous event occurred, where it occurred, the age of the Shroud or the identity of the victim. Nor would carbon dating by itself answer how the Shroud's images were encoded. Carbon dating the Shroud by itself, without any context of other testing, was one of the fundamental mistakes made in 1988.

As we saw, C-14 to C-12 testing should be conducted *after* Cl-36 to Cl-35 and Ca-41 to Ca-40 testing. Multi-spectral imaging, molecular microscopy and X-ray fluorescence analysis, along with the study of Robert Rucker's MCNP codes, should all occur before samples are even selected for Cl-36 to Cl-35, Ca-41 to Ca-40 and C-14 to C-12 testing.

People around the world will have even more reason to want additional comprehensive scientific testing performed on the Shroud when they realize the fiasco that took place in the 1980s. This nine year process saw the cancellation of a testing protocol in which the Shroud could have been examined in 26 areas of testing, including radiocarbon dating. Instead only C-14 dating took place. While no one realized it at the time, C-14 dating results conducted in isolation from other testing at the atomic level were doomed from the start. Yet, the carbon dating scientists could have realized that additional scientific testing was needed to accurately ascertain the cloth's age and to answer so many other outstanding questions. They could have come to this realization if their leader hadn't been so egotistical and duplicitous during the entire process, and if the carbon dating scientists hadn't been so inconsiderate and unprofessional in their actions.

14

NINE YEAR CARBON DATING FIASCO

—~—

Dating the Shroud of Turin was a much more challenging and complex problem than the directors of the radiocarbon laboratories ever understood. In fact, accurately dating this unprecedented burial cloth solely by its C-14 to C-12 ratio proved to be an exercise in futility. Worse yet, the cavalier attitudes and lack of knowledge about the Shroud among the directors and the key participating scientists eventually turned this problem into a full-blown fiasco. When by design, they eliminated any participation in the selection of the samples and eliminated concurrent testing of the Shroud in dozens of other scientific areas by the group of scientists (STURP) who knew more about the Shroud than any other scientists in the world, they greatly jeopardized the project. However, when they then failed to share their raw dating data with other scientific institutions in Italy, and prematurely interpreted their questionable raw data into a date — all contrary to what they had agreed to in writing — then any chance of any success coming from this project completely disappeared. In fact, their misleading results were much worse than no results at all.

(Fig. 214) STURP scientists beginning their 120 hour examination of the Shroud in 1978.

The undisputed leader and spokesman of the radiocarbon laboratories was Harry Gove [1] who was the director of the radiocarbon dating laboratory at the University of Rochester. Dr. Gove was also the co-inventor of one of two new methods of carbon dating small samples that were being developed and becoming operational at radiocarbon laboratories in the late 1970s and early 1980s. Prior to the development of these new methods, the long established radiocarbon dating method required the destruction of a cloth sample the approximate size of a handkerchief, thus preventing its utilization on the Shroud. STURP and many others had always wanted the Shroud to be radiocarbon dated and for a whole series of scientific tests to be conducted on this cloth, if possible. In this regard, one of STURP's founding members, John Jackson, had initially contacted Dr. Willard Libby, the inventor of radiocarbon dating as early as 1976.[2]

Following their comprehensive examination of the Shroud in 1978, STURP formed a C-14 committee the next year and contacted scientists at carbon dating laboratories throughout the world, asking them if they would be interested in joining STURP's effort to carbon date the Shroud of Turin. Five of these laboratories initially joined STURP's effort: the Arizona, Brookhaven and Rochester laboratories in the United States, and Oxford and Harwell in the United Kingdom. Thereafter, the radiocarbon laboratory in Zurich, Switzerland and the British Museum joined the effort.[3] This evolved into a project in 1983 in which the British Museum, as coordinator, mailed two control linen samples to the six radiocarbon laboratories for dating. All six laboratories, as well as the British Museum, in its role as coordinator, were participating in an inter-comparison test of the two new techniques for dating small samples as a prelude to dating similar samples from the Shroud of Turin. Their inter-comparison test results were announced at a radiocarbon dating conference held in Trondheim, Norway in 1985. The results were reported in 1986.[4]

Trondheim Report

Five of the six laboratories correctly dated the first sample, but Zurich was approximately one thousand (1,000) years too young on its date. However, *all* five of these laboratories then dated the second sample incorrectly. *All* five of these laboratories erroneously dated this sample at approximately half of its historical age. Yet, this systematic error on the part of *all* the laboratories was hardly discussed or analyzed in the report. Instead, the report states that the second sample was simply replaced because its five radiocarbon dates "suggested that the material was of much more recent date than expected."[5] Incredibly, the very next sentence states, "This was probably erroneous as it turned out."[6] In other words, all five laboratories dating results for sample two were "probably erroneous." Unfortunately, these results were both a prelude to and a *precedent* for what would occur only two years later. In 1988, three radiocarbon dating laboratories correctly dated some control samples, but *all* three participating laboratories would date the Shroud centuries younger than its actual or historical age.

The report does not state when the British Museum or the radiocarbon laboratories even realized their *systematic* mistake with the second sample. What is clear, however, is that the report was published three years after the second sample was mailed to the laboratories, yet it does not indicate that any investigation was undertaken by the British Museum or any of the participating laboratories. The second sample was simply replaced by a third sample. This cavalier attitude with the second sample stands in contrast to the treatment given to the Zurich laboratory, which also dated the third sample incorrectly by 1,000 years, this time being too old, while the other five labs correctly dated the third sample.

In Zurich's case an investigation was undertaken in which a number of follow-up tests and measurements were performed at the Swiss lab using their new and normal pretreatment cleaning methods. These subsequent tests supported the report's conclusion that Zurich's two 1,000 year mistaken dates "must have been due to contamination introduced by a new method of pretreatment."[7] Zurich's dating errors were not even included in the report's final statistical analysis.[8]

The two original samples that were mailed by the British Museum in 1983 were dated a total of 11 times by the six participating laboratories. Stunningly, 6 of 11 samples appear to have been carbon dated incorrectly by half or more of their actual or historical age. All three samples were dated a total of 17 times by the six participating laboratories. Incredibly, 41% of these samples (7 of 17) appear to have been dated incorrectly by half or more of their actual or historical age. In spite of these results, the report concludes that "Overall, there is good agreement between the results obtained and the expected historical dating of the samples, in particular as far as samples 1 and 3 are concerned."[9] However, this is just another way of saying that although the labs were in "good agreement" for sample 2, their results were *not in agreement* with its actual or historical *age*. The report also concludes that, "A coherent series of results can be obtained when *several* laboratories undertake separate blindfold measurements of the same sample."[10] (Italics added) Yet, the results from the second sample showed that *all* the dating laboratories can mistakenly date a sample even when it is dated by five laboratories.

These conclusions actually indicate how unsure the labs and their coordinator were about their accuracy, yet how anxious they were to go ahead and date the Shroud, regardless. Normally, only one laboratory is used when an object is carbon dated. Radiocarbon dating laboratories are extremely expensive to run and are heavily supported by national institutions in each country. Naturally, being one of the laboratories that carbon dated the Shroud would look very impressive in the next round of funding applications. In addition, as admitted by the inventor of the other new method of carbon dating small samples, Garman Harbottle, Brookhaven's director, there would be a certain amount of prestige and fame from dating the Shroud.[11]

More investigation, analysis and dating should have been undertaken by the coordinator and the laboratories that participated in the Trondheim inter-comparison testing and reporting, especially since those tests served as a prelude to dating the Shroud of Turin. Unfortunately, the British Museum, and three of the six radiocarbon laboratories at Arizona, Oxford and Zurich would also cavalierly and erroneously carbon date the Shroud two years after the release of the Trondheim Report. They would be especially careless in analyzing and reporting their data from the Shroud.

Carbon Dating Protocols

In the 1980s, STURP was easily the most knowledgeable and qualified group of scientists in the world to investigate the Shroud. They had not only initiated and conducted the only comprehensive scientific investigation of this burial cloth in 1978, but had continued testing and examining numerous Shroud samples at laboratories and facilities in the United States and Europe. Its members not only cont-

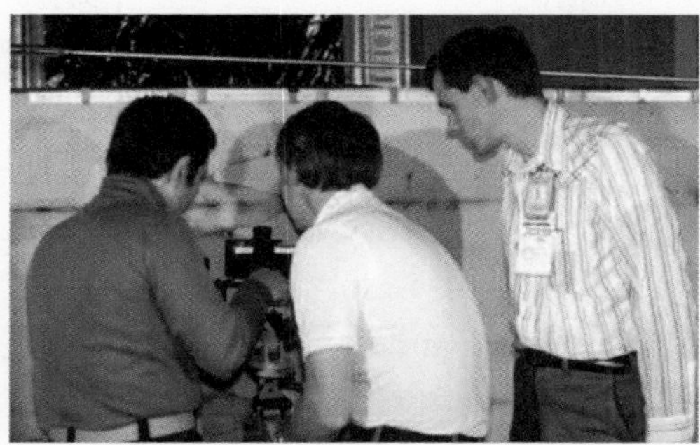

(Fig. 215)

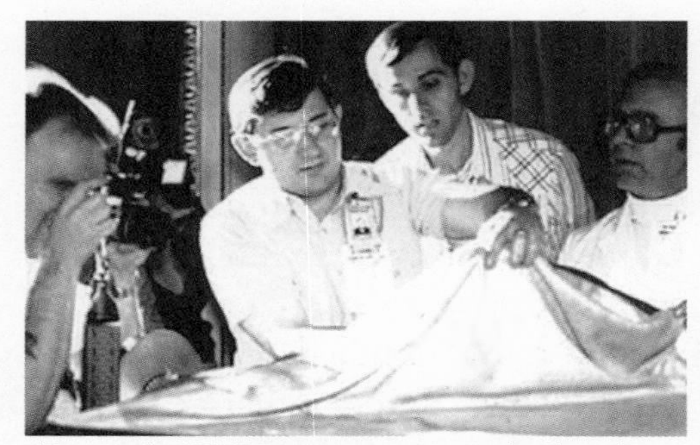

(Figs. 216) STURP scientists examine the Shroud.

inued to share their findings with each other, but with the world at large by publishing dozens of scientific articles in the 1970s and 1980s in peer reviewed scientific journals and other publications.

In 1984, concurrent with the intercomparison experiment, STURP submitted a 177-page scientific proposal describing 26 areas of scientific examination of the Shroud, including dating the cloth by radiocarbon analysis with the dating proposal being approved by the original five carbon dating laboratories.[12] STURP's 25 other areas of scientific examination were designed as a follow up to their previous examination. Their new range of scientific testing was designed to investigate a variety of issues concerning the Shroud's authenticity, conservation and image-formation, as well as its age, origin and history.

Following the announcement of the intercomparison test results in Trondheim in 1985, the six participating laboratories, each of which had joined STURP's carbon dating effort, agreed upon a dating protocol, known as the Trondheim Protocol. Among other things, this carbon dating protocol specifically stated:

- *The good offices of STURP would be used to arrange for appropriate samples to be removed.*

- The British Museum (as coordinator) would obtain written agreement not to reveal results to anyone other than designated officials of the museum.

- The six laboratories would use whatever methods they chose for preparing the samples for measurement, *but detailed descriptions of procedure and how they calculated their mean values and their uncertainties would be carefully recorded.*

- Results would be sent to the Vatican and the Archbishop of Turin. A press release would be issued only when Turin and the Vatican had received the results. (emphasis added)[13]

In 1986, more than twenty individuals with expertise in Shroud matters and/or the field of carbon dating gathered at a conference held in Turin, Italy. This group included several directors of the now seven participating radiocarbon laboratories that met over the course of three days. Despite a number of scientific differences that occasionally became intense, a consensus on a protocol for carbon dating the Shroud was reached, according to the participants. Among other things, this protocol — the Turin Protocol — contained the following points:

- For logistic reasons, samples for carbon dating will be taken from the Shroud *immediately prior to a series of other experiments planned by other groups.*

- *Seven samples* containing a *total of 50 mg* of carbon will be taken from the Shroud.

- Upon removal of the Shroud samples, a date will be chosen for *submission of experimental results from the seven laboratories to the following three analyzing institutions:* the Pontifical Academy of Sciences, the British Museum, and the Meteorological Institute of Turin, "G. Colonnetti." These institutions will keep the results in sealed envelopes until an agreed upon date, at which time they will be opened for statistical analysis.

- *After the analysis of the experimental results by the three analyzing institutions, a meeting will be held in Turin among the three analyzing institutions and representatives of the seven laboratories to discuss the results of the statistical analysis with the objective of deciding the final results of the measurement program.* (emphasis added) [14]

By the time the samples were removed in 1988, destructive undercurrents that had been set in motion by the leader of the carbon dating laboratories had taken their toll on events. In 1987, STURP learned that their testing program in 25 other areas of scientific investigation had been denied. In the same year, the number of labs that would be given samples had also been reduced from seven to three, Arizona, Oxford and Zurich. Yet, in January 1988, a few months before the samples would be removed from the Shroud, the directors of the three remaining laboratories, Michael Tite of the British Museum and Luigi Gonella, the scientific advisor for the Archbishop of Turin decided, among other things that:

- The Shroud samples will be taken from a single site on the main body of the Shroud.

- On completion of their measurements, *the laboratories will send their data for the three samples to the British Museum and the Institute of Meteorology "G. Colonnetti" in Turin for preliminary statistical analysis.*

- The laboratories have agreed not to discuss their results with each other until after they have deposited their data for statistical analysis.

- *A final discussion of the measurement data will be made at a subsequent meeting between representatives of the two above institutions and the three laboratories at which the identity of the three samples will be revealed.*

> The results will form a basis for both a scientific paper and for *communications to the public.* (emphasis added)[15]

Eliminating STURP was the Primary Focus

For years, STURP scientists had no idea why their promising follow-up testing program had been denied, but the denial was no accident. It was the direct result of several years of deceitful and hypocritical conduct on the part of the leader of the carbon dating laboratories — Harry Gove. His underhanded and unscientific activities throughout the Shroud's nine year carbon dating process are described in full in Chapters Eight and Nine of *The Resurrection of the Shroud* and are provided almost completely by Gove himself in his own 1996 book.[16] I repeatedly quote and cite from his book throughout the two chapters in my book. Gove all but brags about his deceitfulness.

Gove admits his deep-seated animosity toward STURP existed even before he called them in 1979 and made any pledges or agreements for his or his colleagues' involvement in the scientific testing of the Shroud. Gove reveals his true feelings and motives toward STURP on the first page that he mentions them in the very first chapter of his book and continues throughout. While discussing the early period of 1978 to 1979, Gove talks about his actual "disdain for those [STURP] scientists."[17] Before he ever called STURP in 1979 to accept its C-14 committee members' presence during the sample's critical removal from the Shroud and during its critical pretreatment and measurement, Gove blatantly reveals his duplicity: "...they [STURP] had good connections in Turin, and could be useful in obtaining a shroud sample for dating — if only they could be prevented from playing any other role."[18] Throughout the book he openly reveals his constant goal " to eliminate STURP once and for all."[19]

(Fig. 217) Harry Gove's constant goal was to eliminate STURP once and for all.

Over the course of nine years, Gove reveals how he, with the help of Michael Tite (the carbon dating coordinator) and the directors of the three carbon dating labs, was able to first eliminate STURP from playing any role in selecting the samples that were to be removed from the Shroud of Turin. STURP had always argued that samples should be removed from several different locations throughout the Shroud. Since they knew the Shroud extensively from first-hand observation of the cloth and its samples, had extensive scientific data from all over the linen and had taken 5,000-7,000 photographs of it in every wavelength and magnification possible, they would have known the best places from which to remove the samples. Avoiding locations that contained other complicating factors that were both visible and invisible, such as scorch marks, water stains, nearby repairs and other miscellaneous but misleading contamination, is always desirable. Regretfully, the people that replaced STURP took only one sample and it was from the worst possible location on the entire cloth.

Gove, the British Museum and the other directors of the radiocarbon laboratories had joined STURP's effort to carbon date the Shroud and had agreed that STURP would arrange for the appropriate samples to be removed for its dating. Yet, eliminating STURP from this role wasn't good enough. Gove and the other directors wanted to eliminate STURP from even being present when the samples were removed.

Giovanni Riggi actually removed the sample in the presence of Gonella and others. However, note that these two men had worked for years with STURP and wanted its expertise and knowledge present. STURP would clearly have been able to influence the locations and numbers of samples selected. Not only would they have influenced these decisions long before the samples were removed, but their scientists would have stepped forward at the time of the actual removal had an incorrect location been chosen. An example of this had occurred previously in 1978 in the confrontation between Dr. Max Frei and one of the STURP scientists: they literally stood face to face, confronting each other over the question of whether Frei would be allowed to put sticky tape on the face of the man in the Shroud and then peel it off. STURP won a ruling from the authorities present, thereby preventing possible harm to the facial image.

(Fig. 218) Max Frei and John Jackson during an impromptu, gentlemanly debate presided over by Luigi Gonella, Scientific Advisor to the Custodian of the Shroud.

If STURP had been allowed to assist in selecting samples from different areas on the cloth, and been present when they were removed from only one site, this would have greatly *aided* in the accuracy of the radiocarbon dating process.

One tactic that Gove used to eliminate STURP from playing any role with the Shroud was through his influence with Professor Carlos Chagas, the president of the Pontifical Academy of Sciences located in Rome (along with his colleague Dr. Vittorio Canuto). According to Gove's book, these men followed almost every suggestion he made. Rarely, if ever, did any of Gove's suggestions deal with substantive scientific matters, however. He focused on falsehoods, maneuvering and procedure. STURP's influence had always been with the custodian of the Shroud, the Archbishop of Turin, Cardinal Anastasio Ballestrero and his scientific advisor, Dr. Luigi Gonella.

Unknown to STURP, Gove wrote Chagas a number of falsehoods about STURP's alleged religious zeal and military mind set, even questioning their sources of support. Gove freely admits while writing to Chagas, "I stated that almost every aspect of the STURP organization was distasteful…."[20] even comparing STURP to the Spanish Inquisition.[21] Chagas, who displayed no knowledge of substantive scientific matters regarding the Shroud or of STURP's work, evidently believed Gove's many lies for he heeded every request by Gove to try to eliminate STURP from playing any role in any future testing of the Shroud.

Whenever it appeared that the authorities in Turin had agreed to STURP's wide range of scientific tests and to radiocarbon dating the Shroud, Gove could get Chagas and Canuto of the Pontifical Academy in Rome to join him in changing the entire focus from reaching such a consensus to insignificant and petty matters such as the location and sponsorship of a meeting in which to finalize the testing protocol. When a meeting to finalize such plans finally was to occur in Turin in 1986 — now sponsored at Gove's insistence by Rome's Pontifical Academy of Science — Chagas followed Gove's incredulous advice not to invite any STURP members to the meeting, while inviting the directors and scientists from the radiocarbon dating laboratories. (Even Gove's traveling companion, Shirley Brignall, received an invitation to this meeting and attended.) Cardinal Ballestreros appealed to the Vatican secretary of state and to the Pope that the Pontifical Academy was taking matters away from him concerning the Shroud.[22]

(Fig. 219) Cardinal Anastasio Ballestrero, Official Custodian of the Shroud and his Scientific Adviser, Prof. Luigi Gonella.

Throughout the carbon dating process, Gove created a number of unnecessary conflicts with Gonella and made repeated unfounded criticisms of him to his carbon dating colleagues and to Chagas and Canuto at the Pontifical Academy of Sciences in Rome. This was all part of his successful strategy to diminish the authority of the Archdiocese of Turin, where STURP had earned enormous respect and influence. Another part of his strategy was for the Pontifical Academy of Sciences in Rome to assert itself into future testing decisions for the Shroud since Chagas and Canuto followed all of Gove's suggestions.

During the 1986-1987 period, when Gove lobbied hardest to undo STURP's planned tests in 25 other areas of scientific inquiry, the directors of the carbon dating laboratories wrote letters to Chagas, and Chagas met personally with the Pope, echoing their complaints about STURP's planned tests. The Archbishop of Turin complained in October 1987 that the directors "...stepped out of the radiocarbon field to oppose research proposals in other fields, with implications on the freedom of research of other scientists and our own research programs for the Shroud conservation ...."[23] In the summer of 1987, a short time before STURP's tests were eliminated, Gove wrote to Chagas his earlier discussed letter containing numerous falsehoods about STURP. In the past, Gove had threatened that the carbon dating laboratories would withdraw without Chagas' continued support; however, this time he *guaranteed* their withdrawal "if STURP participates in the carbon dating enterprise in any way."[24] Gove admitted in his book that the Archbishop's "thinly veiled accusation that we were attempting to prevent STURP from carrying out its scientific investigations was quite correct."[25]

Throughout the prolonged nine year process to radiocarbon date the Shroud, Gove made repeated threats to Chagas that the carbon dating laboratories would withdraw. By the 1980s, STURP and others had scientifically demonstrated that the Shroud could not have been painted, as it had been reputed for centuries. All of the modern scientific, medical, archaeological and historical evidence from the 20th century indicated that the Shroud was the authentic burial garment of the historical Jesus Christ. The only major scientific test remaining at that time was its radiocarbon dating. Until methods to radiocarbon date small samples had been invented by Gove and others in the 1980s, the Shroud could not be scientifically dated. In this sense, Gove and the directors of laboratories who could carbon date small samples had the new owner of the Shroud over the barrel. (In 1983, the Vatican inherited the Shroud when former King Victor Emmanuel II left it to the Holy See in his will — on the condition the cloth remain in Turin, where it had been for over four hundred years.) God only knows what part Gove's guarantee to withdraw the cooperation of the radiocarbon laboratories played in Chagas' thinking, or in his recommendations to the Vatican. God only knows what role Gove's many lies played in Chagas' thinking.

Dr. Gonella, who was intimately involved in the entire infamous carbon dating process, said of the directors:

> At the beginning, when they themselves had asked us to be allowed to examine a sample of the Shroud, they had guaranteed us the utmost seriousness and completeness in the analysis, as well as promising to collaborate with the custodian of the Shroud, the Archbishop of Turin, and with his scientific consultant, the undersigned. Seized however by a feverish desire for celebrity, they began to renege on their promise: no further interdisciplinary investigations; just the carbon-14 test. They even badgered Rome, bringing pressure to bear so that Turin would have to accept their conditions. Through the intervention of Professor Chagas, then president of the Pontifical Academy of Sciences, they set aside the undersigned so that they could do whatever they wanted.[26]

If you think my above characterizations of Gove, or the other directors of the radiocarbon dating labs are incorrect, please read Chapters 8 and 9 of my 2000 book and Gove's 1996 book. I have only given you a flavor of what went on. These sources can tell you many more instances of unprofessional actions. Worse yet, no scientific or credible reasons are apparent for any of Gove's unprofessional and deceitful actions. The only constant that was displayed was Gove's selfish and egotistical desire to not only make radiocarbon dating the center of testing, but the only testing to be done on the Shroud. Equally apparent was his jealous and egotistical desire to eliminate the most knowledgeable group of scientists from any involvement in future testing of the Shroud.

If only Gove, the carbon dating coordinator and the other directors could have spent an equal amount of time and energy on researching this unique cloth with the unprecedented images and blood marks of a crucified corpse wrapped within it that have never been duplicated. If only they had seriously considered that this burial cloth could have wrapped the historical Jesus Christ and that the unprecedented event that caused the man's full-length frontal and dorsal images could have affected the types and amounts of radioactive atoms that were in the Shroud.

Dating the Shroud Samples

In the absence of STURP, but in the presence of others, including the directors of the three participating laboratories and their coordinator, a large piece of cloth, approximately 285 mg., was cut and removed by Giovanni Riggi from the worst location possible on the Shroud. This sample was a full 75 mg. more than the total amount required by all seven original laboratories.[27] The right-hand portion of this large sample was then divided in approximately equal amounts among the three laboratories as indicated in Fig. 220 below. Because the first sample (given to Arizona) was about 10 mg. smaller than Oxford's or Zurich's samples, they were given a second small sample that was approximately 14 mg.

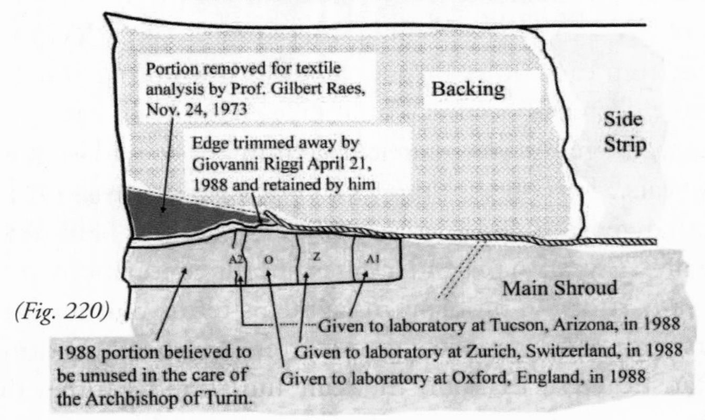

(Fig. 220)

Although the last two protocols called for the laboratories' dating data to be sent to the Pontifical Academy of Sciences and/or the Institute of Meteorology in Italy, as well as to the coordinator for statistical analysis, the labs only sent their data to the British Museum. Furthermore, despite the last two protocols, the British Museum did not share its raw data with either analyzing institution. Instead, the laboratories and the British Museum reworked the data to make it more compatible.

Arizona was the first radiocarbon laboratory to date the Shroud samples, taking eight different measurements on four separate dates in May and June of 1988. The dates of Arizona's measurements and the radiocarbon ages in years are provided below.

May 6	May 12	May 24	June 2
574 +/- 41 years	753 +/- 51 years	676 +/- 59 years	701 +/- 47 years
606 +/- 43 years	632 +/- 49 years	540 +/- 57 years	701 +/- 47 years[28]

Thereafter, Zurich reported five radiocarbon ages and Oxford reported three radiocarbon ages as indicated below:

Zurich	Oxford
733 +/- 61 years	795 +/- 65 years
722 +/- 65 years	730 +/- 45 years
635 +/- 57 years	745 +/- 55 years[29]
639 +/- 45 years	
679 +/- 51 years	

However, these ages provided a serious problem for the British Museum for the radiocarbon ages of the samples ranged from 540 to 795 years old. In this case, these particular ages were considered to be "outliers." A 245 year age range for samples taken less than five centimeters from each other on the same cloth would be too great for an acceptable degree of accuracy or a 95% certainty in age. After all, the radiocarbon ages of these particular samples also had +/- ranges of 57 and 65 years. This means the ages for the very nearby cloth samples could even range from 483 years old to 860 years old. This translates to a range of 377 years or calendar dates of four centuries apart. (I am expressing a part of the problem in laymen's terms that most readers can understand. The statisticians and scientists speak in formulistic terms of statistical tests that only they can understand, yet they disagree about many aspects of the formulas and even on which tests should have been applied to the Shroud data in 1988. The formulas and overall debates essentially address the question of whether the dating data supports an age that can be attributed to the Shroud with a high degree of certainty. In this particular case, the Shroud was dated from 1260-1390, a range of 130 years with a 95% certainty.)

Based on private correspondence with two scientists at Oxford and one at Arizona, chemist Remi Van Haelst states that the British Museum solved this problem by asking Arizona to combine or to essentially average the two radiocarbon sample ages from each of the four above dates in May and June, which Arizona did.[30] The British Museum and the Arizona laboratory thus combined eight age measurements into only four age measurements. This combination was not mentioned in the official *Nature* report containing the Shroud's radiocarbon date of 1260-1390 with a 95% certainty. This report was authored by scientists at the radiocarbon laboratories and the British Museum.

This combination of ages from eight to only four had the effect of eliminating not only Arizona's youngest sample age of 540, but the youngest of all 12 sample ages from all three laboratories. The 540 year age measured on May 24, 1988 was now *eliminated* from the overall data and replaced with a new essentially averaged age of 606 years,[31] a net gain of 66 years for this May 24th date (and the elimination of the lowest outlier for the entire group.) Furthermore, the +/- range for this particular sample's age now was only 35 years instead of +/- 57 years. Arizona's youngest sample age now was 591 years, a lessening of its worst outlier age by 50 years. Incidentally, this new youngest age was acquired by essentially averaging the ages measured on May 6th, thus also eliminating Arizona's and the group's second youngest date as well.[32]

Since all the measured dates acquired by the laboratories go into the formula for ascribing a final age range and a percentage of certainty, the elimination of Arizona's two worst dates helped the entire age grouping of all twelve dates from all three radiocarbon laboratories. The above combining or essentially averaging of each of Arizona's May and June dates also eliminated what would have been the larger group's third youngest age of 632 years measured on May 12th. This was now replaced by a new date of 690 years, which is within the range of ages measured by Zurich and Oxford.[33] (Another 28 +/- years was also eliminated by the last two new dates.)

These are the kinds of ways that the combining — or essentially the averaging of Arizona's ages on all four dates — helped the British Museum and the radiocarbon labs to claim an acceptable degree of certainty for the medieval age attributed to the Shroud with 95% certainty. (While the AMS labs counted the C-14 to C-12 atoms correctly, they failed to consider that something extraneous could have contaminated these samples causing the larger than normal range of ages.) These descriptions are given in laymen's terms and not in the formulistic terms that neither the author nor most readers can understand.[34] If any statements regarding the raw data are incorrect, I invite representatives of the Arizona, Oxford and Zurich radiocarbon laboratories and the British Museum to supply all the raw data from their initial reporting and that figured into the final report, including the locations of the subsamples.

The combination and elimination of radiocarbon ages from Arizona is what Remi Van Haelst is referring to when he reports that the initial raw data was reworked by one of the radiocarbon labs and the British Museum to minimize the outliers and range of dates found in their initial unreported raw data.[35] I believe this combination and elimination of Arizona's initial ages is also part of what statistician Dr. Bryan Walsh is referring to when he reports that the radiocarbon labs masked significant underlying differences in the raw data in their final report.[36] Yet, as pointed out by Dr. Marco Rianni and his scientific colleagues, and by Walsh and Van Haelst, the reworked or combined data in the final report is not even compatible or homogenous,[37] containing ages or dates that still vary more than 200 years over a distance of less than five centimeters on the cloth.

Even though the raw data was reworked and the locations of the twelve subsample dating results (from the laboratories' three samples) were not provided in the report, extensive statistical analysis of the carbon dating results was performed by scientists Walsh, Van Haelst, Riani and his colleagues, who have been able to reveal a number of further important points. Riani and his co-authors ran over 387,000 spatial possibilities and concluded that data from Arizona's second sample was not contained in the final report, which was confirmed by one of the participating scientists at Arizona.[38] Walsh states that his statistical analysis leads to the conclusion that the Shroud subsamples each contained differing levels of C-14.[39] Furthermore, his analysis indicates that a *relationship* exists between the C-14 content within each sample and their location on the cloth.[40]

Walsh's statistical analysis indicates that the C-14 content within the samples increases with their distances from the edge or the bottom of the cloth.[41] The three samples were located about a foot and a half away from the body, running parallel with it around the ankle, as seen in Fig. 221.

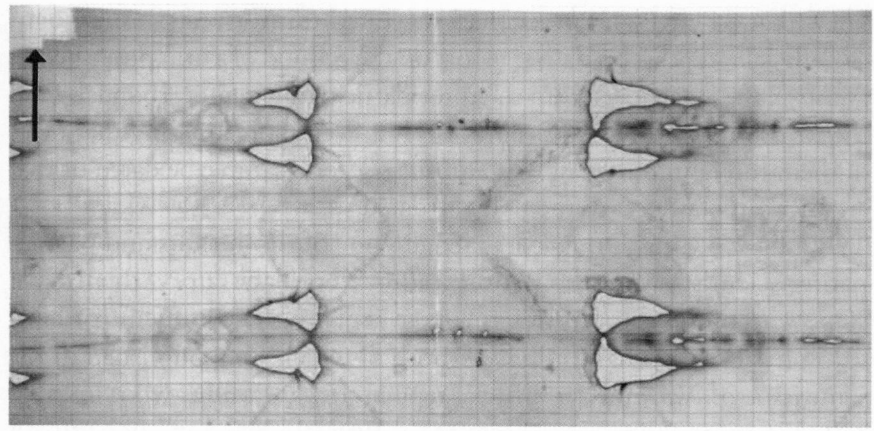

(Fig. 221)

Even this distant set of small parallel samples indirectly indicates the *farther* away the samples were from the body, the *older* they date. Conversely, the *closer* they were to the bulk of the body, the *younger* they date. If Arizona had dated its second sample (labeled A2 in Fig. 220) and its results were included in the report, it may have shed further light on this possible correlation. Moreover, if Arizona's two youngest ages of 540 and 574 weren't completely eliminated from the report, this correlation would have been *more* obvious to everyone who read the report. The combination and elimination of the youngest ages from Arizona's initial reporting may have deceived the authors, as well as the readers of the report.

In light of the range of ages in the initial data and the correlation that the ages or dates became younger as the three samples moved away from the bottom of the cloth and toward the mass of the body (even at a considerable distance from the body), the labs should have dated sample A2. As seen in Fig. 220, A2 is even further away from the bulk of the body and closer to the edge of the cloth than Oxford's sample. The coordinators at the British Museum and the laboratory scientists should also have asked Arizona to date a portion of sample A2, instead of combining their ages and eliminating the youngest ones from the report and their statistical tests.

Furthermore, they should have revealed all the raw data from all three laboratories, explaining that the outliers were too great for a sufficient degree of certainty. They could have requested another Shroud sample from another location on the cloth, preferably near the body. They clearly knew that STURP and others had suggested this for years. Amazingly, the official radiocarbon dating report on the Shroud of Turin published in *Nature* has 21 scientists from the three participating radiocarbon labs and the British Museum listed as authors. It is even more amazing that none of them thought to request that Arizona's second sample be dated or that another sample closer to the body be dated.

In light of this, Dr. Robert Hedges, one of the authors and participating scientists from the Oxford radiocarbon laboratory, then made some very presumptuous statements in reply to the comments of Harvard physicist Thomas Phillips — both of which were published in the same issue of *Nature* as the official report. It was Phillips who first suggested in light of the Gospel accounts of Jesus' resurrection, the Shroud's body images and their general properties, that the body wrapped within it could have irradiated neutrons. He noted this event would have created C-14 atoms (isotopes) whose amounts would "vary in different parts of the Shroud."[42] He also stated that these neutrons would have caused radioactive Cl-36 and Ca-41 atoms (isotopes) in the Shroud. As we saw earlier, such an irradiating event would have invalidated a radiocarbon dating from any location on the Shroud.

Remarkably, the very first sentence of Hedges' reply states, "The processes suggested by Phillips were considered by the participating scientists."[43] In his second sentence he adds, "However, for the reasons given below, the likelihood that they influenced the date in the way proposed is in my view so exceedingly remote that it beggars scientific credulity."[44] Although all of Hedges' reasons were inapplicable or superseded by much more important and relevant evidence – even if they weren't – neither he nor any other scientists at the radiocarbon laboratories have any *scientific* basis to claim that such an event is so incredulous. This is especially true in light of what little overall knowledge they had of the Shroud, and the fact that such an event would have been *consistent* with the *evidence* acquired from this burial cloth.

The failure to consider this evidence or the possibility of such an event lies at the heart of their erroneous dating results and their lack of further consideration and consultation with the analyzing institutions with whom they had agreed to consult. The next chapter will address the fact that scientists must be open to and guided by relevant evidence whenever investigating any area of importance, such as the age or authenticity of the Shroud of Turin.*

Sadly, part of the reason why no deliberation or meeting with any Italian analyzing institutions ever took place to review the laboratories' raw dating data was because the labs' erroneous results were also leaked several times before the official announcement. These leaks not only offended common decency, they were totally inconsistent with all three protocol procedures. The ultimate source for these leaks appeared to be Harry Gove or someone at the Arizona laboratory.[46] Although the University of Rochester did not date a Shroud sample, Doug Donohue, one of the co-directors of the Arizona laboratory, allowed Gove to attend when Arizona dated its sample.

These inconsiderate acts are consistent with all of the labs', their coordinator's and Gove's actions over the entire nine year process of cavalierly carbon dating the Shroud. Their actions over this entire time indicate their primary goal was not to participate in or even allow an interdisciplinary scientific investigation of the Shroud, even if this overall examination would shed additional light on the Shroud's authenticity, age or origin. Nor did they ever seriously consider that in light of its unprecedented body images and all the consistent and remarkable evidence indicating this burial cloth could have wrapped the dead body of Jesus, that they might want some of the help and expertise that was available. They even eliminated the most knowledgeable group of scientists in the world regarding the Shroud, who could have provided insight into the selection of samples and the complicating factors of accurately dating the Shroud. Remember, it was the carbon dating scientists who sought to join STURP's effort to carbon date the Shroud. All of the above actions by Gove, the radiocarbon dating laboratories and their coordinator throughout the entire process shows their unprofessional conduct and lackadaisical effort. It's as if their primary goal was to occupy center stage and to have all the focus be upon themselves.

***I don't mean for Hedges' above remarks to be the last word on him. At the 1986 Turin workshop, he spoke in favor of removing samples from more than one location on the Shroud.[45] In "Hedges Replies," he also explains that because C-14 also converts from N-14, and is much more abundant than C-13, that a much, much lower dose of neutrons would be needed than proposed by Phillips.**

Having acquired center stage at last, when the Shroud's date was officially announced at press conferences in Turin and in the British Museum, its date was smugly provided by Edward Hall, the Director of Oxford's radiocarbon dating laboratory, and Michael Tite with an exclamation point as seen in Fig. 222.

On the same date, Edward Hall also dismissed the Shroud during a televised interview as "a load of rubbish."[47]

Archaeologist William Meacham has utilized radiocarbon dating in his profession even longer than he has been an independent Shroud expert for well over three decades.* At the International Scientific Symposium held in conjunction with the Shroud's millennial exhibition

(Fig. 222) Edward Hall, director of the Oxford University carbon dating laboratory, and Michael Tite of the British Museum as they announce the Shroud's 1988 carbon dating results.

in 2000, Meacham called the 1988 Shroud dating a "fiasco." In 1987, he warned that the major problem with the entire C-14 dating issue was, "The labs seem to have put themselves in charge of the entire dating operation."[48]

Over Gove's objection, Gonella had to lobby Chagas just to secure an invitation for Meacham to attend the 1986 carbon dating meeting in Turin. When Gove and Chagas openly stated at that meeting that STURP's larger scientific testing should not even be considered until the radiocarbon dating was completed, Meacham provided perspective. He said the objective was to scientifically investigate the Shroud. He drew the analogy that an archaeologist would not abandon all of the critical investigative activities associated with an excavation or archaeological dig because of the results of one carbon date. He argued that all of the planned tests concerning the Shroud's authenticity, the cause of

---

***His article "The Authentication of the Turin Shroud: An issue in Archaeological Epistemology" Current Anthropology 24:3 (June 1983) is still one of the best ever written.**

the images, its conservation, age and origin would have relevance regardless of what the carbon dating results were. Meacham said they should go forward. Consistent with Gove's real goal of eliminating STURP and their investigation of the Shroud, he noted that Meacham "should never have been allowed to play a role in the workshop."[49]

Meacham also advised that samples should be removed from as many locations on the Shroud as possible, as an archaeologist attempts to do at a field site. He argued that taking samples from various locations would give greater credibility to the carbon dating of the Shroud. Meacham recalled that "Alan Adler* and I urged, pleaded, cajoled and literally begged . . . for at least two sites on the Shroud to be sampled."[50] Instead of taking these comments at face value, Gove states that these comments "seemed remarkably inappropriate as applied to the Shroud."[51]

Gove couldn't have been more wrong. Meacham's comments were *especially* appropriate for a cloth that is more than 14 feet long and 3-1/2 feet wide; has a documented, varied and unknown history, as well as wrapped a body to whom an unknown or unprecedented event occurred. Just on the basis that the Shroud could have been the burial cloth of Jesus, you would have thought that Gove would have wanted to attach as much credibility to its carbon dating as was feasible, and also date another sample from another location. Meacham stated, "A major divergence of views occurred over the sampling strategy. Strangely, the ^{14}C specialists insisted on having splits of the same single sample. It appeared as if they wanted above all else to achieve harmonious results amongst themselves, as opposed to any results that might indicate a variation of the Shroud's radiocarbon content."[52] Meacham noted, "Unfortunately, the blame for

(Fig. 223) Harry Gove during 1986 carbon dating meeting in Turin. Directors Paul Damon of Arizona and Willy Woelfli of Zurich are on Gove's right. Gove's companion, Shirley Brignall, is on his left. She got to attend the conference while almost all of STURP's scientists did not at Gove's behest.

***Chemist Alan Adler was the only STURP scientist who got to attend the workshop meeting, and this was only through the intervention of Prof. Luigi Gonella.**

this [C-14] fiasco lies mainly on the shoulders of the extremely overconfident, overbearing and haughty attitudes on the part of most of the C-14 lab directors who were involved."[53]

A few chapters back we described at length the forms of atomic testing that must be carried out on the Shroud of Turin. This testing should be undertaken, not only to accurately assess whether the Shroud's medieval dating was accurate, but to calculate its real age. This book has detailed at length all the reasons to legitimately question the Shroud's purported medieval origin. Suffice it to say that a medieval origin is inconsistent with the test results of thousands of other scientific tests, including the recent dating by three new scientific methods in 2013.

Testing samples at the atomic level is not a new area of research. It has been occurring for more than eighty years since the discovery of neutrons in 1932. Radiocarbon dating has existed since the 1950s and AMS measurements have existed since the 1980s. Measuring the amounts of C-14 atoms within material is simply one form of testing at the atomic level, as is measuring the presence and the amounts of Cl-36 and Ca-41 atoms. The reader should understand that the Shroud has already been tested at the atomic level. Because it was tested in such a cavalier and incomplete manner at every phase throughout the entire carbon dating process (and was likely dated erroneously), atomic testing must now be undertaken to demonstrate whether the Shroud linen and blood marks are from medieval times or the first century. This testing could also prove that a miraculous event occurred to the dead body wrapped within this cloth, as well as when, where and to whom it occurred.

Ironically, the carbon dating fiasco could actually have been beneficial in some ways despite the duplicitous and inexcusable behavior from some of the radiocarbon community. Ideas and hypotheses about neutron and/or particle radiation emanating from the man in the Shroud started emerging in order to explain the very aberrant radiocarbon date ascribed to the cloth. In what could possibly be a providential result, this atomic testing could detect millions of items of evidence in Shroud cloth, blood, and charred materials, and in the limestone from Jesus' reputed burial tomb, which would, as we've emphasized before, confirm that Jesus' passion, crucifixion, death, burial and resurrection were actual events in history.

15

HUMANITY'S RIGHT

During the last round of scientifically testing the Shroud in the 1980s, a small group of scientists (who knew very little about the overall subject), in effect, took over the entire project. They not only excluded other scientists who knew much more about the total subject, they excluded the worldwide public from the information that would have been acquired from 25 other areas of testing. What raw data the radiocarbon scientists and their coordinator did acquire was not shared with other analyzing institutions, as had been agreed. Instead, they masked its significant underlying differences by combining and eliminating the youngest two ages from the reported data. Then they ignored the remaining data indicating that each contiguous sample had noticeably different amounts of C-14 content, which correlated with their distances from the edge of the cloth. Worse yet, they next attributed a date to the cloth based on very limited information that was not only erroneous, but was misleading. This inadvertently caused people all over the world to disregard, or never even be aware of, the unprecedented and incomparable evidence on this cloth.

We should learn from these mistakes. We should attempt to acquire as much scientific information from as many proven testing techniques and technologies as possible. For all the reasons stated earlier in the book, we should not conduct radiocarbon dating independently of other radioactive testing of the Shroud. As we have learned, testing the Shroud and its samples at the atomic and molecular levels could prove that particle radiation emanated from the length, width and depth of the dead body wrapped within it. This testing cannot only prove that a miraculous event happened to the dead body in this burial cloth, but *when* it happened and *where* it happened. This testing could not only completely refute the Shroud's 1988 C-14 dating, but reveal the real age of the cloth and its blood marks and demonstrate *why* the 1988 dating was in error. This and other testing could also confirm that the tortured, crucified victim within this burial cloth, to whom this miraculous event occurred, was the historical Jesus Christ. This and other testing could also confirm that this miraculous event caused the unique frontal and dorsal body images to develop over time and that naturalistic and artistic methods did not cause these images, blood marks or affect its medieval C-14 dating. These tests could conceivably resolve every outstanding issue relating to the Shroud.

In light of all the new and unprecedented evidence that we have reviewed — its objectivity and independence, the variety of fields that it came from, that most of it is unfakable, and that the detailed similarities of all the events and circumstances that occurred to the man in the Shroud also occurred to the historical Jesus — most people would agree that testing the Shroud at the atomic and molecular levels (and by other developed and proven scientific methods) is clearly warranted.

The worldwide public's lack of knowledge of this overall evidence, combined with their knowledge of its aberrant, controversial C-14 dating, once dampened any enthusiasm for the Shroud or for testing it further. However, people in the 21st century can now readily acquire access to and obtain perspective and understanding of the overall evidence. They can also look forward to the development and perfection of even newer technology to further test the cloth, blood and charred material on the Shroud and from the limestone within Jesus' reputed burial tomb(s).

There are only a few absolute, universal conditions of life that are recognized by all mankind. For example, while there could be one Creator for all of us, we don't universally agree that there is. Even among those who agree that there is one creator, we don't agree on who this Creator is. There is really only one absolute, universal condition of life that we all acknowledge: that *every single one* of us is going to die. This is the most inherent and fundamental certainty facing every member of the human race.

If a miraculous event really happened to the dead corpse wrapped in the Shroud, which is consistent in every way with the resurrection of the historical Jesus Christ, then this event would be relevant to every one alive. Some or all of the unprecedented evidence revealed in this book is new to most of us and to humanity as a whole. For the first time in history, all of humanity can learn of new, comprehensive and independent evidence that relates to every one of us and addresses our most fundamental problem of all. This evidence is real and unforgeable and could even provide a solution for our inescapable and inevitable deaths.

As we have learned, atomic and molecular testing could easily confirm that a miraculous event occurred to the dead, crucified body wrapped in the Shroud, and that the body emitted particle radiation as it suddenly disappeared. This miraculous event, as well as the entire sequence of events that occurred to this man, were captured and encoded in the cloth. Atomic and molecular testing could easily add *millions* of items of evidence onto thousands of other items of evidence confirming that this miraculous event caused both the unique body images and the blood marks on this cloth. These tests can confirm the actual age of the burial cloth and its blood marks, when and where the miraculous event and the sequence of events occurred, and that they happened to the historical Jesus Christ. Thousands upon thousands of items of evidence on this burial cloth would also corroborate in every way the accounts of these events in the most attested and reliable sources of ancient history. These sources also state that before his death Jesus predicted on numerous occasions that he would be resurrected after he was killed and that all of these occurrences were prophesized centuries earlier in the Old Testament. The Gospels tell us that Jesus also resurrected several people from the dead. Moreover, Jesus is recorded to have said on numerous occasions that, if among other things, we believe in his resurrection — that we, too, could have life after our certain deaths.*

As human beings we have the most inherent right of all to this new and sophisticated evidence. No government, institution or organization of any kind is going to die a *physical* death. They do not require this information for themselves and have no right whatsoever to limit or deny it from *any* human being. They have a duty to guarantee access

---

***This was most distinctly said by Jesus before he raised Lazarus from the dead. "I am the resurrection and the life. Those who believe in me, even though they die, will live, and everyone who lives and believes in me will never die." (John 11:25, 26).**

to this information by every human being within their borders — it is a basic human right. *God* has left this evidence for all of humanity.

Interestingly, the skin color of the man in the Shroud (who appears to be in his 30s), is not apparent on any of the photos, not even on the highly resolved positive images. His grayish skin color on these images resembles the color of his beard, mustache, eyebrows and the blood marks on his body. Because of the light/dark reversal of the straw-colored negative body images (and their reddish blood stains), they all appear lighter or whiter on the positive image (photographic negative), but they are actually darker. The man's hair and skin would have been browner or darker than the way they appear on the photographic negative (positive). The man's skin color would have actually resembled the tan, brown or darker shades of skin found throughout most of the world in Central and South America, Africa, the Middle East, Asia and Southeast Asia.

The Realm of Science

Regardless of how impressive the overall evidence already is or how unanimous and overwhelming it could become, there will be some scientists and others who will be against testing the Shroud. Most of the scientists that I have heard voice these objections were not familiar with the overall evidence on the cloth. They object to testing whether radiation or particle radiation emanated from the body of the man in the Shroud. Their common objections are listed below. Although these scientists were either unfamiliar with the overall evidence or are in a minority of those who are familiar with it, I want to address these concerns since this unprecedented evidence has been left for every human being to see.

The common concerns that these scientists have expressed are that:

- No plausible physical mechanism has been proposed to explain how the resurrection was accompanied by a significant neutron flux.
- Particle or other radiation cannot emanate from a body, let alone a dead body.
- Unless we can reproduce an event in the laboratory, we can't argue for its occurrence.
- Miracles or the resurrection are beyond the realm of science.

A common response to all of these arguments is that we're not trying to duplicate the resurrection; we're investigating whether it occurred this one particular time in history. The resurrection of the historical Jesus Christ is not a recurring event, but neither is any other event in history. There will be only one Battle of Gettysburg or Hastings, only one assassination of Abraham Lincoln or Julius Caesar. Certainly other battles and assassinations have occurred, but these did not have the identical impact and influence upon history or future generations. If you tried to recreate any of these events in a laboratory, or on the actual fields of battle or in the original Roman Senate, it would be *impossible* for many reasons. Yet, this does not prevent us from acquiring relevant evidence of important historical matters by scientific analysis. While such evidence will not always provide absolute proof, it will shed light on or even be dispositive on a number of questions. For example, attempts have been made to gather evidence to prove if Yasser Arafat was poisoned. Different forms of scientific analysis have also been utilized to determine which bullets struck which victims in the Kennedy assassination, and at what angles or directions the bullets were traveling. Scientific analysis is conducted every day on past events that relate to the public welfare or other matters of importance.

Scientists have spent decades studying the Big Bang or the creation of the universe. The fact that this event obviously cannot be duplicated in a laboratory, yet was obviously an important occurrence, does not preclude its scientific investigation. This event was arguably more supernatural or miraculous than the resurrection of Jesus Christ. Not only was all matter and all energy within our universe created from a tiny speck according to this theory, but so was all of space and time. Yet, the fact this event could be considered the most supernatural of all does not preclude its scientific investigation. And, although we cannot provide a plausible physical mechanism or explanation how all of this matter, energy, space and time were formed within and released from such a tiny speck in a momentous explosion — we are clearly not precluded from investigating this occurrence. Since we can scientifically investigate the occurrence of the Big Bang, or the creation of our universe, then surely we can investigate the occurrence of the resurrection. The fact that the resurrection arguably has much more *relevance* to peoples' lives than the Big Bang provides even more reason to investigate it.

Thankfully, science has never had a known or fixed realm. If a predetermined definition of a known "realm of science" existed that prevented scientists from inquiring and investigating into various areas, then most of our advances in science and medicine would never have occurred. The objective evidence acquired from an investigation has always determined the worthiness of scientific inquiries. The key question has always been whether or not a particular matter was within our *realm of investigation*, not whether it was within a predetermined "realm of science."

Instead of a fixed or static realm, the realm of scientific inquiry, knowledge, and even its boundaries has expanded in all areas of modern science throughout the relatively *few* centuries it has existed in the course of human history. Yet, the more we inquire and learn, the more we realize how very little we actually know. We know very little, if anything, about 95% of our universe, which scientists think consists of invisible dark energy and dark matter. We also know very little about the 5% of visible matter/energy in our universe, most of which is well beyond our reach or inspection. We don't even know how many dimensions there are or when they operate. We know of three spatial dimensions and we understand some concepts about what is called the fourth dimension of time, yet the fourth dimension could easily contain more attributes than we currently understand.

Since science indicates that 95% of the universe is comprised of invisible, dark energy and matter, contains quantum strangeness and hidden dimensions, and that all of the matter, energy, space and time in the universe were formed within and then released from a tiny speck in a momentous explosion, why would it have been impossible for the resurrection of the historical Jesus Christ to have occurred? Unless scientists can absolutely prove that neither God nor a Godly power exists, they can't absolutely and arbitrarily rule out an event that is attributed to this power in numerous historical sources.

When the Shroud's controversial 1988 radiocarbon dating was published in the scientific journal *Nature*,[1] a letter from Thomas Phillips of the High Energy Physics Laboratory at Harvard University was also published in the same issue. In the letter, Dr. Phillips noted that the body images on the Shroud had not been duplicated and had qualities "indicating that the body radiated light and/or heat."[2] Phillips asserted that the body may have radiated neutrons, which would have increased the C-14 content within the linen Shroud thereby making it appear much younger than its actual age.

Dr. Robert Hedges of the Oxford University radiocarbon laboratory, who participated in the Shroud's controversial radiocarbon dating process, replied in turn to Dr. Phillips. His reply appeared on the same page as Phillip's letter. Dr. Hedges pointed out that a much lower dose of neutron radiation — a thousand times less than what Phillips proposed — would have been sufficient to create the number of C-14 isotopes within the Shroud to account for its medieval C-14 dating.[3] Hedges even admitted that "the processes suggested by Phillips were considered by the participating [radiocarbon] laboratories."[4] Hedges had also acknowledged after the radiocarbon dates were announced in 1988 that a "sufficient level of neutrons from radiation on the Shroud would invalidate the radiocarbon date which we obtained."[5]

Despite these considerations and admissions, Hedges dismissed the possibility that such an event could have influenced the Shroud's medieval radiocarbon date.[6] And, the *first* reason he asserted was that "No plausible physical mechanism has been proposed to explain how the resurrection was accompanied by a significant neutron flux."[7] That wasn't even the point! The point was that if the Shroud was irradiated with a neutron flux, it would have invalidated the 1988 radiocarbon dating, and it was the laboratories' job to accurately date this cloth. Neutrons were discovered by scientists in 1932, so it is conceivable the Shroud could have been irradiated by neutrons before 1988.

I happen to agree that because of the cloth's unprecedented and unduplicated images, their 130 corresponding blood marks, and the numerous similarities between the passion, crucifixion, death and burial of the historical Jesus Christ with the victim in the Shroud, that the resurrection is the most likely explanation for the neutron flux. But, it is completely ludicrous to state that you have to provide the physical mechanism for the neutron flux or the resurrection in order to consider their occurrence. If the resurrection was accompanied by a neutron flux, it neither happened in the 20th century nor was it caused by humans. It could *only* have been caused by God and no human could possibly provide a *plausible* physical mechanism. Hedges just arbitrarily applied an artificial Catch-22 to the questions of whether the Shroud was neutron irradiated or whether the body was the source of the radiation.

Neither Copernicus, Galileo nor any of the generations of scientists have had to first provide a plausible physical mechanism for the creation of the universe before they studied its features and component parts. Likewise, scientists can study whether certain events took place, or whether various kinds of matter or energy exists — with or without plausible physical mechanisms for their occurrence. Scientists can often provide better explanations for such questions after they confirm the existence or occurrence of such an event, matter or energy, or the various attributes thereof. If Dr. Timothy Jull of the University of Arizona, who also participated in the Shroud's 1988 radiocarbon dating, can use part of their remaining C-14 samples to test a seemingly invisible reweave hypothesis, he can also donate part of their remaining Shroud sample for Cl-36 to Cl-35 testing once this technique is perfected.

Of course, we're not trying to preclude any further testing of artistic or naturalistic methods, even though these methods have been tested since the beginning of the 20th century and have all failed to duplicate the Shroud's body images, their blood marks and other remarkable features. We just want to *include* testing for unprecedented events that are consistent with very attested historical accounts and are also indicated by extensive scientific, medical and archaeological evidence.

Countless experiments in natural laboratories have also been conducted long before and after the time of Christ. As we have read, millions of people have been bloodied and/or buried under a variety of circumstances and covered with shrouds, blankets, sheets, shirts, jackets, soldiers' uniforms, bandages etc.; yet none have left any images approaching the full-length, frontal and dorsal images on the Shroud or their 130 corresponding blood marks. If one insists that all of the Shroud's mutually exclusive image features could *only* have occurred naturally or artistically, and that radiation or particle radiation emanating from the body wrapped within it cannot also be scientifically tested, despite *all* the evidence indicating such an event occurred — then this reflects the thinking of a naturalist (or possibly a humanist), but it does not reflect the mind of a scientist.

Over 150,000 people die *every day*. They have the right to know whether there is unfakable scientific, medical and archaeological evidence that particle radiation emanated from the dead body of Jesus Christ as it disappeared — and if this event occurred after he had been beaten, scourged, crowned with thorns, carried a heavy rough object across his shoulders, was crucified, killed and buried — all according to the detailed descriptions and circumstances in the most accurate and attested sources of ancient history.

There are approximately 7.3 billion people in the world, but that number is just a snapshot of those alive today. More than 350,000 people are born every day. This is a turnover of over 500,000 people every day.[8] If you include all of those who will die along with all of those born in the generations to come, the total number of people who are *certain* to die over the course of time will be so much greater than the original 7.3 billion number. Billions upon billions of people could be fundamentally and eternally affected by this new and unprecedented evidence.

One way or the other, every single one of us will literally test Jesus' promise in full when we die. If it is possible, why not scientifically investigate the critical part of this promise while we are still alive on earth? If new scientific evidence could be obtained that related to the authenticity of Jesus' burial cloth and his physical resurrection, then it would be relevant to every person who will *ever* live, and thus die. If God and Jesus Christ left extensive evidence of Jesus' physical suffering, crucifixion, death, burial and resurrection in recorded history, then corroborating scientific and medical evidence of the same events could also have been left on Jesus' burial cloth. Since an extensive and surprising amount of scientific and medical evidence has already been acquired from earlier *accidental* discoveries and *limited* inquiries of the unique body images and blood marks on the Shroud of Turin, then we should obviously continue to investigate and acquire all additional evidence that may be available to confirm the occurrence of all of these events.[9]

We have *everything* to gain and nothing to lose by developing atomic and molecular testing techniques and uniquely applying them to linen, blood and charred material from the Shroud and to limestone from Jesus' reputed burial tomb(s). If these test results conclude that the Shroud is from medieval times (as its C-14 dating indicates), then we are no worse off than we are now. But, if this evidence confirms that the passion, crucifixion, death, burial and resurrection of Jesus Christ were actual events in history, then its implications are mind boggling.

Very, very few scientists today even grasp the extent or the uniqueness of the total evidence derived from the Shroud. Because of the C-14 dating results, there has been no real need for scientists to be concerned with the Shroud. Some also avoid the Shroud thinking that there could be some vague conflict with the natural laws of science or between science and religion, which they prefer to evade.

If the Shroud is medieval, there is no real need to be concerned with it. Conversely, if the Shroud is Jesus' burial cloth there is every need for scientists (and every human being) to be concerned with the unfakable evidence that has been and can be derived from this cloth and the critical events that happened to Jesus. The ultimate goal of science, of course, is truth. If a miraculous event occurred to the dead body of Jesus Christ, it would not change the physical laws of science. It would just mean that God can perform physical events that humans cannot. However, proof of these events would be the closest that humans on earth have ever come to ultimate or eternal truth.

Science knows that physical bodies go into rigor mortis and then decompose after death. Scientists do not have to worry that the natural physical laws are going to change in this regard, even if God miraculously intervened on this one occasion and resurrected Jesus from the dead. The uniqueness of this event should give scientists an incentive to discover whether such a miraculous event occurred. Scientists should have even more motivation (or even curiosity) to investigate in light of the existing evidence that radiation emanated from the dead body in the Shroud as it disappeared, encoding the most unique, unforgeable images in history. They should also want to investigate the similar sequence of critical events that happened to Jesus and the man in the Shroud, under all the same circumstances, as recorded in the most attested and reliable sources of ancient history.

These scientists would have further incentive to investigate such a miraculous event since these same sources record that on numerous occasions Jesus predicted God would resurrect him from death after this very sequence of events occurred to him. Since these same sources record that Jesus raised people from the dead and said that anyone who believed in him and his resurrection could also have life after death, then even further incentive exists to investigate whether there is tangible *physical* evidence of the passion, crucifixion, death, burial and resurrection of the historical Jesus Christ.

While we know what happens to our physical bodies after death, science has not conducted any experiments or tested whether a person's soul continues to exist. (Think of a soul as an individual's mind, if you will, but without a physical body.) A person on earth who acknowledges that this sequence of events occurred to Jesus Christ can acquire lasting benefits from them. The Gospels and New Testament consistently state that *because* this sequence of events occurred to Jesus, we, too, can have eternal life, despite our sins. We are saved from the consequences of death *because* Jesus' own suffering and death paid for our sins, allowing us this opportunity for eternal life, if we choose. It is a free gift that has already been paid for. If our souls continued to exist in heaven with Jesus Christ, it would not violate any principles of science any more than it would violate the principles of finance, and would be consistent with the most authenticated sources of history.

After his resurrection, Jesus demonstrated that he had a real physical body. He walked, talked, ate with the apostles and allowed them to touch his body. He could do all the physical things that he did before his death, but could also suddenly disappear and reappear as a matter of will. Those who have been saved will *also* get new physical bodies on the Day of Judgment, as will Christians who are living at the time. Jesus' power or ability to suddenly disappear and appear does not alter our laws of science. It merely means that only God or the Son of God has such power to do this. Whether Jesus did this by transitioning to and from an alternate dimensionality (not violating the laws of science as they are currently understood), is actually a secondary or tertiary point. The main point is that only the power of God can do these things. If the saved have such abilities after the Day of Judgment, they would have acquired this opportunity and ability from the suffering, death and resurrection, as well as the power of Jesus.*

If scientific testing confirms that a miraculous event happened to the dead body of Jesus Christ that is consistent with his resurrection (while embedding his various blood marks and encoding his full-length body images over time), this would not conflict with the Gospels or alter the laws of science. We on earth would still be totally incapable of performing such an act, even though we could all become the *beneficiaries* of this miraculous event. There is no conflict between science and God. The lack of such a conflict is actually a secondary or tertiary point when compared to the critical matters that the new and overall evidence addresses.

If this miraculous event occurred, it would clearly indicate there is a God, that he intentionally intervened in history, that Jesus Christ was the recipient of this power, and that God and Jesus Christ are the source of ultimate power or truth, as they can overcome the consequences of death and the laws of nature or science. The same historical sources that are already confirmed by extensive and unforgeable evidence, whose critical events were prophesized centuries before they occurred, and that would become the most thoroughly attested accounts in history, consistently tell us that Jesus was sent to save all of humanity. This is perhaps most succinctly stated in John 3:16-17. "For God so loved the world that he gave his only Son, that whoever believes in him should not perish but have eternal life. For God sent the Son into the world, not to condemn the world, but that the world might be saved through him."[10]

We stand at a unique moment in history. Science and mankind have an opportunity to benefit many billions of people in this century alone. In light of the inexcusable and unprofessional conduct by some in the radiocarbon community, and their aberrant, inaccurate results, science

*As to Jesus' latent power to resurrect from his own death and the source of this power, he states, "I lay down my life, that I may take it again. No one takes it from me, but I lay it down of my own accord. I have power to lay it down, and I have power to take it again; this charge I have received from my Father. (John 10:17-18)

has an extra incentive and *duty* to correctly acquire *all* the relevant radioactive ratios from the Shroud's samples and from the limestone in Jesus' reputed burial tombs. In light of the earlier actions that restricted the evidence acquired from the Shroud, science has a duty to obtain all the relevant evidence and to confirm all the events that happened to the extensively wounded, crucified dead man wrapped in this burial cloth.

While science cannot 100% absolutely guarantee what happened to this man, they can be 95-99% certain on most matters. For example, a person could say we don't have 100% absolute proof the victim was a man because his hands were folded over his groin. Yet, you can still easily conclude this was a man on the basis of his beard, mustache, lack of breasts and the contours and features of an adult male visible on the front and back of his body. A person is free to conclude that since there is not 100% proof that this is a man that they do not have to accept even this point, and they don't. Yet, from all the evidence discussed previously regarding this body, the wounds and blood marks and all of their intricate alignments, reactions, formations and flows, and the identifications of human hemoglobin, human albumin, human DNA, human immunoglobulins and human whole blood serum, it would be preposterous to claim these images and features came from a painting, statue, bas-relief, a woman or a transgender.

You can always refuse to look at the evidence or to believe it. It is your own free choice and no one can make you think a certain way. If a person is going to argue any position on any subject to another person or to the public, he has the burden of proving his point(s) with a preponderance of the evidence. This is not a bad idea for guiding an individual's own thoughts and conclusions for he alone will bear the benefits and consequences of his own thoughts and conclusions.

Scientists may never be able to duplicate all the unique, mutually exclusive body image features, 130 blood marks and off-image properties throughout both sides of a burial cloth.[11] Yet, this does not preclude them from examining the Shroud at the molecular and atomic levels, and by other new methods, and to conclude on the basis of these and all other test results that a miraculous event occurred to the dead body wrapped in the cloth. They don't have to be able to perform a miracle to conclude that one occurred based on millions of items of unfakable evidence that would exist to support this conclusion.

Particle radiation appears to have emanated from the length, width and depth of the dead body wrapped within this burial cloth as it disappeared and caused the unprecedented human body images and their numerous blood marks, along with so many other unfakable features. This process even encoded unforgeable evidence of the entire series of wounds, injuries and events that occurred to this man, all of their surrounding circumstances and the identity of the victim. Modern scientists, Shroud experts and the owners and custodians of this burial cloth, thus have the highest obligation to investigate and reveal to every human being who will ever live, and thus die – whether this miraculous event actually occurred.

Eternity — the last great objective desired by most individuals on earth — could become the next great frontier and discovery for all of humanity.

YOU CAN HELP CHANGE THE WORLD

You can not only advance the research and development of the new scientific testing that we have discussed in this book, you can play a vital role in the publication and dissemination of this information to the worldwide public.

Please send your donations to:

Test The Shroud Foundation
122 South Central Avenue
Eureka, MO 63025

OR

Make an online donation at www.testtheshroud.org

APPENDIX A

Laboratory Requirements and Procedures to Capture and Measure Natural and Radioactive Chlorine in Linen

This Appendix is not just written for chemists and scientists. It is also presented for the scientific-minded reader, enthusiast and potential donor, so that all can better understand in layman's terms what is involved in achieving the very realistic goal of testing whether the Shroud linen cloth was irradiated with neutrons and the amount of particle radiation that each sample received.

The scientific goal is to capture all the chlorine that is in the neutron irradiated linen sample and to accurately measure its Cl-36/Cl-35 ratio. Although measuring this ratio in very solid objects such as rocks has been performed before, these are not routine measurements for milligram sized samples. Furthermore, they have never been performed before on less substantial objects such as linen containing trace amounts of chlorine. Fortunately, as described below, the experimental scientists will be able to know whether all the organic and radioactive chlorine is being captured and measured from their neutron irradiated control samples.[1]

Before scientists measure the organic chlorine in linen they must first wash and remove any inorganic chloride from the linen sample and save the wash water if further testing is suggested. The reason for washing the linen is that any surface inorganic chlorine is most likely there from contamination, such as sweat or saliva. Because chlorine from a person's hands and breath will be acquired by the linen sample, masks and gloves will need to be worn during all procedures.

When scientists measure the inorganic chlorine in rocks, for example, they first grind the rock up and dissolve it with nitric acid. The chlorine is collected by adding silver nitrate to form an insoluble precipitate (a solid substance separated from a solution by a chemical reaction) consisting of silver chloride that drops to the bottom of the container. Unfortunately, the amount of chlorine in most samples is so small that not enough silver chloride forms to precipitate at the bottom. Thus, chlorine is added to increase the amount of precipitate and carry the small amount of the sample's silver chloride to the bottom. This added chlorine is called a "carrier." The amount of added chlorine relative to the chlorine in the sample must be known so that the following procedures can measure the actual Cl-36/Cl-35 ratio of the sample, free from the influence of the carrier chlorine. To accomplish this, the carrier chlorine uses chlorine that is 99.66% chlorine-35; chorine in nature is 75.76% chlorine-35 and 24.24% chlorine-37.

This silver chloride precipitate can now be measured in an accelerator mass spectrometer (AMS), which measures the Cl-36/Cl-35 ratio and the Cl-37/Cl-35 ratio of the silver chloride precipitate. From these two ratios it is possible to compute the true Cl-36/Cl-35 ratio of the linen sample. That is, the effect of the added chlorine in the carrier is removed and the true Cl-36/Cl-35 ratio of the sample can be determined accurately.

The use of the 99.66% Cl-35 carrier method will not work if the amount of chorine in the sample is too small because these small amounts of chlorine become overwhelmed by the carrier. This causes normal experimental errors to become greatly magnified as the sample size decreases. Thus, it was determined that it was not possible to obtain a reliable Cl-36/Cl-35 ratio from very small linen samples when enriched chlorine is used as a carrier. However, for the last few years, Dr. Arthur C. Lind has recommended that a bromide salt be used as a carrier instead of a chloride salt when trying to measure the Cl-36/Cl-35 ratio within a small linen sample. Using a bromide salt as a carrier instead of a chloride salt will result in a precipitate containing silver bromine and silver chloride. Since the AMS is set up to only measure chlorine, there is no need to correct for the added silver bromide in the precipitate. This eliminates the errors when correcting for the added chloride-containing carrier. An accurate Cl-36/Cl-35 ratio should be measurable from the precipitate that results from a solution in which bromine was added.[2]

Before bromine can even be added, however, a solution containing all of the chlorine from the irradiated control sample must be first obtained. This effort actually begins with the selection of a laboratory and the technicians' apparel. Many of these requirements stem from the fact that sulfur is present in air and looks like chlorine to an AMS. A laboratory in which air goes through a filter that filters out sulfur dioxide is required. A laboratory with double doors and an air filtration system by which filtered air comes into the lab, while air with sulfur dioxide goes out through the roof is required. In the small outer room, scientists would put on masks, gloves, booties and outer garments. The linen samples and flask containers are also thoroughly cleaned of all traces of sulfur or chlorine in this room before proceeding into the main laboratory.

In the main laboratory, the total chlorine content of the neutron irradiated control linen samples will be acquired by the oxygen flask combustion method. This method entails completely burning the control sample in a flask with a solution of sodium hydroxide in the bottom. If the control sample is completely burned, all of its chlorine will be captured in a gas within the flask. If the flask containing the solution of sodium hydroxide is slowly swung back and forth, or around, all of its gas will go into the solution. When the chlorine gas goes into this solution, sodium chloride will be in the solution at the bottom of the flask. (Sodium bromide would also have been added to the solution with sodium hydroxide, except it would have been added as the carrier.)

If the gas and the solution were properly mixed, none of the chlorine gas should escape when the scientist opens the container to pour the liquid into another container with a pointed bottom (centrifuge tube). When he adds silver nitrate, it will combine with the sodium bromide carrier and the sodium chloride from the linen to produce a cloudy precipitate containing both silver chloride and silver bromide. When the centrifuge tube swirls everything together, the white precipitate will be forced to the bottom of the v-shaped tube. When the remaining liquid is poured off, the precipitate can be tested in the AMS.

Dr. Lind has spent considerable time also developing a protocol that will purify the precipitate before it is submitted for measurement. He and other scientists associated with the Test the Shroud Foundation would be glad to work with any laboratories or AMS facilities in developing and perfecting these testing methods.

Since the laboratories should first non-destructively measure the indigenous chlorine within its sample;* will know the amount of neutrons that each irradiated sample received; and can ascertain the

---

*Because the chlorine content will change from one location to another within a sample, several representative locations should be measured from each sample and averaged.

exact rates that Cl-36 is created, they will know whether the Cl-36/Cl-35 ratios that are measured within their irradiated samples are accurate. If any of these measured ratios are incorrect, it could also be determined whether the laboratory was failing to capture all the chlorine from the linen sample or whether the AMS was failing to accurately measure these ratios. This could be determined by putting known amounts of Cl-36/Cl-35 within a precipitate and testing whether the AMS accurately measures it. Chemistry laboratories and AMS facilities can work together and develop these testing capabilities to perfection.

These techniques could not only be perfected for accuracy, but they could determine the minimum size linen samples that were needed to accurately measure the Cl-36/Cl-35 and C-14/C-12 ratios in irradiated control samples. These determinations will depend upon the amount of neutrons that each control sample received and their areal densities. Similar estimates of the minimum sample size at each Shroud location could also be made. Nuclear engineer Robert Rucker, who first applied Monte Carlo Neutron Particle analysis to Shroud studies, and other scientists associated with the Test the Shroud Foundation would be glad to work with any laboratories or AMS facilities in any of these ways.

APPENDIX B

Measurements and Calculations Needed to Prove the Shroud was Neutron Irradiated and When it was Irradiated.

First, it must be stated that natural nitrogen (Nitrogen-14) in linen can be converted to radiocarbon (Carbon-14) when neutrons collide with the nitrogen. Thus, if the linen is artificially neutron irradiated, then the added Carbon-14 will make a radiocarbon date measurement of the linen appear to be younger than its actual age. It has been postulated that during Christ's resurrection neutrons were emitted, which caused the Shroud to date to the middle ages. Nuclear engineer, Robert Rucker has made detailed calculations to compute the age changes as a function of location on the Shroud.[1] According to his calculations, the location at the center the man's back has the largest radiocarbon age change; dating into the future to 8400 AD! The procedure that follows describes what measurements are required to determine when the Shroud was neutron irradiated. So as not to unnecessarily remove samples from the Shroud, the procedure is broken into three steps.

Step 1. Proof of neutron irradiation. Neutron radiation would cause radiocarbon measurements of the Shroud linen from the middle of the man's back (or close to this location) to date far into the future. If the Shroud sample from this location does not date to the future, then the following steps should not be performed.

Step 2. This step assumes that the measurements performed in Step 1 date to the future. This assures that the Shroud was neutron irradiated and that the irradiation was large enough for the $^{36}Cl/^{35}Cl$ ratio to be easily measured. Additional samples should be taken from areas also calculated to date to the future. However, it is important that these samples be taken from different areas to make sure they will not all date to the same time. Each of these samples should be processed so that the $^{14}C/^{12}C$ ratio and the $^{36}Cl/^{35}Cl$ ratio can both be obtained from the same sample. The $^{36}Cl/^{35}Cl$ ratio will tell us how much the $^{14}C/^{12}C$

ratio was increased by the neutron irradiation. The measured $^{36}Cl/^{35}Cl$ ratio indicates what neutron fluence was received by the Shroud at that location, but the ratio will not prove when this neutron radiating event occurred.

Step 3. With the data from Step 2, calculations can now be made to determine the age of the Shroud and when it received the neutron irradiation. At least two samples are needed and they should not be identical; it is best if they radiocarbon date to significantly different times. From Robert Rucker's analyses of the wide range of expected radiocarbon dates to be found on the Shroud, it should be easy to obtain such samples.

To aid in developing and visualizing what is involved in making these calculations, the graph below is used. The blue curve represents the decline in radiocarbon (Carbon-14) content in the linen thread used to make the Shroud. It is assumed that the thread was made from a flax plant soon after it was harvested, so that it contained 100 percent Modern Carbon (pMC). In this analysis the slight variations in modern carbon content (caused by variations in sun activity) are neglected, so 100% is used. As time passes, the percent modern carbon decreases as the Carbon-14 decays with a 5730 year half-life.

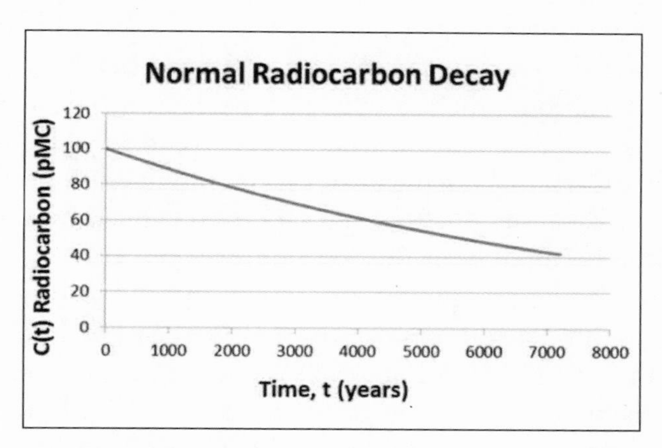

The equation for this decay is given below, where C(t) is the Carbon-14 content for this sample and t is in years.

$$C(t) = 100\% \times (1/2)^{t/5730}$$

To make this clear, this equation shows that at the time when t is 0 years, C(0 years) =100% and when t = 5730 years, C(5730 years) = 50%.

For the Shroud, we don't know for certain when the time is zero but we would like to know when it was. Since we don't know when time equals 0, we use the following equation that just relates the carbon content at any one time, t1, to a later time, t2. This equation is written as follows. This allows the radiocarbon content of a sample to be compared at two arbitrarily different times, so it can be used with samples whose time of origin is not definitely known, as is the case with the Shroud.

$$C(t2) = C(t1) \times (1/2)^{(t2-t1)/5730}$$

We know that when the flax is harvested, the radiocarbon content is very close to 100% and it begins to decay at time zero, t = 0. This flax is processed and woven into linen soon afterward, close to t = 0. The radiocarbon then decays as shown in the graph below as the blue curve. Note that while the time scale is in years, no calendar date is given, so t = 0 can be any calendar year, but it is the year that the flax was harvested and woven. At some unknown time, t1, it is suspected that the linen was neutron irradiated, which caused the radiocarbon content to increase and this is shown below by the sudden increase in red, followed by the normal decay. It seems obvious that the neutron irradiation occurred at t1 during Christ's resurrection, but some may claim it was maliciously done in modern times.

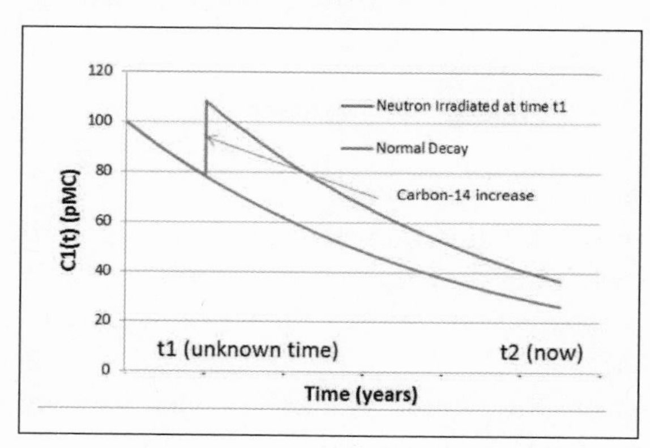

Fortunately, it is possible to determine the Carbon-14 increase at time t1. At time t1, just as the neutron radiation produced Carbon-14 from Nitrogen-14 in the linen, this same neutron radiation also produced Chlorine-36 from Chlorine-35 in the linen. Since Chlorine-36 has a half-life of 301,000 years and has a natural abundance that is practically zero, the amount of Chlorine-36 measured at time t2 is equal to the amount that was produced by the neutron radiation at time t1. From this measured amount of Chlorine-36, the amount of neutron radiation the linen received at time t1 can easily be calculated. Knowing this amount of neutron radiation at time t1, the increased amount of Carbon-14 produced at time t1 can be determined from the concentration of nitrogen in the linen. Note that the concentration will vary from place to place and this will require measurements at different locations to determine an average value and a standard deviation to determine the reliability of the final answer for t2-t1 in the last equation.

Now if we take another linen sample from a different location that received a greater amount of neutron irradiation at time t1, the graph for this sample would look like the one below.

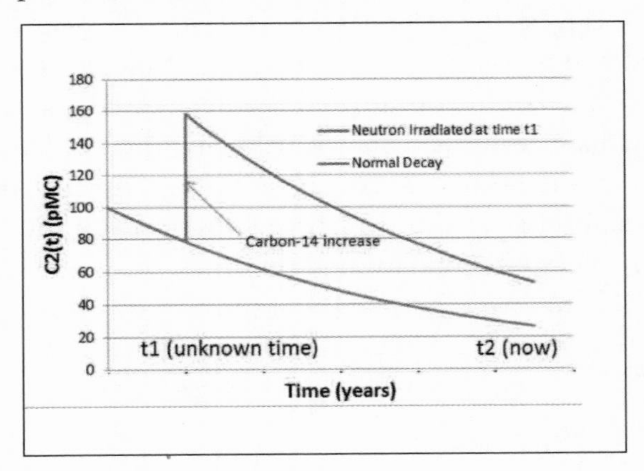

To determine the time t1, when the linen was neutron irradiated, requires that what was shown above in the two graphs for the two samples be put into equation form, as shown below. Each sample has its own equation.

The equation for Sample 1 describes what happens in the graph above for the decline of radiocarbon $C_1(t2)$ after it receives the neutron radiation at time t1.

$$C_1(t2) = [C_0 \text{ x } (1/2)^{-t1/5730} + \Delta C_1] \text{ x } (1/2)^{-(t2-t1)/5730}$$

The first two terms in the square bracket are as follows. Term 1 is C_0, the initial value of the radiocarbon content at an unknown time t1 before the radiation event multiplied by the quantity that accounts for its 5730 year half-life decline of the radiocarbon during the time t1. Term 2 is the added Carbon-14 increase, ΔC_1, at the time of the neutron radiation, t1. The sum of these two are multiplied by the quantity that accounts for the decline of the radiocarbon from time t1 to the time t2, at which time the sample is taken for the radiocarbon and Chlorine-36 analyses.

The equation for Sample 2 is identical to the equation for sample 1, except that it is taken from a different place on the Shroud that is expected to have a significantly different neutron irradiation; hence ΔC_2 is added instead of ΔC_1.

$$C_2(t2)=[C_0 \times (1/2)^{-t1/5730}+\Delta C_2] \times (1/2)^{-(t2-t1)/5730}$$

These two equations can easily be solved for the difference in times between t2 and t1. Since t2 is the present time, knowing the difference between t2 and t1 allows knowledge of time t1, the time the Shroud was neutron irradiated. The intermediate result is obtained by subtracting the two above equations and this result follows.

$$C_2(t2)-C_1(t2) = (\Delta C_2 - \Delta C_1) \times (1/2)^{(t2-t1)/5730}$$

This equation can easily be solved for t2-t1 and the result is

$$t2-t1 = [5730/\log(1/2)] \times \log[(C_2(t2)-C_1(t2))/(\Delta C_2-\Delta C_1)]$$

Remember that this measurement requires that only two samples be taken from areas of the Shroud that Robert Rucker's calculations indicate will have high radiocarbon content and will contain measurably different radiocarbon content. The Carbon-14 contents ($C_2(t2)$ and $C(t2)_1$, along with the two Chlorine-36 contents (which will provide the values of the added radiocarbon, ΔC_2 and ΔC_1), are sufficient to compute t2-t1, or the calendar date of the radiating event. If t2-t1 is close to 2000 years, it will indicate that the Shroud was neutron irradiated at the time of the resurrection.

APPENDIX C

Maillard Reaction

This hypothesis claims that during the Shroud's production its spun fibers were individually moistened with a paste of crude starch as other ancient linen supposedly received. The complete or woven Shroud was then washed in saponaria officinalis or soapwort, a soap-like solution, and then laid out to dry. When the water thus evaporated a thin layer of carbohydrates containing starch or sugar then resulted throughout the surface of the entire Shroud. Rogers claims as the corpse in the Shroud decomposed and putrefied that it gave off amines or amino acids in a gas diffusion that would have reacted with a reducing sugar as a form of non-enzymatic browning or carmelization (Maillard reaction) that resulted in the unique body images on the Shroud of Turin.[1]

There are a number of fundamental problems with this hypothesis. First of all, there is no evidence of a decomposing body on either of the Shroud's full-length body images, yet by definition the body in this image forming hypothesis would have been decomposing. Furthermore, there is no evidence or any decomposition stains at any location on the burial cloth, nor is there any evidence of sugar on the Shroud or the distribution of starch as hypothesized by Rogers. The chemical reaction within the Maillard reaction also will not occur unless the corpse's temperature is 104° F (40° C) or more.[2] Since the dead body in the Shroud first went into rigor mortis while it was in the vertical crucifixion position, it is extremely unlikely this coldblooded corpse would have been at such a high temperature when it was subsequently wrapped in the Shroud and buried in a cool tomb, as in the case of Jesus. A cool temperature will maintain rigor mortis as was the case for the man in the Shroud. Nor is this chemical reaction capable of theoretically occurring on both the inner and outer sides of the burial cloth.[3]

Even assuming every undocumented element of Mr. Rogers' hypothesis, this method would only produce stains on the cloth. This necessarily means it could not contain the three-dimensional and vertically directional information also found throughout the full-length

Shroud body image. This method would not produce negative images that contain highly-resolved, detailed images of a man when they are photographed.

All sorts of diffusion, vapograph, direct contact and various combinations of these and other naturalistic methods have been tested over the course of more than a century, but all have noticeably failed. No signs of staining are present on the Shroud's body image even at the microscopic level where capillary action and the meniscus effect is absent on its encoded fibers. At orifices like the mouth the stains would be the most visible under this decomposing reaction,[4] yet the lips of the man in the Shroud are the smallest resolved body feature on both images.

Rogers only attempted to simulate parts of his method to try to achieve a very few of the less complex features found on the Shroud's images. His results necessarily do not come close to duplicating the images found on the Shroud. To give you an idea how inherently flawed his method is, consider that thousands of Shroud body image fibers located throughout the full-length, frontal and dorsal body images are not only uniformly encoded, but they are encoded 360° around their individual cylindrical surfaces. Even using simulated methods where the underlying assumptions to Rogers' hypothesis are artificially produced, they failed to encode one individual fiber 360° around its cylindrical surface.[5]

Rogers' conclusions how the Shroud's images were formed naturally are similar to his conclusions how the Shroud was invisibly rewoven by hand. Both claims could be thoroughly tested scientifically, but he did not. Instead he just makes the claims without even thoroughly testing, let alone backing his claims up with test results. Rogers has made many similar erroneous and unsubstantiated claims about radiation.

In the case of radiation, however, he gives impressions that it could not have played any part in the Shroud's image formation. He makes very brief and broad statements with very little experimentation to substantiate his claims. What little illustrations or experiments he provides are also incomplete and sketchy with many energies and other information absent. He has also given very general impressions and statements about the effects of radiation that may be applicable to high energy, but not to lower energy radiation.

In a 2005 paper, Rogers illustrates recent microscopic effects of neutron radiation on Egyptian linen cloth and claims to compare these effects to those on the Shroud, yet he doesn't provide comparative illustrations from the Shroud.[6] Moreover, if the Shroud was irradiated with neutrons, it would have happened when the cloth was *new*. Its present microscopic effects would now be 2000 years old. The Egyptian linen was irradiated *after* it was thousands of years old. This approach is backwards. Rogers attempts to rule out all forms of energetic radiation and disparages hypotheses involving it. Yet, he didn't even attempt to irradiate modern linen with lower amounts of energy while observing the microscopic, as well as a variety of other effects at various stages, as the linen artificially aged. The energy, intensity and even the duration of the particle or other accompanying radiation that the Shroud may have received during the image encoding event may never be known for certain. The actual types or the extent of its exposure to air and the corresponding aging effects on the cloth at its various locations throughout its actual history will also never be known for certain. Yet, irradiating modern linen and observing its microscopic and a wide variety of effects as it artificially ages, at least, displays an attempt to understand or investigate the leading radiation hypotheses as they are presented.

Rogers' 2005 paper claims to investigate "the effects produced by different kinds of radiation interacting with flax fibers."[7] Like his unsubstantiated paper claiming the Shroud was invisibly repaired in medieval times, this paper was also published near the time of his death in 2005. In the last sentence of the abstract, Rogers claims that the Shroud image "could not have involved energetic radiation of any kind; photons, electrons, protons, alpha particles, and/or neutrons." Yet, his documentation, illustrations and reasoning do not support these conclusions. In fact, the last sentence of Rogers' paper acknowledges that his conclusions in the abstract were not established and were actually the goals of his limited investigation: "I *believe* that the current evidence *suggests* that all radiation-based hypotheses for image formation *will ultimately* be rejected."[8] (emphasis added)

Rogers also ignored or failed to realize that lower energy radiation is the only agent that can account for the presence or the transference of most, if not all of the body image features from a corpse onto this burial cloth. Rarely does his scientific work with Shroud fibers and threads contain elemental analysis, but instead consists of his own subjective observations seen through a microscope to support broad or incorrect conclusions. In the above and other ways, Rogers also violates the "Scientific Method" that he began asserting others should apply to Shroud research following his brief return to this research two decades after he departed from it.[9]

Many people think that because Rogers was part of STURP that he must have made valuable contributions toward the study of the Shroud and thus give his views credibility without studying them. Yet, they don't realize that it was Ray Rogers who, without authority or permission from any other member of STURP, gave their Shroud samples to Walter McCrone. McCrone was the microscopist who merely examined the Shroud's fibers under a microscope and erroneously concluded its body images had been painted and that paint products had been added to the blood marks. McCrone scooped the world press for a couple of years until several STURP scientists showed up unexpectedly at his home and retrieved their samples (which were returned in horrible condition), and subjected them to elemental and other analyses. McCrone's painting conclusion would not be publicly refuted until STURP's comprehensive, peer reviewed, scientific papers were published beginning in the early 1980s.

Many people also mistakenly referred to or thought of Rogers as Dr. Raymond Rogers. From his first involvement with STURP in the 1970s until his death in 2005, Ray Rogers either held himself out as having a doctorate and/or allowed others to do so without correcting these published falsehoods. However, he never earned a doctorate or Ph.D. degree at any time in his life.

APPENDIX D

Powder Daubing/Pigment Rubbing Techniques

Professor Luigi Garlaschelli has improved a method first advocated by Joe Nickell in which he rubs a powder or pigment mixture within a small sack onto a cloth that has been laid over a human torso (and over a bas-relief for the face). However, only the most prominent features such as the upper parts of the legs, knees, arms, elbows and hands were rubbed. If all the parts of the body and face were rubbed, horizontal distortions would be present on the cloth image and on all resulting images like the photographic negative or the three-dimensional relief. He then removes the cloth and fills pigment by free-hand into the body image.[1] At this point in his description he discusses how the scourge marks and bloodstains were added with a small brush by red-ochre, cinnabar and alizarin pigment suspended in a very diluted solution with water. He then heats his cloth for 3 hours at 215° C and washes it several times with distilled water.

His claims for reproducing certain Shroud properties are mixed in with his descriptions of the powder daubing and pigment rubbing techniques, so it is difficult to tell which techniques, and at what stages, the properties on his cloth are being measured or compared to those on the Shroud. For example, he seems to be saying that the fluorescent characteristics of his cloth match those of the Shroud after he has heated it, but not after he has washed it.[2] Another unanswered question was wouldn't this washing remove Garlaschelli's scourge marks and bloodstains from his cloth?

After stating the most likely explanation for the Shroud image was this rubbing technique, he states, the image was "much more visible . . . shortly after it was made."[3] Later he states "The original pigment came off during the many years of the Shroud's [medieval] history. . . ."[4] This is necessary to account for the fact the Shroud's body is not comprised of pigmentation, but consists of oxidized, dehydrated cellulose. Yet, he offers no evidence of the Shroud's image looking or being described differently during its European history. If the original pigment came off of the Shroud body image, why didn't the very diluted pigments from the blood marks also come off? The Shroud's numerous, yet faint scourge marks and its many bloodstains are still quite visible throughout the cloth today. Is Garlaschelli suggesting the unknown medieval artist washed the Shroud after he encoded his body image with this technique and then added the blood marks? If so, there should be chemical and other traces of this washing on the Shroud, as there clearly are with other water stains on the cloth. His results are better than most attempts, but the precise technique(s) are not given and the stages at which he measures its visual and other properties are not always apparent.

Shroud expert, Dr. Thibault Heimburger, notes that in Garlaschelli's first experiment with dry powdered pigment (red ochre), he attempted to mix solid acids or salts with this pigment to mimic the impurities assumed to be present in the original medieval red ochre, thereby causing the degraded cellulose on the body image. According to Heimburger, this failed when Garlaschelli found that "solid acids or salts without water do not leave any trace in the following artificial aging process."[5] Garlaschelli, therefore, tried another method with 1.2% of sulfuric acid in water mixed with a blue pigment, resulting in a semi-fluid "paste" that he applied onto the cloth. Thereafter, the cloth was artificially aged at 140° C for three hours and then washed to remove the blue pigment. According to Heimburger, "The lack of blue color after heating and washing shows that the resulting image is only due to the action of the diluted acid."[6] I would think that multi-spectral imaging and molecular microscopy could detect differences between Garlaschelli's degraded images and the Shroud's.

Heimburger doubts that dry powder, solid pigment or its impurities could have caused the oxidized, dehydrated chemical reaction found on the Shroud body images. He does not think that the chemical reaction will even occur in the solid phase. He also doubts the pigment or powder would come off all at the same time, so even if the solid phase did cause this effect, it would leave a patchy effect that occurred more in one place than another.[7] In addition, when the powder, paint or the impurities came off, they would fall into the crevices of the cloth. Some of this material or its degrading effects would be detected by multi-spectral imaging and molecular microscopy. According to Heimburger,

Garlaschelli would attempt to duplicate several Shroud features with a thin mixture of water and pigment (slurry) and some with rubbing powder, but did not combine them into a coherent method that a medieval forger could use.

Yet, regardless whether this technique is one coherent method, or whether the powdered or solid substances utilized in this method can produce oxidized, dehydrated cellulose (also consisting of conjugated carbonyls), there are still other problems with this technique.

This method will not encode all of its body image fibers uniformly nor will it encode the individual fibers 360° around their circumference. Further, the body image will not lie on the outer surface or primary cell wall of the fiber as it does with the Shroud's body image fibers.[8] The lightness or darkness of the coloring on the body images will not correlate with the number of uniformly colored fibers within these areas. Garlaschelli concedes that the body image features found at the microscopic level of the Shroud will not be imitated by his method.[9]

True three-dimensional information will not be contained on body images formed by this technique. While Garlaschelli's positive and negative images are good, his three-dimensional images do not contain a variety of relief. (It should also be noted that his three-dimensional relief was taken without the further distorting blood marks on the face.) The three-dimensional relief has either elevated or hollow areas. That is because the images were not encoded in a manner that reflected the original distances between the draped cloth and the underlying body. The intensity of the coloring on this cloth depended directly upon the rubbing technique of the dauber or the free hand application of pigments in additional areas.

For the same reasons, it would be impossible for the frontal body image to be encoded in a vertical straight-line direction from the body to the cloth by these methods. The continuous shading seen on the Shroud's facial image (containing a wider variety of white and gray) is absent from the facial images produced by Garlaschelli's method. While his images are better than many, they lack the high resolution seen on the Shroud. The noses and eyebrows are the same shade, appear connected and artificial. The hair on the head is all the same color, semicircular and looks even more artificial. More than anything the faces lack the obvious human expressions and contours immediately visible on the man in the Shroud.[10]

His full-length images, especially the one made from ochre, lacks the contours and shapes of a human body. These body images do not reflect the presence of rigor mortis, especially the dorsal body images. While it is also to his credit that he produced a dorsal image, it looks nothing like a human body. The back of the head is simply a round circle with no shape or form. The gist of his problem is stated by Garlaschelli when he acknowledges "The [Shroud] imprint itself corresponds to how a real body looks when observed frontally...."[11] However, to avoid the horizontal distortions on his cloth images, Garlaschelli had to "rub the pigment only over the more prominent [body] features."[12] He then improves the body image by adding pigment by free-hand to the cloth after its removal from the body. Yet, only a photographic process in which low energy light or radiation is emitted from a body and captured by the enveloping cloth or film will yield an imprint that "corresponds to how a real body looks."

While it is to his credit that he also attempted to encode some of the odd body image distortions, he missed some and misunderstood others. While he attempts to give an explanation for a couple of these odd features, he doesn't acknowledge that an original forger would *not* have encoded them or allowed them to remain on his final image. He also fails to acknowledge that a medieval forger would not think to encode features whose concepts wouldn't even be heard of for 500-600 years — such as photographic negativity or three-dimensional information encoded on a two-dimensional surface in a vertical straight line direction.

To his credit, he also attempted to encode bloodstains and scourge marks with pigment by using a very diluted suspension of red ochre, cinnabar and alizarin in water that was applied with a small brush.[13] These bloodstains would have none of the many human traits associated with the Shroud's real human blood marks. First of all, these blood marks would not have initially bled from, or formed and flowed on, a human being. Nor could they have subsequently transferred from the body and embedded into the enveloping cloth in this same shapes and forms as when they were on the body, even in the parts of the cloth that weren't touching the body.

Garlaschelli's scourge marks wouldn't have indented centers and upraised edges with serum surrounding borders that are visible only with photographic enlargers and UV fluorescent lighting. His method couldn't encode real human blood flows on the arms or as degraded cellulose, let alone as both. His other blood stains throughout the body images would also lack serum at their borders. Garlaschelli's medieval forger could not have known that postmortem blood and watery fluid would have flowed from the right or *opposite* side of the man's chest from where the blood marks were seen on the Shroud. Thus, he could not have encoded them on the left side so that 500 years later they and the wound would line up superbly on the man's right side on the positive images.

Furthermore, unlike the Shroud's blood marks, Garlaschelli's blood marks do not contain human hemoglobin, human albumin, human DNA, human whole blood serum and human immunoglobulins.

APPENDIX E

Corona Discharge

Although various researchers have directly or implicitly hypothesized about corona discharge since 1900,[1] its principal advocate today is Professor Giulio Fanti.[2] I've never had an ability to understand electricity in general and do not have a great grasp of corona discharge, lightning, piezoelectric effect, electric fields, etc. that are involved with this method. I can actually understand nuclear chemistry better than I can corona discharge.

I do know, however, that Robert Rucker has told me that an electrical discharge could uniformly encode fibers and travel 360° around them, like the encoding process on the Shroud's image fibers. Fanti also indicates that many other Shroud image characteristics can be duplicated on millimeter size linen material.[3] Physicist Arthur Lind explained to me that the Corona Discharge Method will not be able to encode three dimensional information correlated to the body's distance from the cloth or encode a body image with the high resolution that is found on the Shroud.[4]

Fanti acknowledges, as is the case with many other experimental efforts, that the Corona Discharge Method has some problems encoding macroscopic features found on the Shroud's body images.[5] He concedes that some image distortions are not easy to eliminate in his experimental tests and is considering whether ionic wind could play a factor.[6] Ironically, like many other methods, this method also has trouble encoding the Shroud's secondary image distortions or features caused by motion blurs. Like all methods, except the Historically Consistent Hypotheses, the Corona Discharge Method also cannot account for the Shroud's numerous and varied blood marks.

Fanti recognizes that many researchers have "reached the general conclusion that the [Shroud] image was caused by a kind of 'radiation' intended in an ample sense as a phenomenon acting at a distance."[7] After conducting a comprehensive study of most image forming hypotheses, Fanti acknowledges, "hypotheses based on radiation do explain many of the particular characteristics of the image, detected on both macroscopic and microscopic levels."[8]

As we saw in earlier chapters, I think the Historically Consistent Hypothesis can account for almost, if not all of the Shroud's macroscopic and microscopic body image features, along with its unique blood mark, secondary and off-image, features. This hypothesis asserts that particle radiation, consisting primarily of protons and neutrons, emanated from the dead man wrapped in the Shroud. Scientists claim if particle radiation was given off at the atomic level that small fractions of alpha particles and deuterium could also be given off, and perhaps gamma rays or other electromagnetic radiation. Perhaps, some kind of an electrical discharge could have also been given off from the body and is part of the solution to the Shroud's body image features.

Future testing of the Shroud at the atomic and molecular levels, along with other scientific testing, could shed dispositive light on the Shroud's body image formation process as well as an entire range of questions relative to this unprecedented and famous burial cloth.

APPENDIX F

Effects on Burial Cloths by Earthquakes

In 2014, three scientists at the Department of Structural, Geotechnical and Building Engineering, Politecnico di Torino, proposed that the Shroud's medieval radiocarbon dating and its body images could be due to neutron emission coming from the rock structure of the burial tomb during an earthquake.[1] In the abstract they state, "The authors consider the possibility that neutron emissions by earthquakes could have induced the image formation on Shroud linen fibers, t[h]rough thermal neutron capture on nitrogen nuclei, and provided a wrong radiocarbon dating due to an increment in C^{14} content."[2]

Let us first discuss the image forming part of their hypothesis. If the source of the neutrons was the rock structure of the tomb, the flux of neutrons would have been directed to the *outer* sides of the burial cloth. The authors also postulate a secondary release of protons that caused the body images. These protons would have been released from the nitrogen atoms within the cloth that captured some of the neutrons released during the earthquake. However, the release of the secondary protons would not have been limited to *only* the nitrogen atoms on the superficial fibers on the *interior* side of the cloth. To duplicate the Shroud's many body image features by proton (or any) radiation, the source of the radiation can only be the *body* wrapped within this burial cloth. The authors clearly rely on Rinaudo's Protonic Model; however, the protons (and neutrons) within this model clearly radiate *from* the body wrapped within the Shroud.

Since the neutron flux came from outside the burial cloth and the nitrogen atoms are distributed *throughout* the linen, when the small fraction of the neutrons were captured by the nitrogen in the cloth, they would be captured throughout the *thickness* of the cloth. When the very attenuating secondary protons were then released, they would be released uniformly throughout the entire linen cloth. Their release would *not* be limited to only the interior side of the cloth, let alone only the most superficial fibers of the inner side. Neither would there be a direct correlation between the various discolorations on the cloth and their distances from the body wrapped within it. Nor would the discolorations be encoded in straight-line vertical directions between the body and the inner part of the cloth. In fact, *none* of the Shroud's numerous primary and secondary body image features or properties would be encoded by this method. This method would only encode vague formless discoloration throughout the entire cloth. It would be impossible to concentrate or form the Shroud's body images on only the superficial fibers of the inner side of the Shroud by this method.

Another glaring error in the authors' paper is that they have not discussed the density of the neutrons that can be emitted in the limestone from an earthquake.[3] The authors already state that an earthquake of the 9th degree on the Richter scale would be required to produce enough protons and C-14 atoms from nitrogen atoms, yet the earthquake would have to last *at least* fifteen minutes.[4] I witnessed an earthquake once, it was over in a matter of seconds. Nitrogen is only contained in 1/10 of 1 percent of the linen to begin with. Without a discussion of the density of the neutrons that could be emitted in the limestone from an earthquake, the size of the earthquake could be much greater or would have to last even longer. Yet such an earthquake could be large enough and long enough that it would damage or destroy the Shroud, the tomb and the surrounding community.

The authors also speak about an earthquake occurring at 3:00 p.m. (the ninth hour) when Jesus died on the cross. Yet, Jesus' body would not have been wrapped and buried in the Shroud for another few hours, so this earthquake could not have caused Jesus' image or altered the radiocarbon dating of this cloth.

If an earthquake naturally caused the man's images, why haven't other earthquakes caused images on other burial shrouds or blankets, sheets, clothing, etc. in the past? If this earthquake naturally altered the Shroud's radiocarbon dating, why haven't earthquakes had similar effects on other linens or other objects? However, the authors have more problems than their failure to address these questions. One of the authors' problems is that they think they have to have a natural explanation for the production of neutrons. Twice in their paper they state,

"No plausible physical reason has been proposed so far to explain the [neutron or proton] radiation source origin."[5]

This is curious for several reasons. They ignore or do not realize the extensive evidence indicating that radiation emanating from the body of the man in the Shroud can account for all the Shroud's primary and secondary body image features and that particle radiation can also account for the Shroud's medieval radiocarbon dating, the possible coin and flower features, the skeletal features and outer side imaging, the excellent condition of the cloth, the still red color of its blood marks and so many other features. They also fail to account for the man's unique blood marks in their hypothesis and seem to think that only an artistic or naturalistic explanation can be plausible.

The authors also ignore or fail to consider the extensive evidence indicating the man in the Shroud was the historical Jesus Christ, who endured the same sequence of wounds, was crucified, died and buried under all the same circumstances. They fail to acknowledge that such a radiating event from a dead body, whether proposed by Rinaudo, Jackson or myself, would be consistent with Jesus' resurrection. The resurrection would obviously be an act of God, just as the most attested and accurate sources of ancient history clearly confirms. The resurrection would be a "plausible physical reason" to support the extensive, unfakable evidence of such a radiating event. You only need evidence of the *occurrence* of such an event. You need not duplicate such an event yourself.

Instead, the authors propose a naturalistic earthquake that couldn't possibly duplicate the Shroud's body images without violating the laws of science. Nor has such an event been documented to have naturally or consistently affected radiocarbon dating accuracy. They also cite an historical secular source that is problematic[6] for connecting their hypothesis to an inapplicable afternoon earthquake mentioned in one of the Gospels, yet fail to connect all the evidence on the Shroud with even the *possibility* of the resurrection mentioned in all four Gospels.*

---

*After I wrote this appendix, I recently learned that *Meccanica* retracted the Carpinteri paper due to a conflict of interest. I left the appendix in this book, however, in case any readers may have thought the content of the paper retained any validity.

APPENDIX G

The Sudarium of Oviedo

The Sudarium of Oviedo is a linen cloth approximately 84 x 54 centimeters (34" x 21-1/4") that is claimed to have been wrapped around Jesus' face and over the top of his head in a somewhat complicated fashion. It was supposedly first wrapped around and over Jesus' head after his death on the cross. It continued to stay on his head while his body laid face down temporarily in the horizontal position at the foot of the cross. In these vertical and horizontal positions, the cloth supposedly acquired extensive blood stains from the front and back of the man's head, along with pulmonary edema fluid that came out of Jesus' mouth and nose. The sudarium was not removed from Jesus' face and head until he was then laid in his burial shroud. It was then rolled or folded and placed by itself, probably on one of the side benches inside the tomb.

Such a napkin or sudarium is specifically mentioned in John 20:7, when Peter entered Jesus' burial tomb and "saw the linen cloths lying, and the napkin, which had been on [Jesus'] head, not lying with the linen cloths but rolled in a place by itself."[1] The reader should recall that all the synoptic Gospels (Matthew, Mark and Luke) specifically state that Jesus was buried in a sindon or a shroud. John used the word "othonia" or burial cloths in general of all sizes and shapes (see Appendix C of *The Resurrection of the Shroud*). For the purposes of this appendix, the question is not so much whether Peter and John saw a napkin size cloth or a sudarium, but whether the Sudarium of Oviedo is the same napkin that they saw.

The reader should also understand something that I didn't understand when I wrote my first book in 2000. The burial cloths of any size and shape mentioned in John 19:40 or 20:6,7 could be referring to the Shroud and one *or two* head bands or cloths that had been placed over Jesus' head. After the sudarium was taken from Jesus' head and rolled or folded in its own place, Jesus was then laid in a shroud and prepared for burial. The positive images and the three-dimensional reliefs of the man in the Shroud both indicate that a chin band appears to have been wrapped vertically over the man's head and under his chin at the time his images were encoded. This would have been permitted as it would have held the jaw closed (which the horizontally wrapped Sudarium of Oviedo would not have done). This chin band would have remained within the larger enveloped Shroud after the body disappeared and the image encoding process ended. As a result, Peter and John would not have seen this head band when they went into the tomb.

The Sudarium of Oviedo is much bigger than a head band and would have covered the front, back and top of a man's head, as well as his neck.[2] None of the Gospels speak of a cloth that covered Jesus' face and head following his death on the cross and until he was laid in his burial shroud, but some kind of temporary postmortem covering would fit well with John 20:6,7. Like the Shroud, the blood on the Sudarium of Oviedo has also been reported to be human blood of the group AB.[3] Pollen grains from the plant *Gundelia tournefertii* have also been found on Shroud of Turin and the Sudarium of Oviedo.[4] These are insect pollinated species of a thorny plant that grows only in Israel and parts of the Middle East. Furthermore, no pollen has been found on the Sudarium connecting it to Constantinople, France, Italy or Europe.[5] Recent X-ray fluorescence studies performed on the Sudarium indicate that, like the Shroud, dirt is found on the part of the cloth that covered the man's nose and the dirt appears to resemble the calcium found in the limestone in Jerusalem.[6] The Sudarium of Oviedo, however, doesn't begin to have as extensive, unique or unforgeable evidence as the Shroud of Turin.

Like the Shroud, the same series of tests at the atomic and molecular levels could and should be conducted on the Sudarium of Oviedo. Multi-spectral imaging and molecular microscopy could indicate, among other things, whether additional pollens or limestone are on the Sudarium cloth or its blood stains. Fifteen or more scientific tests have been performed on the Shroud's blood. These same tests should be performed, if possible, on some of the very abundant blood on the Sudarium to see if they match the blood results from the Shroud. New DNA testing should be performed on the blood from both cloths.

The plain woven Sudarium was reported to have been radiocarbon dated to A.D. 679 and A.D. 710 by two different laboratories in the 1990s.[7] However, if the Sudarium was folded or rolled in its own place in Jesus' burial tomb, say on one of the side benches, then such a date would actually be *consistent* with the various radiocarbon dates predicted by MCNP analysis.[8] While neutrons would have ricocheted within the entire tomb, more would have landed on the Shroud than the Sudarium. While both cloths would appear to be younger than their actual ages, every part of the Shroud would appear younger because every part of it would have received more neutrons than any part of the Sudarium. The same atomic tests recommended to be performed on cloth and blood samples from the Shroud should be performed on cloth and blood samples from the Sudarium.

Chlorine-36 to Chlorine-35 ratios should first be measured from its cloth and blood samples. These ratios would tell if the Sudarium or its blood had been irradiated with neutrons and the number of neutrons that each sample received. If the Cl-36 atoms in the Sudarium are above their natural infinitesimal levels, this would allow scientists to calculate how many new C-14 atoms were created within its cloth and blood samples. Subsequent C-14 to C-12 measurements of these samples would reveal how many new C-14 atoms remained from the neutron radiating event, and from the time when the flax was first harvested and the blood originally flowed onto the Sudarium.

If *all* the procedures and calculations discussed earlier in this book are followed, scientists could determine when the neutron radiating event occurred to the Sudarium, its actual age and whether it was present when a miraculous radiating event occurred to Jesus' dead body while it was wrapped in a shroud and laid in his reputed burial tomb.

APPENDIX H

The Disappearance of Jesus' Body

By Robert A. Rucker, July 1, 2015

The importance of God's existence cannot be overstated; from its emphasis in "The Great Books of the Western World" to its foundational element in the rise of western civilization to the foundational documents of United States of America. When the United States' "Declaration of Independence" refers to "Nature's God" and says that "all men are created ... by their Creator" and that the authors of the declaration were "appealing to the Supreme Judge of the World ... with a firm Reliance on the Protection of divine Providence", they were not referring to the God of the Muslims, the Hindus, or any other non-Biblical religion. They were referring to the God of the Bible. This is made clear in our Constitution in the last paragraph in the dating of the document. There it says "in the Year of our Lord one thousand seven hundred and Eighty seven". They dated the Constitution by giving the number of years since Jesus' birth, and calling Him "our Lord". The word "Lord" refers to God in the Old Testament, and when applied to Jesus refers to Jesus being God in the flesh. So in dating the Constitution, the authors referred to the deity of Jesus Christ, and then signed their names under it. But our belief in the deity of Jesus is predicated upon the truth of Jesus' resurrection from the dead. For as the Apostle Paul said "If Christ has not been raised, our preaching is useless and so is your faith." (1 Cor. 15:14) So logically, if Jesus was not raised from the dead, then our common morality built on the Bible has no basis in truth, our legal system has no basis in absolute morality, and the foundation of the United States is false and thus illegitimate. It all depends on whether Jesus was actually raised from the dead.

Unfortunately, in our culture, there are many that attack the resurrection of Christ; attempting to convince people that it never happened – that it is not a true historical event. This attack upon Christ's resurrection usually argues that Christ' resurrection violates the laws of science and is therefore impossible. But the "laws of science" are not static things. The laws of science, as they are currently defined, have a long history of development. And our understanding of the laws of science will undoubtedly change in the future as we learn more about the mysteries all around us. Part of the problem is in defining a "miracle" as a "violation of natural law", coupled with the belief that the laws of science are known with absolute certainty and perfection so that they can never change, so that a miracle in this sense is a logical impossibility. In this way, the skeptic does not need to prove that God does not exist; rather, the skeptic merely defines God out of existence. Defining a "miracle" as a "violation of natural law, as we currently understand science" would perhaps be better. But a Christian definition of a miracle would be "An event caused by God that causes awe in the beholder." This definition of a miracle allows for God's existence and for His working in the world. The Christian concept of the laws of nature has always been that God, as the creator and sustainer of the universe, operates on a much higher plane than we do, so that He is able to do what He wants to, without regard to the status of our understanding of the laws of nature, restricted only by His own character and the rules of logic.

But after the importance of Jesus' resurrection is discussed, and the possibility of miracles is discussed, the same question remains: was Jesus actually raised from the dead? The evidence from Scripture clearly affirms that He did rise from the dead, and so in most people's minds the question becomes how could this have happened? The discussion below focuses on only one aspect of this issue, and that is how Jesus' body disappeared from the tomb in the resurrection. This issue will be discussed based on three areas of investigation: 1) the Biblical basis for the nature of Christ's resurrection from the tomb, 2) the theological basis for resurrection throughout the Bible, and 3) the physical basis for how His body could have disappeared from the tomb in the resurrection. One consideration utilized in this third section is the belief that the Shroud of Turin is the authentic burial cloth of Jesus.

Biblical Basis

When Jesus was crucified, a group of believing women was standing near the cross including Jesus' mother Mary (John 19:25-27). Standing near them, near the cross, was the disciple whom Jesus loved, who is best identified as John (John 19:26-27, 35, Ref. 1). The rest of His followers, men as well as women, were standing off watching from a distance (Matt. 27:55, Luke 23:49). Of those who would later be called apostles, apparently only John was standing near the cross.

After Jesus died on the cross, Joseph of Arimathea went to Pilate to ask for the body of Jesus, which Pilate granted. Joseph then "took the body and wrapped it in a clean linen cloth, and laid it in his own new tomb" (Matt. 27:57-60, Mark 15:43-46, Luke 23:50-53, John 19:38). He could not have done these things alone. Nicodemus helped with the burial and he brought a mixture of myrrh and aloes to apply to the body (John 19:39-40). Having stood near the cross during the crucifixion, it seems very natural for John to also have been involved in the burial of Jesus in the tomb. If John was present in the tomb, it is likely that he was the only apostle to be there since the other apostles had fled or were maintaining their distance for fear of the authorities. The role of the women was evidently to watch the men take the body down, transport it from the cross to the tomb, take the body into the tomb, and do the burial (Matt. 27:61, Mark 15:47, Luke 23:55).

In Jewish culture, there were two phases involved in burial of a dead body. The first phase, or primary burial, consisted of wrapping the body in a sheet, placing it on a level area, or shelf, cut into a tomb, and leaving it there until the flesh has rotted away. After the flesh has rotted away, the secondary burial takes place. This involves someone going back into the tomb to place the remaining bones in a stone ossuary, or bone box. This stone box containing the bones is then placed into a niche cut into the wall inside the tomb. The Biblical description of the burial of Jesus' body refers to only the primary burial. There was no secondary burial of Jesus' body due to His resurrection.

After the resurrection, when John and Peter ran to the tomb, John arrived first (John 20:3-5). He bent down outside the tomb to see the linen wrappings inside the tomb. When Peter arrived at the tomb, he went in and saw the linen wrappings that had been around Jesus' body. He also saw the face cloth that had been around Jesus' head lying by itself (John 20:6-7). Then John went into the tomb and he also observed the linen wrappings and the face cloth. John says that when he went inside, "He saw and believed" (John 20:8). What did he believe? The other three times that the word "believe" is used in this chapter (John 20:25, 29, 31) all refer to believing in the bodily resurrection of Jesus. It is certainly best to understand that in John 20:8 John believed that Jesus had risen from the dead, for the context of the following verse makes it clear that John came to believe in the resurrection of Jesus' body even though he still did not understand that the prophesies in the Old Testament required that the Messiah "must rise again from the dead" (John 20:9).

But what did he see in the tomb that convinced him that Jesus had risen from the dead? What he saw in the tomb was the linen wrappings and the face cloth, but what was it about these items that allowed him to logically deduce that Jesus had risen from the dead? Though it is not stated in Scripture, a reasonable explanation can be made for what John saw regarding the linen wrappings that caused him to believe in Jesus' resurrection. John evidently knew how the linen was wrapped around Jesus' body, either from being in the tomb and watching others place the shroud around His body, or from wrapping the shroud around the body himself, or if he was not in the tomb, then from his knowledge of Jewish burial practices. So what John saw was that the linen wrappings had not been disturbed from what they ought to have been, except that there was no longer a body inside the shroud to support it. The linen wrappings had probably collapsed to a large degree, but not totally. The outline of Jesus' body might still have been present on the linen. No one had moved the linen wrappings back to steal or move the body. Jesus had not pushed the linen wrappings back to allow him to get up and walk out of the tomb. When John saw the linen wrappings in their undisturbed partially collapsed condition probably with the outline of His body still present to a degree, he realized that no one had touched the linen wrappings. Jesus' body had simply disappeared from within the shroud. This led John to logically deduce that Jesus' body could not have been moved or stolen, but rather Jesus must have resurrected from the dead.

John came to believe in Jesus' resurrection based on what he saw in the tomb rather than Old Testament prophecies (John 20:9) or Jesus' predictions that He would be raised from the dead (Matt. 12:39-40, 16:4, 21, 17:23, 20:19, 26:32, 27: 40, 63, Mark 8:31, 9:9, Luke 9:22, 24:6-7, John 2:19-22) which the disciples did not yet understand. Initially, when Jesus had made these predictions of His death and resurrection, the disciples had not believed them in a literal sense because they expected the Messiah to be a conquering hero who would rescue Israel from her enemies, i.e. Rome. Jesus often taught in parables and other forms of figurative language, so His disciples hesitated to believe Jesus' predictions about His resurrection literally (Mark 9:9-10). They did not realize that Messiah coming as a conquering hero was to be fulfilled in Christ's second coming. Christ's first coming was to fulfill His role as the Lamb of God, the suffering servant who dies to pay for our sin as prophesied in Isaiah 52:13 to 53:12.

What do the commentaries say about the interpretation that it was the undisturbed collapsed condition of the burial shroud that caused John to believe that Jesus must have resurrected from the dead? Many commentaries do not discuss exactly what John saw that caused him to believe, but those that comment on it take this interpretation. Edwin A. Blum, professor at Dallas Theological Seminary, in "The Bible Knowledge Commentary" says that "John perceived that the missing body and the position of the grave clothes was not due to a robbery. He realized that Jesus had risen from the dead and had gone through the grave clothes." (Ref. 2) Merrill C. Tenney, past dean of the graduate school at Wheaton College, in "The Expositor's Bible Commentary" says that "The disciple saw the meaning of the empty graveclothes and 'believed.' The unique phenomenon of the graveclothes looking as if the body were in them when no body was there undoubtedly recalled Jesus' previous words (cf. John 2:22, 11:25, 16:22)." (Ref. 3) Warren W. Wiersbe, former pastor and prolific author of Christian literature, in his "Bible Exposition Commentary" wrote "What did they see in the tomb? The graveclothes lying on the stone shelf, still wrapped in the shape of the body (John 20:5-7). Jesus had passed through the graveclothes and left them behind as evidence that He was alive. They lay there like an empty cocoon. There was no sign of struggle, the graveclothes were not in disarray." (Ref. 4) John Phillips, radio Bible teacher for the Moody Bible Institute, said in his commentary on the book of John "Jesus had risen from the dead. He had risen right through the grave clothes. Of course! All the clues pointed to that conclusion. Then and there he believed. It was incredibly, gloriously true. Jesus was alive!" (Ref. 5) And William Barclay, Professor of Divinity and Biblical Criticism at the University of Glasgow, in his "Daily Study Bible Series" said "Then something else struck him – the grave-clothes were not disheveled and disarranged. They were lying there still in their folds – that is what the Greek means … The whole point of the description is that the grave-clothes did not look as if they had been put off or taken off; they were lying there in their regular folds as if the body of Jesus had simply evaporated out of them. The sight suddenly penetrated to John's mind; he realized what had happened – and he believed. It was not what he had read in Scripture which convinced him that Jesus had risen; it was what he saw with his own eyes." (Ref. 6)

The New Testament has three other examples of Jesus' post-resurrection body either disappearing or reappearing in an instantaneous miraculous way. After His resurrection, on the road north out of Jerusalem going toward Emmaus, Jesus came up and started walking next to two of His lesser known disciples (Luke 24:13-36). Though He walked and talked with them for quite a while, they did not recognize Him. The exact reason is not given except that "their eyes were prevented from recognizing Him" (Luke 24:16). Of course, Jesus was the last person that they would have expected to see, since they knew that He had been crucified. Possible other naturalistic reasons for their not recognizing Him could include poor lighting as the sun was going down, bad weather such as rain or fog, the use of head coverings to protect against bad weather, or the two disciples simply looking down as they walked along the road, perhaps with their eyes filled with tears. Jesus, though still recognizable, may have looked somewhat different in His post-resurrection body, perhaps more youthful. At any rate, they did not recognize Him until they arrived in Emmaus and He reclined with them at the table. When He took the bread and blessed it in His characteristic manner, and "breaking it, He began giving it to them. And their eyes were opened and they recognized Him; and He vanished from their sight." (Luke 24:30-31, NASB, i.e. New American Standard

Bible) Perhaps they recognized him when they first saw the nail prints in His hands, i.e. wrists, when He broke the bread and offered it to them.

The Greek word translated "vanished" in Luke 24:31 in the NASB is άφαντσς (athantos). In "The NASB Interlinear Greek-English New Testament" by Alfred Marshall, his literal translation for this passage is "and He invisible became from them" so that "athantos" is literally translated as "invisible". The various other translations translate it as either "vanished" or "disappeared":

"He vanished out of their sight." (KJV, RSV)
"He vanished from their sight." (NASB, ESV, NBV, NEB, REB, Phillips, Wuest)
"He had vanished from their sight." (NJB)
"He vanished from them." (Williams, Beck)
"And He vanished (departed invisibly)." (Amplified)
"He disappeared from their sight." (NIV, TEV)
"At that moment he disappeared." (NLT, TLB)
"When they saw who he was, he disappeared." (NCV, Expanded)
"And then he disappeared." (Message)
"but He disappeared." (CEV)

The meaning of this text should be clear. He did not walk swiftly out of the room so as to vanish in the night. If John had wanted to say this, he certainly could have, but he didn't. Jesus simply vanished/disappeared while they were looking straight at Him. It is important to take this at face value in order to properly understand the nature of Jesus' resurrection body.

On two other instances, Jesus reappeared in a miraculous manner; once to the ten disciples (John 20:19, Luke 24:36-37) with Thomas missing, and once to all eleven remaining disciples (John 20:26) with Thomas present. In both cases, the disciples were gathered together in a room, perhaps the same room in which they had shared the last supper with Jesus. The doors were "shut" (John 20:19, 26, NASB) in the sense of being "locked" (John 20:19, 26, NIV, i.e. New International Version) for fear of the authorities who might be looking for them. In both cases, Jesus simply appeared in the midst of them without coming through any entrance into the room.

By looking closely at these four occurrences of the disappearance or reappearance of Jesus' body, several things can be learned.

1. His post-resurrection body had the capability to disappear or reappear without interacting with the atoms surrounding it. He did not need to move the top of the shroud in order to exit from the shroud (John 20:8). The walls and locked doors of the room where the disciples were meeting did not prevent Him from appear ing in the middle of His disciples (John 20:19, 26).
2. The disappearance or reappearance of Jesus' body could be essentially instantaneous. (Luke 24:31, John 20:19, 26)
3. They recognized him (Luke 24:31) to be Jesus. He showed them His hands and His feet (Luke 24:39) as well as His side (John 20:20, 27) where the wounds from His crucifixion could still be seen.
4. Jesus was definitely there in a physical body with flesh and bones, so that He could interact physically with His surroundings. He was not a ghost or a spirit. He told them this (Luke 24:39) and He proved it by inviting them to touch Him and by eating a piece of fish in their presence. (John 20:27, Luke 24:37-43)
5. He had continuity of memory with the time period prior to His crucifixion and resurrection, because He repeated what He had told them while He was with them prior to His death. (Luke 24:44)

The next step in understanding Jesus' disappearances and reappearances after His resurrection is through a systematic study of what the entire Bible teaches about the nature of resurrection. For this, we turn to the study of systematic theology.

Theological Basis

In this section, we will investigate the nature of Jesus' resurrection body from the larger perspective on resurrection as taught in the entire Bible. This involves definitions, a brief defense of resurrection, comparison of resurrection for the righteous verses the ungodly, the necessity of the soul continuing to exist beyond the death of the body, and application of the resurrection body of the righteous to the resurrection body of Jesus.

The reference to the "righteous" as used above is not referring to those who have never sinned, for no one is righteous in this sense (Psalm 143:2, Eccles. 7:20, Rom. 3:10) for Scripture says that all have committed sin (1 Kings 8:46, Is. 59:2, 64:6, Rom. 3:23, 1 John 1:8). The "righteous" as used here refers in a general sense to those who have had their sins forgiven by being declared righteous, i.e. credited with righteousness, by God as a result of responding in faith to God's revelation to them (Gen. 15:6, Rom. 4:3, 9, Gal. 3:6). Specifically, in the New Testament sense, the righteous are those who have been saved as a result of trusting in Christ's death on the cross to pay for their sin (John 3:16, 20:31, Rom. 3:19-28, Gal. 2:16, Eph. 2:8-10). For the purpose here, the terms "the righteous" and "the saved" will be used interchangeably.

The reference to the "ungodly", as used here, does not refer to those who have committed a certain degree of wickedness but rather refers to those that are without God in the sense of not being among the righteous as defined above. So at any point in time, every individual will either be in the class of the righteous or in the class of the ungodly, for either God will have credited righteousness to the individual or He will not have.

In Jewish culture and literature, the concept of resurrection only applies to a person's body; it is the person's body that is brought back to life (Job 19:25-27, NIV), though changed as discussed below. The concept of resurrection never applies to the person's soul. The person's soul is not what is brought back to life in the resurrection because the person's soul does not cease to exist, or go into a state of soul sleep, at the death of the body. For example, the Sadducees, when they tried to trap Jesus regarding His teaching about the resurrection (Matt. 22:23-33) based their argument on the bodily resurrection of a woman who was the wife of seven brothers, one after the other. Their argument makes no sense if the resurrection only applies to their souls coming back to life.

In this example, when the Sadducees, who rejected belief in bodily resurrection, tried to trap Jesus regarding His teaching that people would be bodily raised from the dead (Matt. 22:23-33), Jesus told them "You are in error because you do not know the Scriptures or the power of God." (Matt. 22:29) If we realized "the power of God", then we should reason as Abraham did that "God could even raise the dead" (Gen. 22:5, Heb. 11:19). Paul said that the person who believes that human beings can not be raised from the dead must logically also deny that Jesus was raised from the dead, so that "your faith is futile, you are still in your sins" (1 Cor. 15:12-19). The Bible teaches that not only has Jesus been raised from the dead, but that all will be raised from the dead for judgment by God (Matt. 16:27, Acts 10:42, Rom. 14:10-12, 2 Cor. 5:10). Both the righteous (Is. 26:19, John 6:39, 40, 44, 54, 11:23-26, Rom. 8:11, 1 Cor. 6:14, 2 Cor. 4:14, Phil. 3:10-11) and the ungodly will be raised (Dan. 12:1-2, John 5:26-29, Acts 24:15) at their appointed times (1 Cor. 15:22-23, 1 Thess. 4:14-17, Rev. 20:4-6, 11-15). For each human being, this resurrection is accomplished by each individual's soul being reunited with a resurrected form of his physical body. The purpose of this resurrection for the righteous is for the distribution of rewards for works done out of love for and faith in our savior (Dan. 12:13, Luke 14:13-14, 1 Cor. 3:11-15, 4:5). The purpose of this resurrection for the ungodly is for assigning punishment for ungodly works done out of selfish motives in rejection of our savior (Mt. 25:31-46, Rom. 2:14-16, Rev. 20:11-15).

The nature of the resurrection body, at least for the righteous, is the same as the resurrection body of Jesus. This is suggested in Rom. 8:29, 1 Cor. 15:49 and 1 John 3:2, but Paul clearly states this point in Phil. 3:20-21 by saying "… the Lord Jesus Christ, who, by the power that enables him to bring everything under his control, will transform our lowly bodies so that they will be like his glorious body." In this verse, "our lowly bodies" refers to our present bodies that are to be transformed to be like His "glorious body", which refers to Jesus' resurrection body. Paul also makes this clear by calling Jesus the "first fruits" of those who would be later resurrected (1 Cor. 15:20-23). In these verses Paul draws on a comparison that would be easily understood in their agrarian culture. If you go up to a tree and pluck off the very first fruit and taste it, you will be able to determine what all the rest of the fruit will be like from that tree because all the subsequent fruit will have the same nature as the first fruit. For example, if you determine that the first fruit is an apple, then all the rest of the fruit from that tree will also be apples. And conversely, the nature of the first fruit will be the same as the subsequent fruit. Thus, we can understand the nature of Jesus' resurrection body by understanding the nature of the resurrection bodies of the righteous that will be later resurrected. Jesus' resurrection body will have the same nature as their resurrection bodies.

Much is said in the Bible about the resurrection bodies of those who will be resurrected at Christ's coming. The phrases "post-resurrection body" and "resurrection body" are not used in Scripture. Paul uses several other adjectives to describe the bodies of those who are resurrected, but in all cases, the noun remains the same - it is still a "body". As with Jesus' resurrection body, it will still be recognizable, touchable, and able to interact with its surroundings when in this world because it is a physical body with flesh and bones (Luke 24:39), but the nature of this future resurrection body is not the same as the physical bodies that we currently have. It will undergo a basic transformation, a metamorphosis as described by the new adjectives used by Paul. Consider the central passage on this issue, 1 Cor. 15:35-54, from the NIV:

> " [35]How are the dead raised? With what kind of body will they come? ... [36]What you sow does not come to life unless it dies. [37]When you sow, you do not plant the body that will be, but just a seed ... [38]But God gives it a body as he has determined ... [42]So it will be in the resurrection of the dead. The body that is sown is perishable, it is raised imperishable; [43]it is sown in dishonor, it is raised in glory; it is sown in weakness, it is raised in power; [44]it is sown a natural body, it is raised a spiritual body. ... [46]The spiritual did not come first, but the natural, and after that the spiritual ... [49]And just as we have borne the image of the earthly man, so shall we bear the image of the heavenly man. [50]I declare to you, brothers and sisters, that flesh and blood cannot inherit the kingdom of God, nor does the perishable inherit the imperishable. [51]Listen, I tell you a mystery: We will not all sleep, but we will all be changed—[52]in a flash, in the twinkling of an eye, at the last trumpet. For the trumpet will sound, the dead will be raised imperishable, and we will be changed. [53]For the perishable must clothe itself with the imperishable, and the mortal with immortality. [54]When the perishable has been clothed with the imperishable, and the mortal with immortality, then the saying that is written will come true: "Death has been swallowed up in victory."

It should be noted that this passage was written to the believers in Corinth, so it has application to the resurrection bodies of the righteous. And in verse 51 when it says that "We shall not all sleep", it is not teaching "soul sleep" but is merely using "sleep" as a euphemism for the death of the body, because, as in sleep, the body will awaken again in the resurrection. It is saying that when Christ returns, the saved who are then alive will receive their resurrection bodies even though they are still alive. This is confirmed in 1 Thess. 4:14-17:

> "[14] For we believe that Jesus died and rose again, and so we believe that God will bring with Jesus those who have fallen asleep in him. [15]According to the Lord's word, we tell you that we who are still alive, who are left until the coming of the Lord, will certainly not precede those who have fallen asleep. [16]For the Lord himself will come down from heaven, with a loud command, with the voice of the archangel and with the trumpet call of God, and the dead in Christ will rise first. [17]After that, we who are still alive and are left will be caught up together with them in the clouds to meet the Lord in the air. And so we will be with the Lord forever.

This passage teaches that when Christ returns in His second coming, He will bring with Him the souls of those who have already died ("fallen asleep") while trusting in Him for salvation. We who are still alive on the earth at Christ's second coming will not receive our resurrection bodies before ("not precede") those who have already died ("fallen asleep"), for those who have died trusting "in Christ will rise first", i.e. their souls that Jesus brings with Him at His second coming will be reunited with their resurrected bodies first, i.e. prior to those still alive on the earth. "After that" those living believers on the earth will also receive their resurrection bodies when they are "caught up together with them" already "in the clouds to meet the Lord in the air." The context of 1 Thess. 4:14-17 is the issue of when the believers, both dead and living, will receive their resurrection bodies. In this passage, Paul tells the believers in Thessalonica that they don't have to be concerned about believers that have already died not being resurrected from the dead, in fact they will receive their resurrection bodies slightly before those who are still alive on the earth at Christ's second coming. The event referred to in 1 Thess. 4:17 when the believers that are alive when Christ returns are "caught up … to meet the Lord in the air" "in a flash, in the twinkling of an eye" (1 Cor. 15:52) is called the translation or rapture of the church.

While it is clear that the ungodly will also be resurrected at the appropriate time, the nature of the bodies received by the ungodly in their resurrection is not described in Scripture. Lewis Sperry Chafer, past president of Dallas Theological Seminary said in his 8 volume "Systematic Theology" that "With respect to the nature of the resurrection body of the unsaved in which they "stand" before the great white throne (Rev. 20:12), little may be determined. There can be no doubt about the fact of their resurrection at the time and place divinely appointed". (Ref. 7)

While it is clear that the ungodly will also be resurrected at the appropriate time, the nature of the bodies received by the ungodly in their resurrection is not described in Scripture. Lewis Sperry Chafer, past president of Dallas Theological Seminary said in his 8 volume "Systematic Theology" that "With respect to the nature of the resurrection body of the unsaved in which they "stand" before the great white throne (Rev. 20:12), little may be determined. There can be no doubt about the fact of their resurrection at the time and place divinely appointed". (Ref. 7)

For clarity, the descriptions used in the above passage (1 Cor. 15:35-54) for the resurrection bodies of the righteous are compared in the following table:

Table 1. Comparison of the Dead from 1 Cor. 15:35-54

How the Dead are Buried	How the Dead are Raised
As a seed that dies to give birth to the plant that follows	With a body that results from the body that was buried, but with its nature changed as determined by God
Perishable	Imperishable
In dishonor	In glory
In weakness	In power

As a natural body	As a spiritual body
In mortality	In immortality

According to this passage, when the righteous are raised to life in the resurrection, they will be given a body, as determined by God, that is a spiritual body as opposed to the previous natural body. This is the essential difference, and the other characteristics flow from it. Notice that the text does not say that those that are raised in the resurrection will be raised as a spirit. If it said this, it would mean that they would be raised without a body of flesh and bones according to Luke 24:39. Rather, 1 Cor. 15:35-54 affirms that they will be raised with a body, which must include flesh and bones according to Luke 24:39. But the adjective is changed from "natural" to "spiritual". This transition from a natural body to a spiritual body at the resurrection is what Paul calls "the redemption of our bodies" (Romans 8:23) so that "if the earthly tent we live in is destroyed, we have a building from God, an eternal house in heaven, not built by human hands." (2 Cor. 5:1) In this verse, the "earthly tent" refers to our present natural bodies, and the "building from God" refers to our future spiritual bodies.

According to 1 Cor. 15:35-54, when a person dies the body that is buried is described as a natural body. The adjective "natural" means that his body operates according to natural things or principles – the principles of nature. We call these principles the laws of nature, such as the laws of physics, chemistry, genetics, etc. That is why his natural body is mortal, perishable, and relatively weak and dishonorable. But when a person receives his resurrection body, the body is described as a spiritual body. So what Paul refers to as a "spiritual body" is still a physical body; having the characteristics of a physical body such as weight and volume, and it would still be able to interact with its surroundings so that, for example, it could eat food and be felt by others; but there is a basic change in the nature of the physical body because the adjective has changed from "natural" to "spiritual". The adjective "spiritual" means that his body operates according to spiritual things; where spiritual things are in the ascendency over the natural things – the laws of nature. Spiritual things can refer to God and His spiritual realm or to the spiritual aspect of a person which is called his soul. Both concepts ought to be combined. Wayne Grudem in his "Systematic Theology" (Ref. 8) explains it as follows: "Paul says that the body is raised a "spiritual" body (1 Cor. 15:44). In the Pauline epistles, the word "spiritual" (Gk. Pneumatikos) never means "nonphysical" but rather "consistent with the character and activity of the Holy Spirit" (see, for example, Rom. 1:11, 7:14, 1 Cor. 2:13, 15, 3:1, 14:37, Gal. 6:1, Eph. 5:19). The RSV translation 'It is sown a physical body, it is raised a spiritual body,' is misleading, and a more clear paraphrase would be, 'It is sown a natural body subject to the characteristics and desires of this age, and governed by its own sinful will, but it is raised a spiritual body, completely subject to the will of the Holy Spirit and responsive to the Holy Spirit's guidance.' Such a body is not at all "nonphysical" but it is a physical body raised to the degree of perfection for which God originally intended it."

This distinction between the "natural" and "spiritual" body indicates that in the resurrection, the person's soul in union with the Holy Spirit is in the ascendency over what we would recognize as normal physical limitations resulting from the laws of nature. A person's soul includes his will. This indicates that a person in his resurrection body, by an act of his will in the power and control of the Holy Spirit, will be able to overcome what would otherwise appear to be limitations that are imposed on a natural body by the laws of nature. That is why his spiritual body is immortal (2 Tim. 1:10), imperishable, powerful and glorious. The resurrection/spiritual body in theology is often also called a glorified body (John 7:39, 12:16, 23, Romans 8:30) and the process of being resurrected into a spiritual body is call glorification (Chapter 17 of Ref. 13). The phrase "glorified body" perhaps better communicates that this body still has physicality, though it has been raised to a much higher level of spiritual control so that it better reflects God's glory.

Wayne Grudem (Ref. 8) says: "The fact that our new bodies will be "imperishable" means that they will not wear out or grow old or ever be subject to any kind of sickness or disease. They will be completely healthy and strong forever. Moreover, since the gradual process of aging is part of the process by which our bodies now are subject to "corruption," it is appropriate to think that our resurrection bodies will have no sign of aging, but will have the characteristics of youthful but mature manhood or womanhood forever. ... In these resurrection bodies we will clearly see humanity as God intended it to be."

Henry C. Thiessen in his "Lectures in Systematic Theology" (Ref. 9) says: "In general it may be said that the resurrection body will not be an entirely new creation. If that were the case, it would not be the present body, but another body. But the body which is sown will be raised (1 Cor. 15:43f, 53f). Nor, on the other hand, will the resurrection body necessarily be in every detail composed of the identical particles contained in this body (1 Cor. 15:37f). All that Scripture warrants us in saying, is that the resurrection body will sustain a similar relation to the present body as the wheat in the stalk sustains to the wheat in the ground out of which it grew. An adult has the same body with which he was born, though it has undergone continual change and does not contain the same cells with which it was born. So the resurrection body will be the same body, though its make-up will be changed." This change is what transforms the natural body into the spiritual body. As the seed that is buried grows into the plant that follows, thus showing both continuity and differences, so it is with the dead body that is buried which is raised into the resurrection body that follows, for "we will all be changed" (1 Cor. 15:51). Even Job in the Old Testament (~1900 B.C.) looked forward to this change in his body (Job 14:14-15, 19:25-27).

An issue that needs to be clarified at this point is the relation between the continuity of the soul and the future resurrection of the body. Due to the pervasive influence of naturalism in our culture, people generally believe that when the body dies, the person dies. This means that there is nothing that survives the death of the body, which means that there can be no ultimate divine judgment upon the individual for sin. This destroys three important things: the possibility of justice in the universe, the possibility of an absolute morality, and the possibility of real significance in life for in the final analysis, nothing matters. So the issue of whether mankind has a soul that survives the death of the body is very important. Fortunately, the Bible is very clear that each living person is composed of a soul as well as a body, and that the soul survives the death of the body. The central passage on the survival of the soul after the death of the body is 2 Cor. 5:1-10. In the NIV, this reads:

> [1]For we know that if the earthly tent we live in is destroyed, we have a building from God, an eternal house in heaven, not built by human hands. [2]Meanwhile we groan, longing to be clothed instead with our heavenly dwelling, [3]because when we are clothed, we will not be found naked. [4]For while we are in this tent, we groan and are burdened, because we do not wish to be unclothed but to be clothed instead with our heavenly dwelling, so that what is mortal may be swallowed up by life. [5]Now the one who has fashioned us for this very purpose is God, who has given us the Spirit as a deposit, guaranteeing what is to come.
>
> [6]Therefore we are always confident and know that as long as we are at home in the body we are away from the Lord. [7]For we live by faith, not by sight. [8]We are confident, I say, and would prefer to be away from the body and at home with the Lord. [9]So we make it our goal to please him, whether we are at home in the body or away from it. [10]For we must all appear before the judgment seat of Christ, so that each of us may receive what is due us for the things done while in the body, whether good or bad."

The above passage teaches several things, but the entire passage is based on the fact that the soul continues its existence after the death of the body. In this passage, Paul is writing to the believers in the city of Corinth. It is thus for the believer that the promise is given that when the soul is "away from the body" it is then "at home with the Lord" (2 Cor. 5:6-8). Other verses in the Bible that indicate that the soul survives the death of the body include Ps. 49:15, 73:24-25, Eccl. 12:5-7, Mt. 10:28, 22:31-32, Luke 16:19-31, 23:42-43, 46, Acts 7:59, Phil. 1:21-23, and Rev. 6:9-11.

In the Bible, the term "death" does not refer to cessation of existence but only to separation; in this case separation of the soul from the body. This means that at the death of the body, the soul separates from the body and continues to exist. Millard J Erickson in his "Christian Theology" says "Life and death, according to Scripture, are not to be thought of as existence and nonexistence, but as two different states of existence. Death is simply a transition to a different mode of existence; it is not, as some tend to think, extinction." (Ref. 10) Authors Gordon Lewis and Bruce Demarest in their three volume set "Integrative Theology" say "In Scripture physical death does not mean annihilation but the separation of the spirit from the body (James 2:26). This understanding is derived not from philosophy but from teachings of prophets and apostles and preeminently Jesus Christ – who knows the other side directly. At death a Christian's unresponsive body is buried, decomposes in the grave, and returns to dust. There the body "sleeps" until it is resurrected. The spirit, however, is not sleeping or nonexistent but feels "unclothed" when the body is dismantled like a tent (2 Cor. 5:1)." (Ref. 11)

Upon death of a person's body at this point in time, depending on whether the person is "saved" or not, an individual's soul either goes to be with the Lord in what is called in theology the "intermediate state", or goes to what is called Sheol in the Old Testament or is called Hades in the New Testament to await the great white-throne judgment (Rev. 20:11-15). The intermediate state refers to the state of the soul between the death of the body and the resurrection of the body for those who are saved. Both Sheol and Hades refer to the place/state of departed spirits now occupied by the ungodly (Mt. 22:13, 25:30, 41, 46) though Sheol-Hades is usually understood to not be the same as hell. The point is that for both the righteous and the ungodly, a person's soul continues to exist after the death of his body, so is available to be reunited with a resurrected form of his body at the proper time appointed by God.

Another item that needs to be clarified is 1 Cor. 15:50, which says that "flesh and blood cannot inherit the kingdom of God." The concept is that an unchanged natural body of flesh and blood cannot inherit the kingdom of God; it must be changed from a natural body to a spiritual body, as taught in the following verses. Though the spiritual body is still a physical body made of flesh and bones that is able to interact with its surroundings, the spirit is in the ascendency over the forces of nature in order to prepare it for the kingdom of God, i.e. heaven.

With the above understanding of the Bible's overall teaching on resurrection, we can now return to the central question of the nature of Jesus' resurrection body. Since "Jesus Christ … will transform our lowly bodies so that they will be like his glorious body" (Phil. 3:21) and since Jesus' resurrection body is the "first fruits" (1 Cor. 15:20, 23) of all the saved who would be later resurrected, the nature of Jesus' resurrection body must be essentially the same as the nature of the future resurrection bodies of all the saved. This means that we can apply the right column of Table 1 to the body that Jesus had after the resurrection. Thus His resurrection body was a spiritual body so that it was immortal, imperishable, with great power and great glory. His body was not limited by the natural operation of this universe, i.e. the laws of nature, and could by His spirit, in the exercise of His will, do things that would be totally impossible for us in our natural bodies. We can not do things with our bodies that violate the laws of nature simply through our will. We can not instantaneously vanish in front of people or materialize in the middle of a crowd of people in a locked room. But Jesus did these things due to the nature of the spiritual body that He had after His resurrection.

Charles C. Ryrie (Ref. 12) summarized it this way: "Christ's resurrection body has links with His unresurrected earthly body. People recognized Him (John 20:20), the wounds inflicted by crucifixion were retained (John 20:25-29, Rev. 5:6), He had the capacity though not the need to eat (Luke 24:30-33, 41-43), He breathed on the disciples (John 20:22), and that body had flesh and bones proving that He was not merely a spirit showing itself (Luke 24:39-40). But His resurrection body was different. He could enter closed rooms without opening doors (Luke 24:36, John 20:19), He could appear and disappear at will (Luke 24:15, John 20:19), and apparently He was never limited by physical needs such as sleep or food."

What is the significance of Jesus' resurrection? John F. Walvoord (Ref. 13) answers the question this way: "The resurrection of Christ, of course, is key to all His present work in heaven. Because He is resurrected, people are justified in believing in His death on the cross as the basis of their salvation. His resurrection also guarantees and undergirds all His future works, including returning for His own in the Rapture, blessing all those who are the objects of His saving grace, resurrecting everyone in their proper order, reigning on the throne of David in the Millennium, and ultimately triumphing over the world in delivering the conquered world to God the Father. His resurrection also adds proof to the inspiration of the Bible and constitutes a tremendous fulfillment of prophecy recorded by inspiration of the Holy Spirit. All these factors concerning the resurrection make it clear that the resurrection and translation of Christians at the Rapture will be a tremendous event, an event worthy of our constant expectation. Knowledge of our future life helps to cast light on our present life goals, encouraging us to consider how our actions will be seen from the viewpoint of eternity."

There have always been people who disbelieved in the resurrection (Matt. 22:23, Acts 23:8, 1 Cor. 15:12). In response to the Sadducees, who disbelieved in the resurrection, Jesus said "You are in error because you do not know the Scriptures or the power of God" (Matt. 22:29). While this response is true and adequate to answer the Sadducees, it is not entirely adequate for dealing with the skeptic in this present-day context. Today's skeptic may phrase his objection to the resurrection of Jesus in various ways, for example:

- But this is all crazy talk.
- This is the 21st century. No one believes in that stuff any more.
- I've never seen anyone be resurrected from the dead, have you?
- Belief in the resurrection is not based on a scientific methodology, so there is no reason to believe in it.
- The resurrection is contrary to the laws of nature and is therefore impossible.
- The resurrection of Jesus, or anyone else, can not be real because there is no physical mechanism.

These objections fall into two categories: 1) an attempt to intimidate the believer through ridicule, and 2) a claim that resurrection ought to be rejected because of science. As ambassadors of Christ (2 Cor. 5:20), Jesus calls the believer to "always be prepared to give an answer to everyone who asks you to give the reason for the hope that you have … with gentleness and respect" by "speaking the truth in love" (1 Peter 3:15, Jude 1:3, Eph. 4:15). Christian scholars have responded to attacks against Biblical teachings, such as the above, so that valuable resources are available to assist the Christian in the areas of philosophy, world view analysis, apologetics, and science as it relates to the Bible (Ref. 15 to 29).

From a theological perspective, in general, God's actions are accomplished by the will of the Father, by the action of the Son, and by the power of the Holy Spirit. This applies to both creation and salvation. This also applies to Jesus' resurrection and to the future resurrection of all people. But from a natural and physical perspective, the question still remains as to how Jesus' body could have disappeared from the tomb. In the next section, several possibilities will be discussed with a focus on the best alternative.

Physical Basis

It seems appropriate to begin this section with a discussion of what science is and what distinguishes good science from poor science as it relates to the Shroud of Turin. The basic characteristics of science ought to be the following:

- Science is a search for the truth,
- Using careful observation of repeated experiments,
- With all conditions that affect the results being known and controlled,
- So that a hypothesis can be developed relating the causes to the observed effects,
- Using mathematical equations to the extent possible,
- So that predictions can be made that are testable,
- So that the hypothesis is falsifiable.

The main point of the above list is that science ought to have truth as the highest objective, so that the researcher ought to follow the evidence where it leads apart from preconceived ideas or biases. This issue becomes very important for the Shroud of Turin because it contains a front and back image of a man who appears to have been crucified exactly as the Gospels say that Jesus was crucified. Thus, if God exists so that miracles are possible, then Jesus' resurrection ought to be considered as a possible explanation for the image.

This view of science is consistent with a realization that the laws of science as we currently understand them have been a long time in development, that there will undoubtedly be many new scientific principles and laws discovered in the future, and that the laws of science in our current statement of them may only cover a subset of a much larger reality. An excellent example of this last point is Newtonian physics (classical mechanics) which was accepted for hundreds of years as being universally true, yet was proven by Einstein's theory of relativity to be true only for relatively low speeds and relatively weak gravitational fields.

According to the above definition of science, to scientifically prove that God does not exist would require careful observation of repeated experiments while knowing and controlling all conditions that affect the results of the experiments. Since God has not been scientifically proven to not exist, the unbiased researcher of the Shroud of Turin ought to assume the possibility that God could exist. The application of this is that in our investigation of the Shroud of Turin we ought not to reject the possibility of Jesus' resurrection just because it violates our current understanding of the laws of science. The researcher on the Shroud of Turin who, for whatever reason, believes that God does not exist and incorporates this belief into his scientific methodology by rejecting any possibility of a miracle, defined as a violation of the laws of science as he currently understands them, reveals that he is not being an objective scientist in that he has allowed the philosophical assumptions of his world view to become more important than a search for the truth.

The application of the above view of science to our following discussion on how Jesus' body could have physically disappeared is as follows. Since God could exist and since our current understanding of the laws of science could only be a subset of a larger reality, it will not be required that we understand the mechanism by which his body disappeared. This allows us to proceed in our consideration of the seven processes proposed below in applying our current understanding of the laws of science to judge the pros and cons of each proposed process. It is acknowledged that the discussion below does not propose an experimental program that satisfies all of the seven characteristics of science discussed above, but it hopefully makes progress in understanding the pros and cons of the seven processes proposed below for the disappearance of Jesus' body.

In the previous sections, the nature of Jesus' resurrection body has been investigated by considering the passages in the Bible that discuss Jesus' resurrection and His post-resurrection appearances, and by considering references to resurrection in general throughout scripture. It was concluded that at the resurrection, Jesus' body underwent a basic transformation, a metamorphosis, from what Paul called a natural body to a spiritual body. With this new spiritual body, Jesus could do things that we cannot do. Jesus evidently disappeared from the tomb without going through the burial shroud or the walls of the tomb. Jesus disappeared while the two disciples in Emmaus were watching Him. And on more than one occasion, He appeared in the middle of a room full of His incredulous followers without entering through a door or any other opening into the room. One second He was not there - the next second He was. We must admit that these abilities are beyond the realm of our understanding of what is physically possible. And this indicates that the process by which Jesus' body disappeared from the tomb could also be beyond our understanding of physical reality.

The above discussion forms the basis for a consideration of how His body physically disappeared from the tomb. Consider what this means; a person's body consists of various organs and tissues such as heart, liver, skin, etc. These organs and tissues consist of various types of cells. These cells consist mostly of various organic molecules called proteins. These molecules consist of various atoms, and these atoms consist of neutrons, protons, and electrons. While the constituents of the neutrons and the protons, i.e. quarks, could also be mentioned, it is sufficient to stop at the level of the atoms with their neutrons, protons, and electrons for the considerations here. So when Jesus' body disappeared from the tomb, the atoms, including the neutrons, protons, and electrons in the atoms, had to disappear from the tomb. While understanding that God is the ultimate cause of Jesus' resurrection and thus of the disappearance of Jesus' body from the tomb, it is legitimate to consider various physical mechanisms that God could have employed in the disappearance of Jesus' body from the tomb, such as the following options:

1. The molecules in Jesus' body broke into their constituent atoms which then passed through the shroud and into the walls of the tomb.

2. The atoms in Jesus' body disintegrated into their neutrons, protons, and electrons, which passed through the shroud and into the walls of the tomb.
3. The atoms in Jesus' body disintegrated, with the entire mass of His body being converted into energy – specifically electromagnetic energy such as light, ultraviolet, and X-rays. The photons of this electromagnetic energy penetrated through the shroud and into the walls of the tomb.
4. The atoms in Jesus' body disintegrated, with the entire mass of His body being converted into neutrinos and anti-neutrinos which would have penetrated through the shroud and through the walls of the tomb.
5. Jesus' body was transported out of the shroud and the tomb into some other location in this physical universe by a wormhole.
6. Jesus' body disappeared from inside the shroud by a transition into an alternate dimensionality.
7. Jesus' body disappeared by an unknown mechanism not related to any known physical phenomenon or law of physics.

In considering the above options, the known laws of physics will be used as far as possible. This approach will be used to consider the energy required or released in options 1 to 4. Beyond these limits, we will use what might be called extrapolations of the laws of physics as suggested by current considerations and models in modern physics. This approach will be used for options 5 and 6. Option 7 is beyond the extrapolations from modern physics.

Options 1 to 4 all involve disintegration of the body in some sense. The two main objections to the disintegration of the body are: 1) the large amounts of energy that would either be required as input to the body or that would be released from the body, and 2) how the multiple reappearances and disappearances of the body would take place.

The energy issue will be considered first. In option 1, sufficient energy must be input to the body to break all of the molecular bonds between the atoms to release the individual atoms which then pass through the burial shroud and into the walls of the tomb. The required energy input to the body would be on the order of the energy required to cremate a body. This amount of heat should have left an effect on the burial shroud.

And a significant fraction of the atoms from Jesus' body should have remained on the shroud. Investigation of the Shroud of Turin, which the author believes to be the authentic burial cloth of Jesus, indicates that neither of these is true.

The energy considerations for option 2 are more complex. Some background information will be needed first. In any atom, the electrons, which have a negative electrical charge (-1), circle very rapidly around the very small central mass called the nucleus. The nucleus contains all of the neutrons and protons. The protons have a positive electrical charge (+1), whereas the neutrons have no electrical charge. The mass of the proton and electron relative to the neutron are: Neutron = 1, Proton = 0.99862349, Electron = 0.00054386734, so that on the average about 99.97% of the mass of an atom is located in the nucleus.

When neutrons and protons combine to form a nucleus, as occurred at the beginning of the universe or in a super-nova, energy is released. Since according to Einstein's equation ($E = M \times C^2$) mass can be changed into energy and energy can be changed into mass, the energy released when a nucleus forms results in the nucleus weighing less than the sum of the weights of the neutrons and protons that combined to form the nucleus. Neutrons and protons in a nucleus are called nucleons. The amount that a nucleus weighs less than the sum of its nucleons is called the mass defect. To split a nucleus again into its component neutrons and protons requires that this mass defect must be overcome by putting energy into the nucleus. This energy that is required to split a nucleus into its component neutrons and protons is called the binding energy, so that the binding energy is equal to the energy equivalent of the mass defect according to $E = M \times C^2$. The binding energy that holds the neutrons and protons together in a nucleus can be calculated using Einstein's formula $E = M \times C^2$ where the energy E that is binding the neutrons and protons together in the nucleus is equal to the decrease in the mass M, i.e. the mass defect, times the speed of light C squared. In other words, to break the nucleus of an atom into separate neutrons and protons requires an amount of energy equal to the binding energy of the nucleus. This energy is not released from the atom; rather it must be put into the atom. The neutrons and protons must absorb this amount of energy in the process of being released from the nucleus as separate particles.

If this binding energy for the nucleus is calculated for each element, and the result divided by the number of nucleons (# nucleons = # protons + # neutrons) in the nucleus, and plotted as a function of the number of nucleons for each element, then the result is a plot of the "binding energy per nucleon" as shown in Fig. 1. The element with the highest binding energy per nucleon is iron, because it is at the highest point on this curve. Energy will be released in a nuclear process if the results of the nuclear process move up on this curve to higher values of the binding energy per nucleon. So if elements heavier than iron fission (break down) into lighter elements that are higher on this curve, or if elements lighter than iron fuse (combine) to produce heavier elements that are higher on this curve, then energy will be released in the process. For example, energy will be released in fission if U^{235}, with 92 protons plus 143 neutrons in its nucleus, fissions into two fission products such as Tellurium (Te^{135}) with 52 protons and 83 neutrons in its nucleus and Zirconium (Zr^{97}) with 40 protons and 57 neutrons in its nucleus, with the release of three neutrons in the process. Energy will also be released in fusion if Hydrogen (H^3 called tritium) with 1 proton and 2 neutrons in its nucleus fuses with Helium (He^4) with 2 protons and 2 neutrons in its nucleus to produce Lithium (Li^7) with 3 protons and 4 neutrons in its nucleus.

But option 2 is not related to fission of the heavy elements or fusion of the light elements, which would be represented by moving up on the curve in Figure 1, and thus releasing energy. Option 2 is the nuclei of all of the atoms being broken down into their constituent neutrons and protons. This would be represented on Figure 1 by moving from any element on the curve to the lower left hand point on the curve where the binding energy per nucleon is zero. For any element, such a process would move down on the curve and thus would require that energy be put into the nuclei to separate the neutrons and protons from each other. So this energy would be absorbed by the nuclei rather than being released from it. To break down all of the nuclei into their constituent neutrons and protons would require a huge input of energy to overcome the binding energy of all the neutrons and protons in the nuclei of all the atoms. The amount of energy can be determined by calculating the total binding energy of all the neutrons and protons in the nuclei of all the atoms in the body of Jesus. Based on the figure on the Shroud of Turin, the weight of

Jesus' body is usually estimated to be about 170 to 175 pounds. Assuming a body weight of 170 pounds and the weight fractions for the elements in the body from Ref. 30, the total energy that must be absorbed by the nuclei in the body can be calculated. The calculation is summarized in Table 2.

Figure 1. Binding Energy That Holds the Nucleons Together in the Nucleus

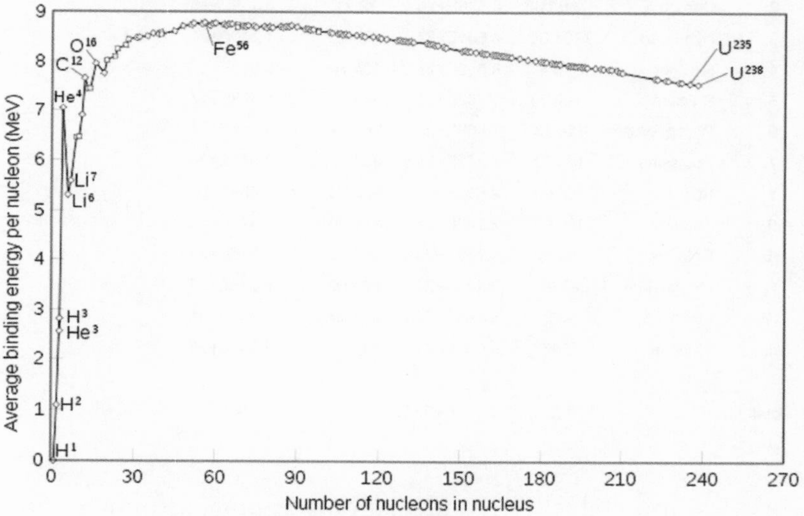

[The graph is from http://physics.ucsd.edu/do-the-math/2012/01/nuclear-fusion/]

Table 2. Calculation of Energy Required to Split All Nuclei into Neutrons & Protons

	Contents of the Body			Binding Energy,	Mev Required to Overcome
	Element	Grams	Atoms	Mev/atom	the Mass Defect
1	Oxygen	47309.89	1.781E+27	127.65	2.273E+29
2	Carbon	17603.68	8.826E+26	92.21	8.139E+28
3	Hydrogen	7701.61	4.601E+27	0.00027	1.240E+24
4	Nitrogen	1980.41	8.515E+25	104.70	8.915E+27
5	Calcium	1100.23	1.653E+25	343.17	5.673E+27
6	Phosphorus	858.18	1.669E+25	262.92	4.387E+27
7	Potassium	154.03	2.372E+24	334.93	7.946E+26
8	Sulfur	154.03	2.893E+24	272.70	7.889E+26
9	Sodium	110.02	2.882E+24	186.56	5.377E+26
10	Chlorine	104.52	1.775E+24	302.79	5.376E+26
11	Magnesium	20.90	5.180E+23	201.02	1.041E+26
12	Iron	4.62	4.983E+22	491.27	2.448E+25
13	Fluorine	2.86	9.068E+22	147.80	1.340E+25
Total		77105.00	7.394E+27		3.305E+29

As shown in Table 2, the calculation was done assuming only the most common 13 elements out of the 59 elements listed in Ref. 30. These 13 elements gave a body weight of 77105 grams (169.988 pounds) so the other 46 elements that were omitted accounted for only 0.012 pounds out of the 170 pound assumed weight of the body. The energy required to be put into the body to separate all the nuclei for these 13 elements into their neutrons and protons is calculated to be 3.305×10^{29} Mev (Million electron volts). This value is equal to 5.30×10^{16} Joules, which is equal to the energy released from 12.7 megatons of TNT. The most powerful nuclear weapon detonated by the United States was the Castle Bravo device set off on March 1, 1954. It released an energy equivalent to 15.0 megatons of TNT. In other words, to split all of the nuclei in Jesus' body into their constituent neutrons and protons would require 84% of the energy of the most powerful nuclear weapon ever detonated by the United States. The amount of energy required to remove the electrons from the atoms would be very insignificant compared to the energy required to split the nuclei into their neutrons and protons. So 5.30×10^{16} Joules would also be the energy required to split all of the atoms into their constituent parts (neutrons, protons, and electrons). This energy would not be released from the body; it would have to be put into the atoms in the body in order to split them into their neutrons, protons, and electrons. In other words, this amount of energy would be absorbed by the neutrons, protons, and electrons in the process of splitting the atoms into their constituent parts. So this energy would be absorbed rather than being given off in an explosion.

Another important objection to option 2 is the number of neutrons and protons that would be emitted from the body. In option 2, all of the atoms in the body are broken apart into their component parts so that all of the neutrons and protons in all of the atoms in the body are emitted from the body. For the estimated weight of the body of 170 pounds, there would be 2.09×10^{28} neutrons and 2.55×10^{28} protons in the body. There would also be the same number of electrons as protons. Calculations with the MCNP (Monte Carlo Neutron Particle) nuclear analysis computer code indicate that if 3.04×10^{18} neutrons are emitted homogeneously in the body, it would cause the C-14 dating for a sample near the feet of the image to be shifted from 30 AD to 1260 AD. (Ref. 31) This number of neutrons is only $3.04 \times 10^{18} / 2.09 \times 10^{28} = 1.45 \times 10^{-10}$ = 0.000000015% of the total number in the body. And from experiments with irradiation of linen with protons (Ref. 32), we know approximately how many protons must have hit the Shroud to cause the discoloration – about 1.2×10^{12} protons/cm^2. Multiplying this number by the approximate area of the front and back image on the shroud (1.1×10^4 cm^2) yields about 1.3×10^{16} protons hitting the shroud. If it is assumed that the protons are emitted homogeneously in the body, and if it is assumed that of the number of protons that reach the surface of the body to exit from it is only 1% of the total (the actual number has not yet been determined), then the number of protons emitted in the body would be $1.3 \times 10^{16} \times 100 = 1.3 \times 10^{18}$, and the fraction of the total number of protons in the body that must hit the shroud would be

$1.3 \times 10^{18} / 2.55 \times 10^{28} = 5.1 \times 10^{-11} = 0.0000000051\%$. So the evidence is very much against 100% of the neutrons and protons being released from the body. If 100% of the protons were emitted from the body it would damage the Shroud beyond recognition. And if 100% of the neutrons were emitted from the body it would have caused a dramatically larger effect on the C14 dating. And if 100% of the neutrons and protons were releasedfrom the body, when they reached the shroud and the limestone in the tomb, they would be captured by the nuclei in the atoms. This would release a small fraction, probably a few percent, of the total energy (5.30×10^{16}Joules calculated above) that was absorbed by the neutrons and protons when they were initially separated from the nuclei. But a few percent of 5.30×10^{16} Joules (12.7 megatons of TNT) is far more than enough energy to pulverize and vaporize the shroud and the tomb. So option 2 is unrealistic for multiple reasons.

Option 3 converts the entire 170 pound mass of Jesus' body into energy. If this were to happen, the amount of energy that would result can again be calculated from Einstein's equation $E = M \times C^2$, where the energy E is in Joules, the mass M is in kg, and the speed of light = 2.9979×10^8 m/s (meters per second). The estimated 170 pound weight of Jesus' body = 77.11 kg, so the energy released would be $77.11 \times (2.9979 \times 10^8)$ squared = 6.93×10^{18} Joules = the energy released from 1655 megatons of TNT, which is 110 times more energy than the largest nuclear weapon detonated by the United States. Since under the assumptions of option 3, this energy is converted into electromagnetic energy such as light, ultraviolet, and X-rays, it would be deposited in the materials around the body. With this amount of energy being released in this way, it would vaporize the shroud, the tomb, and Jerusalem, and probably kill most people in Israel. These events obviously did not happen, so option 3 must be rejected.

In option 4, the atoms in Jesus' body disintegrated, with the entire mass of His body being converted into neutrinos and anti-neutrinos. In option 3, the energy from the mass of the body was assumed to be converted into electromagnetic energy such as light, ultraviolet, and X-rays which would be deposited in the surrounding materials, and as a result would vaporize them. To avoid this problem, option 4 assumes that the energy is converted into neutrinos and anti-neutrinos. These particles are believed to have a very small mass but it is so small that as yet it has not been accurately measured. If they have mass, then they would travel at a velocity which is just below the speed of light. These particles have no charge so interact with matter to only an extremely slight extent. It is estimated that about 65 billion (6.5×10^{10}) neutrinos per second from the sun pass through every square centimeter on the earth, including through our bodies, yet we don't even notice them. The extremely low interaction probability of these particles means that the energy resulting from the disappearance of the body is not deposited in the surrounding materials, so that the shroud, tomb, and city of Jerusalem would not be vaporized. In fact, the interaction of neutrinos and anti-neutrinos with matter is so minimal, that there would probably be no noticeable effect when this huge energy release took place within the shroud.

One author that takes the view that the mass of Jesus' body was converted into neutrinos and anti-neutrinos is Frank J. Tipler, professor of mathematical physics at Tulane University. He believes (Ref. 33) that in the resurrection of Jesus, His "Dematerialization can be accomplished by electroweak quantum tunneling ... This would convert all the matter in Jesus' body into neutrinos". Tipler admits that the problem with this is the probability that all atoms in Jesus' body (about 6.7×10^{27}atoms) would undergo "electroweak quantum tunneling" simultaneously, which he defines as within the same 0.01 second, when the probability of a single atom dematerializing in this short time interval is about 10^{-100}. The probability of all atoms in Jesus' body undergoing this process within the same 0.01 second is beyond the realm of credibility, given any commonly held view of probability theory. And this is without consideration of the multiple dematerializations and rematerializations that must have happened to explain Jesus' post-resurrection appearances.

In review of the options involving disintegration of Jesus' body, option 1 should be rejected because of lack of evidence on the Shroud of Turin, option 2 should be rejected because of the huge amount of energy that must be put into the body without any evidence of it being left on the Shroud, option 3 should be rejected because the Shroud, the tomb, and Jerusalem were not vaporized, and the mechanism behind option 4 should be rejected due to probability considerations. Another sufficient reason to reject all concepts of disintegration is that it leaves you with no body. If the body disintegrates, then it no longer exists. So to explain Jesus' appearances after the resurrection, of which there are at least 10 (Mt. 28:9-10, 16-20, Luke 24:13-32, 36-49, John 20:11-21:25, Acts 1:3-12, 1 Cor. 15:3-7), the body has to be either recreated out of new atoms for each appearance, or it has to be recomposed from the original materials. This would involve bringing the atoms back together that had scattered into the shroud and the limestone walls of the tomb (option 1), or bringing the neutrons, protons, and electrons back together (option 2), or bringing the photons of light, UV, and X-rays back together (option 3), or bringing the neutrinos and anti-neutrinos back together (option 4). From this consideration, it should be obvious that a far preferable option would be for Jesus' body to continuously exist after the resurrection so that it doesn't need to be recreated or recomposed for each appearance after the resurrection. Two options have been suggested to satisfy this requirement: a wormhole (option 5) and an alternate dimensionality (option 6).

In option 5, Jesus' body was transported out of the shroud into some other location in this physical universe by a wormhole. The term "wormhole" is used for a topological curving of our four dimensional reality (3 space dimensions + 1 time dimension) known as space-time in Einstein's relativity theory. There is some theoretical basis for believing that a wormhole could result when the topological curving in space-time is sufficient to construct a passageway (tunnel, shortcut) from one point in space-time to another point in space-time.

The concept of a wormhole is readily recognized by many in our culture through Star Trek and other science fiction TV shows and movies. Wormholes are sometimes used in such science fiction programs to create new possibilities for the plot in the program, and because a wormhole is considered to be a theoretical possibility by physicists based on Einstein's equations on general relativity. In our current understanding of physics, it is not possible for anything to travel faster than the speed of light in a vacuum. But in Star Trek, the Enterprise is able to travel many orders of magnitude faster than the speed of light because the ship has what is called warp drive. Powered by annihilation of matter and anti-matter in the warp drive engines on the Enterprise, the warp drive creates a bubble of normal space-time immediately around the Enterprise, but outside this bubble creates a curvature (bending, warp) in the fabric of space-time, i.e. in the space-time field, which moves with the Enterprise, thus allowing the Enterprise to exceed the speed of light and sometimes even travel through time. In contrast to warp drive on the fictional Enterprise, according to relativity theorists a wormhole could be a naturally occurring phenomenon and thus a self-sustaining warp in the fabric of space-time. A sufficiently strong warping in the fabric of space-time to create a wormhole is normally associated with a black hole, but there are evidently other theoretical possibilities to create a wormhole. In this discussion it should be emphasized that there is no experimental evidence for the existence of any wormhole, and that things don't have to happen or even be physically possible just because they are mathematical solutions to equations in certain theories in physics and mathematics. For example, 2 – 3 = -1, but if I have two apples on the table, I can't take three apples from them.

So in option 5, a wormhole would have to be created inside the Shroud as it covers the body inside the tomb. The wormhole would have to transport the body to another location and/or time in this physical universe, according to current theory. The wormhole would have to transport the body back for the first post-resurrection appearance and then away again when the appearance is over. This process would have to repeat itself for each of the ten post-resurrection appearances. And each time, there would have to be nothing else transported in either direction. For example, when the wormhole initially transports the body out of the shroud, it would have to transport no part of the shroud away, because we still have the entire shroud as the Shroud of Turin. And when the wormhole transports Jesus' body into the midst of His disciples in the upper room, His disciples are not injured in any way by the wormhole and there is no indication of anything else being deposited in the room along with Jesus' body. A wormhole occurring as a natural phenomenon would not behave this way, and a wormhole resulting from a black hole would have destroyed anything that falls into it, including Jesus' body. As with disintegration, this would again leave us with no body to return for the post-resurrection appearances. Even if God is appropriated as creating the wormhole each time, the complexity of option 5 and the many miracles that would be required to make it work argues strongly against it.

In option 6, Jesus' body disappeared from inside the shroud by a transition into an alternate dimensionality. Some background is necessary at this point to help the reader understand the concepts being discussed. The reality in which we exist appears to consist of three space dimensions and one time dimension. The three space dimensions can be visualized in a rectangular room by two walls and the floor. These three surfaces meet in a point in the lower corner of the room. This point in the lower corner of the room can be designated as the origin, or starting point, for what is called a coordinate system. Coming out from this point is three lines where the two walls and the floor intersect each other. Where one wall and the floor intersect can be called the x-axis. Where the other wall and floor intersect can be called the y-axis. And where the two walls intersect can be called the z-axis. Thus the point in the lower corner of the room along with the two walls and floor that go through it is a coordinate system composed of an x, y, and z-axis going through the origin (point in the corner). The concept of a coordinate system is useful because it allows us to locate every other point in the universe by specifying its distance from the origin along the x, y, and z-axis of the coordinate system.

It will now be useful to develop a nomenclature to help us abbreviate these concepts. A dimensionality consisting of a certain number of dimensions will be designated by the capital letter "D". Following the "D" will be parentheses which contain within them the names of the dimensions. For example, in our world with three space dimensions, the dimensionality would be designated as D(x, y, z). But there is also a time element in our reality. At one point in time, the location of the tip of my pencil can be specified by the x, y, and z coordinates in our coordinate system. But the tip of my pencil is at different locations (x, y, and z coordinates) at different times. We actually need four dimensions (x, y, z, and t) to specify where the tip of my pencil is located at any time. So our reality consists not only of three space dimensions but also a time dimension. The dimensionality of our reality is thus designated as D(x, y, z, t), where four values (x, y, z, and t) are required to locate the tip of my pencil, or any other object, in our reality. How these concepts can be related to option 6 and a "transition into an alternate dimensionality" can be illustrated by the following story.

> Once upon a time, there was a man named Mr. Dotman who lived in a country named Lineland. Mr. Dotman was very small; in fact he had no dimension (length, width, or height) at all. He was just a dot. But he didn't feel bad about himself for where he lived; everyone was only a dot, having no dimension at all. Now the country of Lineland was a strange place too, for it had only one dimension in space, where everyone lived on the x-axis and everyone was convinced that this was all that existed. But they could move along the x-axis, which was important. For this meant that their reality also had a dimension of time.
>
> Now Mr. Dotman loved to tickle his imagination by reading strange sounding things with strange looking symbols, and so one day he found himself reading about strange seeming thoughts by a strange sounding man named Einstein. And as a result, our Mr. Dotman realized that he was living in a universe of one spatial dimension and one time dimension so that Lineland could be designated as a dimensionality of two dimensions, D(x, t). And when he realized this, strange seeming thoughts started bouncing around in his zero dimensional brain as to why this might be so. But upon discussing it with his friends in Lineland, they thought him rather strange, and for their own protection and profit they started behaving in a most unfriendly manner. Now Mr. Dotman thought their response was very strange, and before long Mr. Dotman found himself confined to prison. He was only allowed to move between x = -1.0 and x = 1.0 on the x-axis by two dot guards stationed at these two points who watched Mr. Dotman at all times lest he attempt an escape. Now the situation for Mr. Dotman was far worse than it might at first appear for he by nature was a very positive person; far preferring to be positive rather than negative, but that is a different story.

Now the advantage for one being in jail is that one has plenty of time to focus one's thoughts on what is important, and so it was for Mr. Dotman. As he meditated upon his confinement along the x-axis, and upon Mr. Einstein's strange ideas, it occurred to him that the dimensionality of the universe in which he lived, D(x, t), might, just possibly, be a part, a subset, a simplification of a larger reality, a higher dimensionality with a y-axis as well, D(x, y, t), where he was not only confined along the x-axis by the prison in which he meditated, but also, by the nature of things, was confined along the y-axis at y = 0. So that this larger reality, this higher dimensionality with three dimensions, D(x, y, t), for him and his ex-friends only appeared to be a two dimensional reality, D(x, t), because of their inability to perceive the third dimension because of their restriction to y = 0.

But with two dot guards on either side of him, looking straight at him, what could Mr. Dotman do? And then the answer suddenly occurred to him. The two dot guards, who were looking straight at him, were looking straight on the x-axis at him. It was their world view of only one spatial dimension that prevented them from even considering the possibility of a larger reality composed of a higher dimensionality. And it was then, when this thought came to completion in his zero dimensional brain that he turned his zero dimensional body to look off the x-axis. And then he saw it – the y-axis. And he wondered why had he not seen it before? It had evidently been there all the time, but he had never even looked for it before.

And then Mr. Dotman did something that had never been done before in all of Lineland. He moved from y = 0 up to y =1, and when he did, the two dot guards, who lived on the x-axis at y = 0, screamed and ran away, for Mr. Dotman had simply vanished from before their eyes as they were looking right at him on the x-axis. And when the guards reported to the king of Lineland what had happened, the king bribed the guards to say that Mr. Dotman's friends had come at night and stolen him from the prison while they were asleep. And this fooled most of the people of Lineland, for in their world view, they were convinced that they lived in a reality of only one spatial dimension, so that it was impossible for a person to move off of the xaxis because it contradicted their very advanced science.

Mr. Dotman now found himself in a new and strange place; in a country called Planeland, because the x and y-axis that he found there defined a broad plane on which the inhabitants lived. And he made new friends, and these new friends renamed him because Dotman no longer seemed to describe him, so they renamed him Mr. Squareman. This was only partially because of his new found shape, but also because his silly ideas seemed to limit their freedom in their frivolity and vanity. So before long, newly named Mr. Squareman again found himself in prison with guards at (x, y) coordinates of (-1, 0), (1, 0), (0, 1), and (0, -1) who watched him very carefully lest he attempt an escape. But this time he knew what to do; for again he suspected that the reality of two spatial dimensions and one time dimension, D(x, y, t), in which he now lived was only a subset of a higher dimensionality, D(x, y, z, t). And as he turned his two dimensional body to look off the x-y plane, he quickly spotted the z-axis. And then Mr. Squareman did something that had never been done before in all of Planeland. He moved from z = 0 up to z =1, and when he did, the four guards, who lived on the x-y plane at z = 0, screamed and ran away, for Mr. Squareman had simply vanished from before their eyes as they were looking right at him in the x-y plane. And when the guards reported to the king of Planeland what had happened, the king bribed the guards to say that Mr. Squareman's friends had come at night and stolen him from the prison while they were asleep. And this fooled most of the people of Planeland, for in their world view, they were convinced that they lived in a reality of only two spatial dimensions, so that it was impossible for a person to move off of their x-y plane because it contradicted their very advanced science.

And he now found himself in a country called earth, where again they gave him a new name, and where everyone was convinced that reality consisted of only three spatial dimensions and one time dimension, so that their universe had a dimensionality of four dimensions, $D(x, y, z, t)$. And very quickly he again found himself in prison because his ideas seemed to limit people's frivolity and vanity. But this prison was deeper and darker for they had killed him – but only his body. And they placed his body in a prison and placed many guards outside with orders to watch very carefully lest anything strange should happen. But again he knew what to do, for the supposed four dimensional reality, $D(x, y, z, t)$, of the country called earth was actually only a subset of a higher dimensionality containing 10 dimensions, $D(x, x', x'', y, y', y'', z, z', z'', t)$, where the primed and double-primed dimensions were wrapped so tightly around the x, y, and z-axis that they were not even noticed by the people living in the land. So when he moved his body from $x' = y' = z' = 0$ over to $x' = y' = z' = 1$, the guards, who were confined to the x-y-z space at $x' = y' = z' = 0$, screamed and ran away for his body had simply vanished from before their eyes as they were looking right at it. And when the guards reported to the king what had happened, the king bribed the guards to say that his friends had come at night and stolen his body from the prison while they were asleep. And this fooled most of the people of the land called earth, for in their world view, they were convinced that they lived in a reality of only three spatial dimensions and one time dimension, $D(x, y, z, t)$, so that it was impossible for a person to move out of their x-y-z space because it contradicted their very advanced science. The End

The main purpose of the above story is to illustrate what it means to make a transition into an alternate dimensionality. Mr. Dotman initially lived in Lineland where everyone thought that the entire universe consisted of only one spatial dimension and one time dimension, for a total of two dimensions, $D(x, t)$, but it actually had three dimensions, $D(x, y, t)$, except that everyone was limited to $y = 0$ by their world view which informed them that the x-axis was all that there was. Mr. Dotman discovered that the y-axis could be found if one only looked for it, and that by an act of his will he could go from $y = 0$ to $y = 1$ and thus escape from being imprisoned on the x-axis. His "transition to an alternate (or higher) dimensionality" consisted in going from what everyone thought was a universe of two dimensions, $D(x, t)$, to a universe of three dimensions, $D(x, y, t)$, although it was a universe of three dimensions, $D(x, y, t)$, all along.

The same can be said for him in Planeland. Everyone in Planeland thought that the entire universe consisted of only two spatial dimensions and one time dimension, for a total of three dimensions, $D(x, y, t)$, but it actually had four dimensions, $D(x, y, z, t)$, except that everyone was limited to $z = 0$ by their world view which informed them that the x-y plane was all that there was. Mr. Dotman, now renamed Mr. Squareman, discovered that the z-axis could be found if he only looked for it, and that by an act of his will he could go from $z = 0$ to $z = 1$ and thus escape from being imprisoned on the x-y plane. His "transition to an alternate (or higher) dimensionality" consisted in going from what everyone thought was a universe of three dimensions, $D(x, y, t)$, to a universe of four dimensions, $D(x, y, z, t)$, although it was actually a universe of four dimensions, $D(x, y, z, t)$, all along.

The same can be said for him in the country called earth, where everyone thought that the entire universe consisted of only three spatial dimensions and one time dimension, for a total of four dimensions, $D(x, y, z, t)$. But in the story the universe actually had ten dimensions, $D(x, x', x'', y, y', y'', z, z', z'', t)$, except that the primed and double-primed dimensions were not perceived by people. So everyone's world view was defective. They actually existed in a much higher dimensionality of 10 dimensions rather than the 4 dimensions that they perceived. Mr. Dotman, now under his new name, for an unexplained reason had the power to find the primed dimensions and by an act of his will transitioned to them, i.e. he went from $x' = y' = z' = 0$ to $x' = y' = z' = 1$. To the guards, this "transition to an alternate dimensionality" was observed as a disappearance or vanishing from their reality because they were confined to $x' = y' = z' = 0$. So his "transition to an alternate (or higher) dimensionality" consisted in going from what everyone thought was a universe of four dimensions, $D(x, y, z, t)$, to a universe of ten dimensions, $D(x, x', x'', y, y', y'', z, z', z'', t)$, although it was actually a universe of ten dimensions all along. A universe of ten dimensions was used because some "string theories" in modern physics postulate this dimensionality.

In the story, the body of Mr. Dotman never disintegrated and it never ceased to exist. In transitioning to an alternate dimensionality, his body was always located somewhere and he maintained continuous consciousness. The transition to the alternate dimensionality did not cause any energy release and did not injure anyone. And the transition to the alternate dimensionality was brought about by Mr. Dotman in an act of his will. Presumably, though not utilized in the story, Mr. Dotman could have transitioned back into jail if he had wanted to do so. Thus, a "transition into an alternate dimensionality" can be compared to a person in a room with a closed door, who opens the door, goes into the next room, and closes the door behind him. When he exits the first room and enters the second room, his body does not need to disintegrate and there is no huge explosion. He is simply in the next room and can come back into the first room anytime he wants to.

In option 6, Jesus' body disappears from inside the shroud by a transition into an alternate dimensionality. If we apply the characteristics of such a transition as presented in the above story to Jesus' resurrection there are many similarities. In Jesus' resurrection, there is no evidence in Scripture that His body disintegrated, or that it ceased to exist, or that it caused a huge energy release or that it injured anyone. The only evidence in Scripture is that His body was no longer in the burial shroud in the tomb as if it had disappeared from within the wrapped burial shroud. After His resurrection, He also had the ability to reappear and disappear and did so at least ten different times in His post-resurrection appearances (Mt. 28:9-10, 16-20, Luke 24:13-32, 36-49, John 20:11-21:25, Acts 1:3-12, 1 Cor. 15:3-7). There is no evidence in scripture that His reappearances and disappearances involved a significant energy release, no one was injured, and these passages read as though His reappearances and disappearances are occurring simply as the result of an act of His will. This is most consistent with his body continuing to exist between His post-resurrection appearances. If this is so then His body never disintegrated and it never ceased to exist; it was always somewhere though not in our dimensionality, D(x, y, z, t). The concept that Jesus' resurrection and His post-resurrection appearances are transitions to and from and alternate dimensionality seems to fit the evidence very nicely.

There is a question whether there could have been a significant energy release due to His resurrection from the tomb because an earthquake is noted in Mt. 28:2. While this earthquake is called great (KJV, ESV, NLT), severe (NASB), or violent (NIV), there is no indication in Scripture that it was felt elsewhere in the city and there is no archeological evidence that the city of Jerusalem was destroyed in about 30 to 33 AD. For these reasons, some have suggested that Matthew's reference to a violent earthquake should either be understood figuratively in some sense, or is simply not true. But there is good reason to reject these views and take the reference to an earthquake in Mt. 28:2 at face value. If a violent earthquake occurred due to shifting rock layers deep in the earth, then its destructive effect would have occurred over a very large area, but this earthquake was very localized at the surface, not deep under ground, because it was due to an event in the tomb, i.e. Jesus' resurrection. For an earthquake that has its point of origin in the tomb, the energy released would take the form of compression or P waves and shear or S waves in the rock and soil that would spread out in a half-sphere in the ground below and around the tomb. An insignificant amount of energy would be transferred to the air above the ground. As the compression and shear waves in the ground spread out in a half-sphere, their strength would decrease as the inverse square of the distance from the tomb, i.e. $(1/r)^2$ where r = distance of the observer from the tomb. There would also be surface waves (Love waves and Rayleigh waves) that would spread out in a circle on the surface so that their strength would decrease as the inverse of the distance of the observer from the tomb, i.e. $(1/r)$. So what might be described as a violent earthquake by the women or guards near the tomb (Mt. 28:2) might not even be felt by most people in Jerusalem. This being the case, the reference to the earthquake in Mt. 28:2 should be understood literally to refer to a real event.

The text of Mt. 28:2 is not clear whether the earthquake occurred at the moment of Jesus' resurrection or whether His resurrection was earlier, and it is not absolutely clear whether the angel caused the earthquake in the process of rolling away the stone or whether it was caused by Jesus' resurrection. Matthew 28:2 in the NIV only says "There was a violent earthquake, for an angel of the Lord came down from heaven and, going to the tomb, rolled back the stone and sat on it." The preposition is translated as "for" or "because" in most translations (NASB, NIV, ESV, NLT, TLB, RSV, NBV, NJB, Amplified, Phillips, Williams, and Wuest). These translations imply that the earthquake was caused by the angel in the process of coming down from heaven and rolling away the stone. But the "for" is not included in some translations (NEB, REB, NCV, Beck, TEV, and The Message). And in The Expanded Bible, the "for" is omitted from the main reading but inserted in brackets with an "L" to indicate it is the literal translation of a word that is present in the Greek. The "for" in English is the translation of the Greek word "γαρ" which is pronounced "gar". According to "The NIV Exhaustive Concordance" (Ref. 34), this Greek word occurs 1040 times in the New Testament. In the NIV, it is translated "for" 502 times, "because" 85 times, "but" ten times, "and" nine times, "now" seven times, "indeed" six times, "in fact" or "since" four times each, and "after all" or "why" three times each. For 31 other occurrences of "γαρ", it is translated in 27 other ways. But most significantly, for 376 occurrences of "γαρ" in the New Testament, which is 36% of the total, it is not translated at all, being entirely omitted in the English. As in any translation, the way in which a word ought to be translated, or whether it is simply to be omitted in the translation, is determined by what makes sense in the context of what comes before and after it.

Several points need consideration regarding the context of the earthquake in Mt. 28:2. It is unlikely that the mere appearance of the angel at Jesus' tomb caused the earthquake referred to in Matthew 28:2, since the scriptures never associate an earthquake with the appearance of any other angel. And it seems unlikely that the angel rolling away the stone would require a violent earthquake to accomplish it. So it seems more likely that the earthquake was the result of Jesus' resurrection. If the disappearance of Jesus' body in the resurrection involved a transition of

His body into an alternate dimensionality, then such a transition might cause oscillations or ripples in the space-time field or continuum that defines our dimensionality, like ripples on the surface of a lake that result when a stone is thrown into the water. Such an oscillating warp or perturbation in the space-time field of our dimensionality could have caused a general shaking of the ground around the tomb, as waves on the water would cause a toy boat to move up and down. But there is no evidence for an earthquake occurring at any of the ten post-resurrection appearances, which argues against the concept that a transition into an alternate dimensionality would necessarily cause an earthquake. In Jesus' resurrection, besides his body disappearing, at least two other important things happened – his body was glorified, i.e. his physical body was changed into a spiritual body, and his soul was reunited with his body to make it alive again. So while it seems more likely that the earthquake was caused by Jesus' resurrection than by the angel moving the stone, we must admit that we have no basis for even speculating how enough energy could be released to cause a violent earthquake around the tomb.

If the earthquake in Mt. 28:2 was due to Jesus' resurrection, as argued for above, then the "γαρ" in Mt. 28:2 could be translated as "and" or possibly as "now" but more likely just left out of the translation to eliminate the implication that the earthquake was caused by the angel rolling the stone away. This is done in the Today's English Version (TEV, also called Good News for Modern Man): "Suddenly there was a strong earthquake; an angel of the Lord came down from heaven, rolled away the stone, and sat on it." The commentaries usually recognize that the earthquake was associated with the angel coming to roll away the stone, but usually do not say that the earthquake was caused by the angel rolling away the stone. For example, "The Bible Knowledge Commentary" for Mt. 28:2 says "There was, however, a violent earthquake associated with an angel coming from heaven and rolling away the stone from the door of the tomb." This would only imply that the earthquake occurred at very close to the same time as the angel coming and rolling away the stone. The conclusion of this matter is that the earthquake could have been caused be Jesus' resurrection, and probably was.

The conclusion of our consideration of option 6 is that the Scriptural evidence regarding Jesus' resurrection and post-resurrection appearances fits nicely with the concept that He was transitioning to and from an alternate dimensionality. But don't we know that our universe only has three spatial dimensions and one time dimension so that the dimensionality of our universe is D(x, y, z, t)? And isn't science so well based that there is no controversy remaining? This may appear to be so to the layman, but in the realm of modern physics, very strange things often seem to be necessary. For example, quantum mechanics has been proven to be true when applied to very small things and general relativity theory has been proven to be true when applied to very large things but they can't both be entirely true because the basic assumptions of these two theories are inherently contradictory: quantum mechanics assumes that gravitational attraction results from the exchange of virtual particles whereas relativity theory assumes that gravitational attraction results from the bending of the fabric of space-time. Another example is the wave-particle duality of light, where light acts like a wave and a particle (photon) at the same time, because if individual photons are allowed, one at a time, to go through either one or the other of two slits in a barrier they can accumulate into an interference pattern behind the barrier, which is characteristic of a wave. The same has been found to be true of subatomic particles such as electrons. Another example is that the universe not only consists of matter and energy that can be seen, but it must also contain what is called dark matter and dark energy that can not be seen. Dark matter was originally hypothesized to explain the fact that the arms of galaxies are usually not all wrapped upon themselves as they ought to be given the age of the galaxies. This dark matter interacts with normal matter through gravitation, but it does not give off light, it does not reflect light, and it does not scatter or absorb light from behind it, so that it can not be seen. And the dark energy, though invisible, must pervade "empty" space so as to cause the acceleration in the expansion of the universe that is observed. It is now believed that there is about 5.5 times as much dark matter as there is ordinary visible matter in the universe, and that dark matter is composed of some as yet unknown subatomic particle. It is believed that dark matter plus dark energy constitute about 95% of the total mass–energy content of the universe.

Many problems remain in our understanding of physics. How can force be exerted across space? Can the four forces in nature be unified into a single force that displays itself in four different ways? Can all the sub-atomic particles be unified into a single understanding (usually called the "standard model") of matter? Why does matter have mass and weight? Why is there so much more matter than antimatter in the universe? Is there a single unified theory that explains all of these questions? In an attempt to develop a unified "theory of everything" which resolves the conflict between quantum mechanics and general relativity theory, some theorists in modern physics have developed various forms of what is called string theory, also sometimes called superstring theory. These theories say that everything is made up of extremely small one-dimensional loops called "strings" which are like infinitely thin rubber bands whose vibration modes account for all the characteristics of matter and energy. These string theorists develop their theories based on geometrical considerations of higher dimensionalities of more than four dimensions D(x, y, z, t). In his book The Elegant Universe (Ref. 35), Brian Greene said "Most of us take for granted that our universe has three spatial dimensions. But this is not so according to string theory, which claims that our universe has many more dimensions than meet the eye--dimensions that are tightly curled into the folded fabric of the cosmos." As conceived by various string theorists, these higher dimensionalities are composed of from 10 to 26 dimensions. In our story of Mr. Dotman in Lineland, the story utilized a dimensionality of 10 dimensions, with six of the dimensions wrapped so tightly around the x, y, and z-axis that they were not observable, as one type of string theory postulates.

Oxford mathematician Roger Penrose has developed an alternative to the various forms of string theory in an attempt to better understand reality. He bases his theory on the use of complex numbers, whereas string theory is based on real numbers. A complex number is of the form A+Bi (A plus B times i) where A and B are real numbers but i is the square root of -1 (which can have no real solution so is called "imaginary"), so that A is the real part of the complex number A+Bi and B times i is the imaginary part of the complex number. With this shift from real numbers to complex numbers in his theory, he can produce a mathematical concept of what he calls a twistor, which has no mass, is 20 orders of magnitude (factors of 10) smaller than a nuclear particle, is shaped like a doughnut or piece of rope with a single twist as it circles a central axis, and can travel at the speed of light in the direction of its axis. According to Penrose, the fundamental subatomic particles are made up of combinations of a small number of various types of twistors, the continuous fabric of space and time in Einstein's equations is replaced by a space of twistors with each twistor being the smallest possible unit so that space-time (now called twistor space) is not continuous, i.e. is not infinitely divisible, but is composed of points with each point being a twistor, and the four basic forces (gravity, electromagnetic, weak nuclear, and strong nuclear) may one day be shown to be the result of various types of deformations in twistor space. Penrose explains twistor theory in Chapter 33 of his book "The Road to Reality" (Ref. 36).

There are 5 different variations on current String Theory, which, according to Penrose, may all include facets of ultimate reality, but nobody really knows. In addition to the various string theories and twistor theory, there is something called "11-dimensional supergravity" theory (Ref. 37). However, none of the above theories answers all of the questions or solves all of the problems. Thus, at the deepest level of our understanding of how things operate in our reality, there are many unknowns and unresolved issues and conflicts, so that our perceived dimensionality D(x, y, z, t) may be only a subset of a higher order dimensionality. This might allow for Jesus' body to disappear from within the shroud by a transition into an alternate dimensionality as option 6 postulates.

In review, considerations for options 1 to 4 involved using the known laws of physics to determine the energy input to or output from the body under different schemes of disintegration, with the conclusion that all schemes of disintegration ought to be rejected. Considerations for options 5 and 6 involved what might be called extrapolations of theories of physics, with the conclusion that the involvement of wormholes ought to be rejected but that the characteristics of Jesus' resurrection and post-resurrection appearances agree with the characteristics of transitions between alternate dimensionalities. The only other option that needs to be considered is option 7 which says that Jesus' body disappeared by an unknown mechanism not related to any known physical phenomenon or law of physics. Due to the unknown nature of this option, it is not possible to consider its pros and the cons. But it is wise to take a position of humility regarding our current state of science and admit that there are probably many new concepts, principles, and theories that are yet to be discovered in science. This means that the eye-witness testimony of Jesus' resurrection and post-resurrection appearances ought to be taken at face value, and thus recognize that these are real historical events. They ought not to be rejected just because they contradict our current understanding of science.

Conclusion

The methodology by which Jesus' body disappeared from the tomb in the resurrection has been considered from the Biblical references to His resurrection, from the theology of resurrection throughout the Bible, and from a consideration of the physical basis of seven options for how Jesus' body might have disappeared from the tomb. It was concluded that at the resurrection, Jesus' body underwent a basic transformation, a metamorphosis, from what Paul called a natural body to a spiritual body. In His resurrected state of this spiritual body, spiritual things such as his soul and the Holy Spirit were in the ascendency over the limitations of our physical reality, so that by an exercise of His will He could do things that might appear to contradict the laws of nature as we now understand them. Consideration of the physical basis for the disappearance of the body from the tomb has led us to reject concepts that involve disintegration of the body or the action of one or more wormholes. But it has led us to endorse the concept that Jesus' resurrection and post-resurrection appearances are consistent with transitions between alternate dimensionalities, with each transition being the result of an act of His will. It is also possible that Jesus' resurrection and post-resurrection appearances were the result of an unknown mechanism not related to any known physical phenomenon or law of physics, though the merits of this option could not be judged due to its vagueness.

Also discussed under option 2, based on computer calculations, was that if a very small fraction (1.45×10^{-10}) of the number of neutrons in the body were emitted in the body, it would explain the shift in the radiocarbon dating for the Shroud of Turin from about 30 AD to an uncorrected value of 1260 AD. And based on experiments, if a very small fraction (very approximately 5×10^{-11}) of the number of protons in the body were emitted from the body, it would have caused a discoloration on the linen similar to that in the image on the Shroud of Turin. If Jesus' body disappeared from the tomb by a transition into an alternate dimensionality, or by another unknown mechanism, then there is no reason to reject the possibility that these very small fractions of the neutrons and protons that were in His body could have been emitted in the process of His disappearance, thus explaining the shift in the C^{14} date and the appearance of the image on the shroud.

Email comments or suggestions to robertarucker@yahoo.com.

APPENDIX I

Old Testament Prophecies of New Testament Events

Even my close friend, in whom I trusted;
Who ate my bread,
Has lifted up his heel against me. (Ps. 41.9)

And while He was still speaking, Behold, Judas, one of the twelve, came up, accompanied by a great multitude with swords and clubs, from the chief priests and elders of the people. Now he who was betraying Him gave them a sign, saying "Whomever I shall Kiss, He is the one, seize Him." (Matt. 26:47, 48. See also 27:3,5,7. See also Mark 14:43-47; Luke 22:47, 48; John 18:2,3)

And I said to them, "If it is good in your sight, give me my wages; but if not, never mind!" So they weighed out thirty shekels of silver, as my wages. Then the Lord said to me, "Throw it to the potter, that magnificent price at which I was valued by them." So I took the thirty shekels of silver and threw them to the potter in the house of the Lord. (Zech. 11:12, 13.)

Then when Judas, who had betrayed Him, saw He had been condemned, he felt remorse and returned the thirty pieces of silver to the chief priests and elders. . . . And he threw the pieces of silver into the sanctuary and departed.. . .And they counseled together and with the money bought the Potter's Field as a burial place for strangers. (Matt. 27:3, 5, 7. See also Matt. 26:15, Mark 14:10, 11.)

Malicious witnesses rise up;
They ask me of things that I do not know. (Ps. 35:11)

Now the chief priests and the whole Council kept trying to obtain false testimony against Jesus, in order that they might put Him to death; they did not find any, even though many false witnesses came forward... (Matt. 26:59, 60. See also Mark 14:56-59.)

He was oppressed and He was afflicted.
Yet he did not open His mouth. (Is. 53:7)

And while He was being accused by the chief priests and elders. He made no answer. (Matt. 27:12. See also Mark 14:60-61.)

I gave My back to those who strike Me. And My cheeks so those who pluck out the beard; I did not cover My face from humiliation and spitting. (Is. 50:6)	Then they spat in His face and beat Him with their fists; and others slapped Him. (Matt. 26:67. See also Mark 14:65; 15-19; Luke 22:63; John 18:22.) ...and after having Jesus scourged, he delivered Him over to be crucified. Mark 15:15. See also Matt. 27:36; John 19:1.)
. . . They pierced my hands and my feet. I can count all my bones. They look, they stare at me. (Ps. 22:16-17)	And when they came to the placed called The Skull, there they crucified Him. . . . And the people stood by, looking on . . . (Luke 23:33, 35)
They divide my garments among them. And for my clothing, they cast lots. (Ps. 22:18)	And they crucified Him, and divided up his garments among themselves, casting lots for them, to decide what each should take. (Mark 15:24. See also Matt. 27:35; Luke 23:34; John 19:23-24.)
My loved ones and my friends stand aloof from my plague; And my kinsmen stand afar off. (Ps. 38:11)	And all His acquaintances and the women who accompanied Him from Galilee, were standing at a distance, seeing these things. (Luke 23:49. See also Matt. 27:55-56; Mark 15:40.)
All who see me sneer at me; They separate with the lip, they wag the head, saying, "Commit yourself to the Lord; let Him deliver him; Let Him rescue him, because He delights in him." (Ps. 22:7-8, See also Ps. 109:25)	And those passing by were hurling abuse at Him, wagging their heads, and saying, ". . . if You are the Son of God, come down from the cross." In the same way the chief priests also, along with the scribes and elders, were mocking Him and saying, ". . . He trusts in God; Let Him deliver Him now, if He takes pleasure in Him; for He said, "I am the Son of God" (Matt 27:39-43)
They also gave me gall for my food And for my thirst they gave me vinegar to drink. (Ps. 69:21)	They gave him wine to drink mingled with gall; and after tasting it, He was unwilling to drink. . . . And immediately one of them ran and taking a sponge, he filled it with sour wine [vinegar], and put it on a reed, and gave Him a drink. (Matt. 27:34, 48. See also Mark 15:23, 36; Luke 23:36; John 19:28-30.)

... Because He poured out Himself to death, And was numbered with the transgressors. (Is. 53:12)	At that time two robbers were crucified with Him, one on the right and one on the left. (Matt. 27:38. See also Mark 15:27; Luke 23:47-48.)
"And it will come about in that day," declares the Lord God, "That I shall make the sun go down at noon And make the earth dark in broad daylight." (Amos 8:9)	And when the sixth hour [noon] had come, darkness fell over the whole land until the ninth hour. (Mark 15:33. See also Matt. 27:45; Luke 23:44-45.)
My God, my God, why hast Thou forsaken me? (Ps. 22:1)	And at the ninth hour Jesus cried out with a loud voice, "Eloi, Eloi lama sabachthani?" which is translated, "My God, my God, why has Thou forsaken Me?" (Mark 15:34. See also Matt. 27:46.)
Into Thy hand I commit my spirit. (Ps. 31:5)	And Jesus, crying out with a loud voice, said, "Father, into Thy hands I commit My spirit." And having said this, He breathed His last. (Luke 23:46. See also Matt. 27:50.)
He keeps all his bones; Not one of them is broken. (Ps. 34:20)	but coming to Jesus, when they saw that He was already dead, they did not break His legs. (John 19:33)
... [T]hey will look on Me whom they have pierced... (Zech. 12:10)	but one of the soldiers pierced His side with a spear. (John 19:34)
I am poured out like water. And all my bones are out of joint; My heart is like wax; It is melted within me. (Ps. 22:14)	... and immediately there came out blood and water. (John 19:34)

"And I will pour out on the house of David and on the inhabitants of Jerusalem, the Spirit of grace and of supplication, so that they will look on Me whom they have pierced; and they will mourn for Him, as one mourns for an only son, and they will weep bitterly over Him, like the bitter weeping over a first-born. (Zech. 12:10)

Now when the centurion saw what had happened, he began praising God, saying, "Certainly this man was innocent." And all the multitudes who came together for this spectacle, when they observed what had happened, began to return, beating their breasts. (Luke 23:47-48)

But he was pierced through for our transgressions,
He was crushed for our iniquities;
The chastening for our well-being fell upon Him,
And by His scourging we are healed. (Is. 53:5)

And He himself bore our sins in His body on the cross, that we might die to sin and live to righteousness; for by His wounds you were healed. (1Peter 2:24)

...Yet He Himself bore the sin of many,
And interceded for the transgressors. (Is. 53:12)

For this is My blood for the covenant, which is poured out for many for forgiveness of sins. (Matt. 26:28. See also Mark 10:45; John 3:17)

His grave was assigned with wicked men,
Yet He was with a rich man in His death ... (Is. 53:9)

... there came a rich man from Arimathea, named Joseph, who ... went to Pilate and asked for the body of Jesus ... And Joseph took the body and wrapped it in a clean linen cloth, and laid it in his own new tomb, which he had hewn out in the rock ... (Matt. 27:57-60)

For Thou wilt not abandon my soul to Sheol;
Neither wilt Thou allow Thy Holy One to undergo decay. (Ps. 16:10)

... He was neither abandoned to Hades, nor did His flesh suffer decay. (Acts 2:31)

But the Lord was pleased
To crush Him, putting Him to grief;
If He would render Himself as a
guilt offering,
He will see His offspring, He will
prolong His days,
And the good pleasure of the Lord will
prosper His hand. (Is. 53:10)

. . . Him who raised Jesus our Lord from the dead. He who was delivered up because of our transgressions, and was raised because of our justification. (Romans 4:24-25)

Arise, shine; for your light has come,
and the glory of the Lord has risen
upon you.
For behold, darkness will cover the
earth,
And deep darkness the peoples;
But the Lord will rise upon you,
And His glory will appear upon you.
And nations will come to your light,
And kings to the brightness of
your rising. (IS. 60:1-3.)

And He is the radiance of His glory and the exact representation of His nature . . . (Heb. 1:3) These are in accordance with the working of the strength of His might which He brought about in Christ, when He raised Him from the dead, and seated Him at His right hand in heavenly places, far above all rule and authority and power and dominion, and every name that is named, not only in this age, but also in the one to come. (Eph. 1:19-21)

New American Standard Bible

ENDNOTES

PREFACE

1. Taken from B. McDonald, Commentary, "The Book Churchill Needed Most," The Philadelphia Trumpet, April 2010, p. 35.

CHAPTER ONE

1. P. Vignon, *The Shroud of Christ* 1902 (New Hyde Park, NY: University Books, 1970). Paul Vignon was one of the greatest Shroud researchers of all. He and anatomist Yves Delage led the first scientific investigation of the Shroud in 1900-02 based on Secondo Pia's photographic plates. He and Delage were the first scientists to assert because the Shroud contains faint, full-length, negative images that display high resolution and other details invisible to the naked eye — with left-right, light-dark reversal when it is photographed — that a medieval painter could not have encoded such features on a cloth. Vignon proposed the first naturalistic explanation for the Shroud's images, but later acknowledged failings inherent in this method. Vignon was also the first to note the similarities between the Shroud as seen with the naked eye and many centuries-old depictions of Jesus. From the time Vignon first saw Pia's amazing photographs from 1898, until his death in 1943, he not only studied the Shroud, but initiated new areas of inquiry into the subject.

2. J. H. Heller, *Report on the Shroud of Turin* (New York: Houghton Mifflin, 1983), pp. 38-39; J. P. Jackson, E. J. Jumper, B. Mottern, and K. E. Stevenson, "The Three-dimensional Image on Jesus Burial Cloth," *Proceedings of the 1977 United States Conference of Research on the Shroud of Turin* (Albuquerque, N.M.: Holy Shroud Guild, Mar. 1977), 74-94; J. P. Jackson, E. J. Jumper and W. R. Ercoline, "Correlation of Image Intensity on the Turin Shroud with the 3-D Structure of a Human Body Shape," *Applied Optics* 23.14 (July 1984): 2244-2270; and J. P. Jackson, E. J. Jumper, and W. R. Ercoline, "Three Dimensional Characteristics of the Shroud Image," *IEEE 1982 Proceedings of the International Conference on Cybernetics and Society* October 1982): 559-575.

3. Heller, *Report,* pp. 38-39; Jackson et al., "Correlation," "Three Dimensional Characteristics," and "The Three-dimensional Image,"; C. Avis, D. Lynn, J. Lorre, S. Lavoie, J. Clark, E. Armstrong, and J. Addington, "Image Processing of the Shroud of Turin," *IEEE 1982 Proceedings of the International Conference on Cybernetics and Society* (October 1982): 554-558; G. Tamburelli, "Some Results in the Processing of the Holy Shroud of Turin," *IEEE Transactions on Pattern Analysis and Machine Intelligence* PAMI-3.6 (November 1981): 670-676; G. Tamburelli, "Reading the Shroud, Called the Fifth Gospel, with the Aid of the Computer," *Shroud Spectrum International* 15 (June 1985): 3-6.

4. L. Gonella, "Scientific Investigation of the Shroud of Turin: Problems, Results and Methodological Lessons," in *Turin Shroud – Image of Christ?* (Hong Kong: Cosmos Printing Press Ltd. 1987), pp. 29- 40, 31.

5. Heller, *Report*; K. E. Stevenson and G. R. Habermas, *Verdict on the Shroud* (Ann Arbor, Mich.: Servant Books, 1981); Gonella, "Scientific Investigation"; L. A. Schwalbe and R. N. Rogers, "Physics and Chemistry of the Shroud of Turin," *Analytica Chimica Acta* 135 (1982): 3-49; Don Lynn interview in *Silent Witness,* produced by David Rolfe, 1978, 45:48-49:20; and personal communications with Don Lynn and Jean Lorre (NASA and STURP scientists), June 2, 1984 and March 16, 1999.

6. Don Lynn interview in *Silent Witness*; Personal communications with Don Lynn, August 21, 1989, and Jean Lorre, March 16, 1999.

7. J. P. Jackson, "The Vertical Alignment of the Frontal Image," *Shroud Spectrum International* 32/33 (1989): 3-26; see also Jackson et al., "Three-Dimensional Image," p. 83; and W. R. Ercoline, R. C. Downs, Jr., and J. P. Jackson, "Examination of the Turin Shroud for Image Distortions," *IEEE 1982 Proceedings of the International Conference on Cybernetics and Society* (October 1982): 576-579.

8. Jackson, "Vertical Alignment"; Jackson et al., "Three Dimensional Image," p. 83; and Ercoline et al., "Examination for Image Distortions."

9. As quoted in W. McDonald, "Science and the Shroud, *The World and I* (Oct. 1986): pp. 420-428, 426.

10. E. J. Jumper, "Considerations of Molecular Diffusion and Radiation as an Image Formation Process on the Shroud," *Proceedings of the 1977 United States Conference of Research on the Shroud of Turin* (Albuquerque, NM: Holy Shroud Guild, March 1977), pp. 182-189; J. P. Jackson, "A Problem of Resolution Posed by the Existence of a Three-Dimensional Image on the Shroud of Turin," in *Proceedings* 223-233; Jackson et al., "Three-Dimensional Image," p. 91

11. E. J. Jumper, A. D. Adler, J. P. Jackson, S. F. Pellicori, J. H. Heller, and J. R. Druzik, "A Comprehensive Examination of the Various Stains and Images on the Shroud of Turin," *ACS Advances in Chemistry No. 205 Archaeological Chemistry III*, J. B. Lambert, ed. American Chemical Society (1984), pp. 447-476; Heller, Report; Schwalbe and Rogers, "Physics and Chemistry"; J. H. Heller and A. D. Adler, "A Chemical Investigation of the Shroud of Turin," *Can. Soc. Forens. Sci. J.* 14.3 (1981): 81-103; Gonella, "Scientific Investigation"; S. F. Pellicori and M. Evans, "The Shroud of Turin Through the Microscope," *Archaeology* (January/February 1981): 32-43; W. C. McCrone and C. Skirius "Light Microscopical Study of the Turin 'Shroud,' I," *Microscope* 28 (1980): 105-113; and W. C. McCrone,

"Light Microscopical Study of the Turin 'Shroud,' II," *Microscope* 28 (1980): 115-128.

12. Some scientists refer to this outer layer as the primary cell wall. G. Fanti, J. A. Botella, P. Di Lazzaro, T. Heimburger, R. Schneider, N. Svensson, "Microscopic and Macroscopic characteristics of the Shroud of Turin image superficiality," *Journal of Imaging Science and Technology*, 54:4, 2010

13. Pellicori and Evans, "The Shroud Through the Microscope," *Archaeology* (January/February 1981): 32-43; Jumper et al., "Comprehensive Examination;" V.D. Miller and S. F. Pellicori, "Ultraviolet Fluorescence Photography of the Shroud of Turin," *Journal of Biological Photography*, 49.3 (July 1981).

14. Jumper et al., "A Comprehensive Examination"; Schwalbe and Rogers, "Physics and Chemistry"; Heller and Adler, "A Chemical Investigation;" Miller and Pellicori, "Ultraviolet Fluorescence."

15. Jumper et al., "Comprehensive Examination"; Pellicori and Evans, "The Shroud Through the Microscope"; S. F. Pellicori, "Spectral Properties of the Shroud of Turin," *Applied Optics* 19.12 (1980): 1913-1920; G. G. Gray, "Determination and Significance of Activation Energy in Permanence Tests," in *Preservation of Paper and Textiles of Historic and Artistic Value,* Advances in Chemistry Series 164 (American Chemical Society, Washington, D.C., 1977), in S. F. Pellicori, "Spectral Properties" (1980); S. Pellicori and R. A. Chandos, "Portable unit permits UV/vis study of Shroud," *Industrial Research & Development*, February, 1981, Vol. 23:186-189; John P. Jackson, Eugene Arthurs, Larry A. Schwalbe, Ronald M. Sega, David E. Windisch, William H. Long, and Eddy A. Stappaerts, "Infrared laser heating for studies of cellulose degradation," *Applied Optics* 27 (1988): 3937-3943; J. Rinaudo, "Protonic Model of Image Formation on the Shroud of Turin," *Third International Congress on the Shroud of Turin*, Turin, Italy, June 5-7, 1998, p. 2; J. Rinaudo, "A Sign for Our Time," *Shroud Sources Newsletter*, May/June 1996, pp. 2-4; Miller and Pellicori, "Ultraviolet Fluorescence."

16. Jumper et al., "Comprehensive Examination"; Heller and Adler, "A Chemical Investigation"; Miller and Pellicori, "Ultraviolet Fluorescence"; P. Di Lazzaro and D. Murra, "Shroud like coloration of linen, conservation measures and perception of patterns onto the Shroud of Turin," *2014 Workshop on Advances in the Turin Shroud Investigation* (ATSI 2014), Bari, Italy, Sep 4-5, http:// www.shs-conferences.org/articles/shsconf/pdf/2015/02/shsconf_atsi2014_00004.pdf; P. Di Lazzaro, D. Murra, A. Santoni, and G. Baldacchini, "Sub-micrometer coloration depth of linens by vacuum ultraviolet radiation," *Proceedings of the International Workshop on the Scientific Approach to the Acheiropoietos Images*, Turin, Italy, May 4-6, 2010, pp. 3-18.

17. This important episode was acquired from chapter two of J. Walsh, *The Shroud*, (New York: Random House, 1963). Many other important, well-described episodes and summaries of investigation into the Shroud can be found in this book.

18. How could a medieval forger possibly paint or otherwise encode in a vertical straight-line direction from all parts on the full-length and width of the supine body, up to all parts of the cloth draped over it, regardless if the cloth was sloping upward, downward or relatively flat?

How could the body image be encoded through the spaces between the cloth and the reclined body (for example, the spaces between the tip of the nose and both sides of the cheeks)? Some naturalistic methods hypothesize that various substances such as perspiration, urea, body fluids, myrrh, aloes, etc. were on the body, or on the cloth in various amounts, and acted in combination with each other, or with fermentation products or decomposition processes, that somehow encoded the Shroud's unique body images. Yet, none of the above or other molecules, compounds, fluids or vapors will act or diffuse in a vertical straight-line direction through the spaces between the cloth and the body. They're isotropic and would travel in all directions. (No traces of myrrh or aloes or the presence of perspiration, urea, bodily fluids* or other sensitizing materials were detected on the body images or the linen cloth. While it is theoretically possible that such molecules or material could have decomposed or oxidized over time, some residual traces of the responsible elements that formed the body images should still be present. No signs of decomposition, which would have started two or three days after death, are present on the body images or the entire linen cloth. As we will see in the next chapter, the body images were formed while the body was in rigor mortis, which would have postponed decomposition staining.)

How could a medieval forger possibly paint or otherwise encode every part of the length and width of the vague, frontal body image in proportion to the original distances they were from the underlying body, regardless whether the cloth was in contact with the body or not? How could he artistically or naturally encode light/dark distance information without encoding any of the fibers more intensely? How could he, instead, encode a greater *number* of colored fibers at every location representing different distances from the cloth?

How could a medieval artist using any naturalistic or artistic method encode only the topmost two to three fibers of the threads? How could he encode them individually? How could he encode them the same uniform color? How could he encode them 360° around each fiber? How could he encode only the outer parts of each fiber while leaving the original white color in the middle of each encoded fiber? A medieval artist in the 1350's would have needed a microscope, but it wouldn't be invented until 1595. The finest material available to medieval painters for brushes was horse hair, yet a single horse hair is thicker than an individual fiber. How could an artist,

---

***Except those associated with blood marks throughout the frontal and dorsal body images (along with a watery fluid that emerged from a wound on the right side of the chest).**

regardless of his technique, encode any of these things without getting extremely close to his work, yet he couldn't see the body images unless he stood about ten feet away from the cloth? (Please keep in mind that our artist would also have to transfer onto cloth both images that were so low in contrast they faded into the background when he stood within six to ten feet of it.) A forger could not intricately encode all of the above features throughout the thousands of body image fibers on the full-length frontal and dorsal images on the Shroud using any artistic or naturalistic method.

How could a forger encode images without using any pigments, powder, or materials of any kind? How could the straw-yellow color of his body images be comprised of double-bonded carbon and oxygen atoms that used to be single-bonded to each other along with hydrogen? He would need a process that caused the single-bonded atoms to break apart throughout both full-length images simultaneously (in direct proportion to their distance from the body, only on the outer layers, 360° around each individual superficial fiber etc.) with two of these atoms double-binding over a period of time. How could he see to encode his body images when they developed over time? As will be seen in later chapters, the Shroud's body images were not first observed or mentioned until hundreds of years after they were encoded.

Why would a forger encode with left-right and light-dark reversal? The latter quality, found on the entire cloth, helps visibility by providing dark contrast for the much lighter, positive body images. How would a medieval forger know that light-dark reversal(s) of all coloration on the entire cloth would be found on photographic negatives, when this technology would not be invented for 500 years? Further, how could he encode vague, indistinct body images on a linen cloth that would turn into resolved images of a human being when seen on a photographic negative 500 years later? The shading and contrasts within both full-length images are very precise and subtle. How could he check his work? How could he get it so precisely correct everywhere without checking his work? Why would he use such a technique when no one else could see his superb artistry? How could he even anticipate, let alone encode, light-dark contrast and high resolution imagery found within photographic technology 500 years later?

In a similar vein, why would a medieval forger encode three-dimensional and vertically directional information onto a vague negative image on cloth when no one could see it? How could he encode the features without ever checking his work? How could he anticipate, let alone encode, three-dimensional and vertically directional information that would be illustrated by computer imaging technology 600 years later?

CHAPTER TWO

1. J. H. Heller, *Report on the Shroud of Turin* (New York: Houghton Mifflin, 1983); K. E. Stevenson and G. R. Habermas, *Verdict on the Shroud* (Ann Arbor, Mich.: Servant Books, 1981); and J. H. Heller and A. D. Adler, "A Chemical Investigation of the Shroud of Turin," *Can. Soc. Forens. Sci. J.*, Vol. 14, No. 3 (1981): 81-103, p. 92; Niels Svensson and Thibault Heimburger, "Forensic aspects and blood chemistry of the Turin Shroud Man," *Scientific Research and Essays* Vol. 7(29), pp. 2513-

2525, 30 July 2012, http://www.academicjournals.orgljournal/sre/edition/30 July, 2012. This is in addition to the forensic judgment of scores of pathologists, anatomists and physicians that have examined the blood marks before and after the revealing details acquired from STURP's investigation in 1978.

2. C. Goldoni, "The Shroud of Turin and the bilirubin blood stains," *Proceedings of the 2008 Columbus International Conference*, ed. G. Fanti, Aug. 14-17, 2008, http://www.ohioshroudconference.com/papers/p04.pdf; P. Barbet, Les Cing Plaies du Christ (The Five Wounds of Christ), 1935, SEI, Turin, 1940; P. Barbet, *A Doctor at Calvary* (New York: J. P. Kenedy & Sons, 1953). P. Barbet, *A Doctor at Calvary* (Garden City, New York: Doubleday Image Books, 1963). Although the last two books are the same, I have cited the second or paperback edition in all subsequent endnotes because this edition is more accessible for the reader.

3. W. Meacham, "The Authentication of the Turin Shroud: An Issue in Archaeological Epistemology," *Current Anthropology* 24.3, June 1983, p. 285; P. Barbet, *A Doctor at Calvary* [100-120]; Stevenson and Habermas, *Verdict on the Shroud* (Ann Arbor, Mich.: Servant Books, 1981) [90-120]; I. Wilson, *The Shroud of Turin*, rev. ed. (Garden City, N.Y.: Doubleday Image Books, 1979), chapter 3, [90-120]; Zugibe, *The Cross and the Shroud*, rev. ed. (New York: Paragon House, 1988); [100-120]; F. C. Tribbe, *Portrait of Jesus?* (New York: Stein and Day/Publishers, 1983), chapter 6; [120]; P.L.B. Bollone, "A Pathologist Observes the Shroud of Turin," in *La Sindone E La Scienze,* Acts of the International Congress of Sindonology, Second Edition, Turin, Centro Internazionale diSindonologia, Edizoni Paoline, 1978 [over 100]; and M. Straiton, "Evidence That the Body was Placed in the Holy Shroud after Death Had Occurred," in *La Sindone Scienza E Fede,* Acts of the II National Congress of Sindonology in Bologna, 1981, CLUEB, 1983. Italian Scientists Giulio Fanti and Barbara Faccini count as many as 372 bloodstains. "New image processing of the Turin Shroud scourge marks," *International Workshop on the Scientific Approach to the Archeiropoietos Images*, Frascati, Italy, May 4-6, 2010, pp. 47-54.

4. Barbet, *Calvary*; R. Bucklin, "Viewpoint of a Forensic Pathologist," *Shroud Spectrum International* I.5 (December 1982): 3-10; Don Lynn (image enhancement specialist and STURP scientist), personal communication, 1985; and Vernon Miller, STURP photographer, personal communications, December 1997.

5. Meacham, "Authentication"; I. Wilson, *Shroud of Turin and The Mysterious Shroud* (Garden City, NY: Doubleday, 1986); T. Humber, *The Sacred Shroud* (New York: Pocket Books, 1978); and R. K. Wilcox, *Shroud* (New York: Bantam Books, 1979); F Manservigi and E. Morini, "The Hypothesis About the Roman Flagrum: Some clarifications," *Shroud of Turin: The Controversial Intersection of Faith and Science, International Conference*, St. Louis, MO, October 9–12, 2014, http://www.shroud.com/pdfs/stlmanservigipaper.pdf.

6. Dr. P. Scotti, as cited by Dr. Alan Adler in "The Turin Shroud Lecture," Department of Chemistry, Queen Mary College, London, July 20, 1984; Adler, "Chemical Investigation on the Shroud of Turin in *The Mystery of the Shroud of Turin Interdisciplinary Symposium* video, Elizabethtown, Penn.: Elizabethtown College, February 15, 1986.

7. Adler, "Lecture" and "Chemical Investigations"; Heller, *Report*, p. 185-6; Heller and Adler, "A Chemical Investigation"; Adler, as stated in T. Case, *The Shroud of Turin and the C-14 Dating Fiasco*, (Cincinnati, OH: White Horse Press, 1996); Jumper, as stated in Adler, "Chemical Investigations"; Gonella, "Scientific Investigation of the Shroud of Turin: Problems, Results and Methodological Lessons," in *Turin Shroud – Image of Christ?* (Hong Kong: Cosmos Printing Press Ltd., 1987), pp. 29-40; and V. D. Miller and S. F. Pellicori, "Ultraviolet Fluorescence Photography of the Shroud of Turin," *Journal of Biological Photography* 49.3 (July 1981)."

8. Adler, "Lecture" and "Chemical Investigations"; and personal communication with Alan Adler, July 18, 1985.

9. Barbet, *Calvary*; D. Willis, cited in Wilson, *The Shroud of Turin*; G. Judica-Cordiglia, cited in G. Caselli, "Ascertainments of Modern Medicine on the Imprints of the Holy Sindon," in *La Santa Sindone;* R. W. Hynek, *True Likeness* (London and New York: Sheed and Ward, 1951); Zubige, *The Cross and the Shroud*; and G. R. Lavoie, "The Medical Aspects of the Shroud of Turin as Seen by a Practicing Physician," in *The Mystery of the Shroud of Turin an Interdisciplinary Symposium*, video, Elizabethtown, Penn.: (Elizabethtown College, February 15, 1986).

10. Barbet, *Calvary*, pp. 21, 91; Bucklin, "Viewpoint of a Forensic Pathologist"; R. Bucklin, in *The Silent Witness*; directed by David W. Rolfe, produced by Screenpro Films (London), 1978; G. Judica-Cordiglia, cited in Caselli, "Ascertainments"; Hynek, *True Likeness*; and P.L.B. Bollone, "A Pathologist Observes the Shroud of Turin," in *La Sindone E La Scienze*, Acts of the International Congress of Sindonology, Second Edition, Turin, Centro Internazionale di Sindonologia, Edizoni Paoline, 1978.

11. Bucklin, "Viewpoint of a Forensic Pathologist"; and Heller, *Report*; See also S. F. Pellicori and M. Evans, "The Shroud of Turin Through the Microscope," *Archaeology* (January/February 1981): 32-43.

12. S. Rodante, "The Coronation of Thorns to the Light of the Shroud," *Shroud Spectrum International* 1 (December 1981): 4-24, Translated and reprinted from *Sindon* 24 (Oct. 1976).

13. Bucklin, "Viewpoint of a Forensic Pathologist"; R. Bucklin and J. Gambescia, in Shroud of Turin Research Project *Update* 3.1 (Fall 1981); Rodante, "The Coronation of Thorns"; Barbet, *Calvary*; P. Barbet, "Proof of the Authenticity of the Shroud in

the Bloodstains" (1950), *Shroud Spectrum International* 1 (December 1981), excerpt reprinted from Sindon 14-15, (December 1970): p. 31; Caselli, "Ascertainments"; Bollone, "A Pathologist Observes"; J. M. Cameron, "The Pathologist and the Shroud: in *Face to Face with the Turin Shroud*, ed. P. Jennings (London: Mayhaus-McCrimmon, 1978), 57-59; Zugibe, *The Cross and the Shroud*; and D. Willis, cited in Wilson, *Shroud of Turin*.

14. Rodante, "Coronation of Thorns," and Caselli, "Ascertainments."

15. Rodante, "Coronation of Thorns." Dr. Rodante's work was greatly influenced by that of Dr. Caselli, who had thoroughly studied all of the wounds previously.

16. S. Rodante, "Coronation of Thorns," p. 8.

17. S. Rodante, "Coronation of Thorns."

18. Ibid. F. La Cava, *Cristo illustrate dalla scienze medica* (Naples: D'Auria, 1953); and Zugibe, The Cross and the Shroud, pp. 24-27.

19. Bucklin, "Viewpoint of a Forensic Pathologist"; Bucklin, "An Autopsy on the Man of the Shroud." *Acts du Illeme symposium scientifique International* – Nice, 1997, Centre International d' Etudes surle Linceul de Turin, pp. 99-101; Barbet, *Calvary*; Bollone, "A Pathologist Observes"; Wilson, *The Shroud of Turin*, and E. W. Massey, "An Interpretation of the Hand and Arm Markings of the Shroud of Turin," *The Hand* 12.1 (1980).

20. John 20:25, 27.

21. According to A. O'Rahilly, *The Crucified* (Dublin, Ireland: Kingdom Books, 1985), p. 137: "In Greek the word *cheir* usually means not hand but arm. So Homer, *Iliad* xi.252; Hesiod, *Theog.* 150; Euripides, *Iph.* 1404; etc. In Xenophon, *Anab* i.5,8, it means wrist (bracelets on their wrists). Hippocrates uses *akre cheir* for forearm. Arm is probably the meaning in Heb. 12:12."; Tribbe, *Portrait of Jesus?*, F. T. Zugibe, *The Cross and the Shroud* also cites Zorell's *Lexicon Herbraicum et Armaicum* and Lidell-Scott's *Greek-English Lexicon*, 7th ed. (New York: Harper Brothers, 1883).

22. V. Tzaferis, "Crucifixion – The Archaeological Evidence," *Biblical Archaeology Review* 11 (January/February 1985): 44-53, p. 52.

23. Tzaferis, "Crucifixion"; and N. Haas, "Anthropological Observations on the Skeletal Remains from Giv'at Ha-mivtar," *Israel Exploration Journal* 20.1/2 (1970): 8-59.

24. John 19:36; Psalms 34:20. Jesus is interpreted as the Passover lamb who takes away the sins of the world (John 1:29; 1 Cor 5:7; 1 Pet 1:19) and thus further prophesized in the Old Testament (Exod. 12:46; Num 9:12) according to John.

25. Barbet, *Calvary*. Alternatively, Dr. Zugibe suggests the nail could have passed through the wrist "through a space created by four other carpal bones. . . where "the trunk of the median nerve would most likely be damaged by this path." F. Zugibe, *The Cross and the Shroud*, p. 63, and "Pierre Barbet Revisited," http://www.shroud.com/zugibe.htm, p. 7.

26. It would not seem significant if you saw a crucifixion victim's thumbs pointed toward his palms while he was nailed on the cross by his wrists. The absence of the thumbs can only be appreciated when you view the top of the hands and the wrists. Most Roman crucifixion victims were left hanging on the cross after their death to be scavenged by wild animals and birds and then thrown onto a common heap. Only if a person received an immediate burial and/or had his hands folded down to his body and his picture taken, like the man in the Shroud, could the absence of his thumbs likely even be appreciated.

27. Bucklin, "Viewpoint of a Forensic Pathologist," and "An Autopsy," p. 100; Barbet, *Calvary*; Willis, cited in Wilson, *Shroud of Turin*; Caselli, "Ascertainments"; Bollone, "A Pathologist Observes"; Hynek, *True Likeness*; and Cameron, "The Pathologist."

28. Barbet, *Calvary*, p. 98; Willis, cited in Wilson, *Shroud of Turin*; Cameron, cited in Wilson. *The Mysterious Shroud*; and E. A. Wuenschel, *Self-Portrait of Christ* (Esopus, N.Y.: Holy Shroud Guild,1957), p. 34.

29. G. Judica-Cordiglia, cited in Wilson, *Shroud of Turin*, p 39, and in Barbet, *Calvary*, p. 97; Lavoie, "The Medical Aspects"; and Barbet, *Calvary*.

30. Miller and Pellicori, "Ultraviolet Fluorescent Photography"; and Heller, *Report*.

31. Heller, *Report*, pp. 112, 152; S. F. Pellicori, "Ultraviolet Fluorescent Photography"; S. F. Pellicori and M. Evans, "The Shroud of Turin Through the Microscope," *Archaeology* (January/February 1981): 32-43; J. Jackson, cited in C. Murphy, "Shreds of Evidence," *Harper's* 263.1578 (November 1981): 42-65. See also Wilson, *The Blood and the Shroud* (New York: The Free Press, 1998), p. 95.

32. Cameron, "The Pathologist," p. 58; personal communication with St. Louis medical examiner Dr. Mary Case, April 4, 1984; R. Bucklin, "Postmortem Changes and the Shroud of Turin," *Shroud Spectrum International* 14 (March 1985): 3-6, and "The Legal and Medical Aspects of the Trial and Death of Christ," *Medicine, Science and the Law* 10.1 (1970); Barbet, *Calvary*, p. 159; and Lavoie, "Medical Aspects."

33. Barbet, *Calvary*, p. 135; Caselli, "Ascertainments"; Dr. L. Gedda as stated in Caselli, "Ascertainments"; and Cameron, "The Pathologist."

34. Barbet, *Calvary*, p. 135. Several Italian scientists have recently concluded that the man in the Shroud suffered, among other things, a dislocation of the humerus on the right side and lowering of the shoulder caused by a violent blunt trauma from behind. M. Bevilacqua, G. Fanti, M. D'Arienzo, R. De Caro, "Do we really need new medical information about the Turin Shroud?" *Injury*, vol. 45, issue 2, pps. 460-464, February, 2014.

35. Miller and Pellicori, "Ultraviolet Fluorescent Photography"; A. Battaglini, "Considerations on the Feet of the Man in the Shroud." *Shroud Spectrum International* 9 (December 1983): p. 6; Vignon, as stated in an excerpt from *Shroud Spectrum International*, No. 9, p. 6; and Barbet, *Calvary*.

36. Bucklin, "Viewpoint of a Forensic Pathologist"; Barbet, *Calvary*; Battaglini, "Considerations"; Judica-Cordiglia, excerpt in *Shroud Spectrum International* 9, (December 1983): 6; Caselli; "Ascertainments"; Bollone, "A Pathologist Observes"; Lavoie, "Medical Aspects"; and Vignon, "The Problem of the Holy Shroud," *Scientific American* 93.163 (1937): 162-64.

37. Bucklin, "Viewpoint of a Forensic Pathologist," p. 8.

38. Barbet, *Calvary*, pp. 125, 183; Caselli, "Ascertainments"; Judica-Cordiglia, excerpted in *Shroud Spectrum International* 9 (December 1983): 6; and Vignon, "Problem."

39. Barbet, *Calvary*, pp. 121, 125-26; Caselli, "Ascertainments"; D. Willis, "Did He Die on the Cross?" *Ampleforth Journal* 74 (1969); Hynek, *True Likeness*; Straiton, "Evidence That the Body"; and Vignon, "Problem."

40. Bucklin, "Viewpoint of a Forensic Pathologist," p. 8, and "An Autopsy," p. 99; Barbet, *Calvary*, p. 144; Caselli; "Ascertainments"; Willis, "Did He Die?"; Hynek, *True Likeness*; and Bollone, "A Pathologist Observes."

41. Meacham, "Authentication," p. 290; Wilson, *Shroud of Turin* and *The Mysterious Shroud*; Barbet, *Calvary*; Humber, *The Sacred Shroud*; and Wilcox, *Shroud*.

42. A. F. Sava, "The Wound in the Side of Christ," *Catholic Biblical Quarterly* 19 (1957): 343-347; A. F. Sava, "The Blood and Water from the Side of Christ," *American Ecclesiastical Review* 138 (1958): 341-45; Bucklin, "Viewpoint of a Forensic Pathologist"; Barbet, *Calvary*; Judica-Cordiglia, cited in Bucklin, supra; and L. L. Gomez, "Legal Medical Study of the Wound on the Side of Christ," in *La Santa Sin-*

done, Nelle Richerche Moderne, Acts of the First National Congress of Studies, Turin 1939.

43. Bucklin, "Viewpoint of a Forensic Pathologist," pp. 8, 9.

44. See for example Vignon, *The Shroud of Christ*, 1902; Barbet, Les Cinq Plaies du Christ (The Five Wounds of Christ), 1935; Caselli, "Ascertainments"; 1939; Barbet, *Calvary*, 1953; Bulst, *The Shroud of Turin* (Milwaukee: Bruce, 1957); Wuenschel, *Self-Portrait of Christ*, 1957; D. Willis, "Did He Die on the Cross?" (1969); and "False Prophet and the Holy Shroud," 1970; Bucklin, "The Legal and Medical Aspects," 1970; Rodante, "Coronation of Thorns," 1976.

45. Barbet, *Calvary*, p. 16.

46. Ibid. p. 29; Yves Delage proposed testing the cloth in 1902, as did scientists at the Sorbonne through Baron Manno, the former president of the 1898 exhibition. Scientists and physicians through Vignon also requested such tests. The 1939 Congress of the Italian Commission and the 1950 International Congress at Rome further passed resolutions to this effect. Wuenschel, *Self Portrait*; Caselli, "Ascertainments"; and Walsh, *The Shroud* (New York: Random House, 1963).

47. Miller and Pellicori, "Ultraviolet Fluorescent Photography," pp. 76-83; and K. F. Weaver, "The Mystery of the Shroud," *National Geographic* 157.6 (June 1980); A. D. Adler, "The Origin and Nature of Blood on the Turin Shroud," [Excerpts from lecture of the Dept. of Anatomy, Univ. of Hong Kong, March 3, 1986] in *Turin Shroud – Image of Christ?*, pp. 57-59.

48. Heller and Adler, "A Chemical Investigation"; Adler, "Lecture" and "Chemical Investigations"; Heller, *Report*; Jumper, Adler, Jackson, Pellicori, Heller and Druzik, "A Comprehensive Examination of the Various Stains and Images on the Shroud of Turin," in *ACS Advances in Chemistry No. 205 Archaeological Chemistry III,* American Chemical Society (1984).

49. A. D. Adler, "Updating Recent Studies on the Shroud of Turin," *Archaeological Chemistry: Organic, Inorganic, and Biochemical Analyses*, Mary Virginia Orna, ed. American Chemical Society (1996), pp. 223-228.

50. A. D. Adler, "The Nature of the Body Images on the Shroud of Turin," *Shroud of Turin International Conference*, Richmond, Va., June 18-20, 1999, p. 2, http://www.shroud.com/; see also Adler, "The Nature and Origin," p. 57.

51. S. F. Pellicori, "Spectral Properties of the Shroud of Turin," *Applied Optics* 19.12 (1980): 1913-20.

52. Heller, *Report*; and Adler, "Lecture" and Chemical Investigations."

53. P. L. B. Bollone and A. Gaglio "Technical Immune-Enzymatic Application of the Shroud Drawings," presented at Third National Meeting of Studies on the Shroud, October 13-14, 1984; P. L. B. Bollone, M. Jorio, and A. L. Massaro, "Defining the Blood Group Identified on the Shroud," in *La Sindone Scienza E. Fede*, p. 178; P. L. B. Bollone, M. Jorio, and A. L. Massaro, "Identification of the Traces of Human Blood on the Shroud," *Shroud Spectrum International* 6 (March 1983): 3-6; and P. L. B. Bollone and A. Gaglio, "Demonstration of Blood, Aloes and Myrrh on the Holy Shroud with Immunofluorescence Techniques," *Shroud Spectrum International* 13 (December 1984). See also Adler, "Lecture" and "Chemical Investigations."

54. *The Mysterious Man of the Shroud*, directed by Terry Landeau, CBS documentary, aired April 1, 1997; L. A. Garza-Valdes, *The DNA of God?* (London: Hodder & Stoughton, 1998), pp. 41-42.

55. Heller, *Report*, p. 210.

56. Barbet, *Calvary*; Caselli, "Ascertainments"; Straiton, "Evidence That the Body"; Hynek, *True Likeness*; Lavoie, "Medical Aspects"; and R. Bucklin, "An Autopsy," p. 99. Experiments with volunteers hanging in a crucifixion position have also produced similar results. See experiments of artist/scientist Isabel Piczek in Wilson, *The Blood and the Shroud*, p. 23, and those of Dr. Frederick Zugibe in *The Cross and the Shroud*, p. 108.

57. Barbet, *Calvary*; W. D. Edwards, E. J. Gabel, F. E. Hosmer. "On the Physical Death of Jesus Christ," *JAMA* 255.11, (March 21, 1986); Hynek, *True Likeness*; Wuenschel, *Self-Portrait*; and Tzaferis, "Crucifixion."

58. Bucklin, "Postmortem Changes," "Legal and Medical Aspects," "Viewpoint of a Forensic Pathologist," "A Pathologist's Viewpoint," and "An Autopsy"; Caselli, "Ascertainments"; Cameron, "The Pathologist"; Hynek, *True Likeness* and "The Real Cause of Crucifixion Death and Cadaveric Rigidity," in *La Santa Sindone*; Masera "The Work of Medical Jurisprudence on the Marks of the Holy Shroud," in *La Santa Sindone*; Straiton, "Evidence That the Body"; Zugibe, *The Cross and the Shroud*; Bollone, "A Pathologist Observes"; Barbet, *Calvary*; Bruckner, "Some Observations on the Medical Aspects of the Shroud of Turin," *Linacre Quarterly* 54 (May 1987): 58-62; Lavoie, "The Medical Aspects"; Dr. William Drake, chief of Pathology, Missouri Baptist Hospital, personal communications in 1985 and 1986.

59. Bucklin, "Postmortem Changes"; personal communications with pathologist Dr. William Drake, 1986-86; personal communication with medical examiner Dr. Mary Case, April 4, 1984; personal communication with pathologist Dr. Dan McKeel in April 1984; Zugibe, *The Cross and the Shroud*, p. 131; and Brucker, "Some Observations."

60. Caselli, "Ascertainments"; Bucklin, "Postmortem Changes"; Brucker, "Some Observations"; Zugibe, *The Cross and the Shroud*; and personal communication with medical examiner Dr. Mary Case, April 4, 1984.

61. Bucklin, "Postmortem Changes," pp. 3-6, "Legal and Medical Aspects," and "An Autopsy," pp. 99-100; Barbet, *Calvary*, p. 122; personal communications with pathologist Dr. William Drake, 1985- 86; Bollone, "The Pathologist Observes"; Straiton, "Evidence That the Body"; Hynek, *True Likeness*; Brucker, "Some Observations"; Zugibe, *The Cross and the Shroud*; Lavoie, "Medical Aspects"; and Caselli, "Ascertainments."

62. Caselli, "Ascertainments"; personal communications with pathologist Dr. William Drake, 1985-86; Hynek, *True Likeness*; Barbet, *Calvary*, p. 87; Brucker, "Some Observations"; and Lavoie, "Medical Aspects."

63. Bucklin, "Postmortem Changes"; Hynek, *True Likeness*; Straiton, "Evidence That the Body"; and Zugibe, *The Cross and the Shroud*.

64. Barbet, *Calvary*; Caselli, "Ascertainments"; Straiton, "Evidence That the Body"; Hynek, *True Likeness*; and Lavoie, "The Medical Aspects."

65. Caselli, "Ascertainments"; personal communications with pathologist Dr. William Drake, 1985-86; Cameron, "The Pathologist"; Hynek, *True Likeness*; Straiton, "Evidence That the Body"; and Barbet, *Calvary*.

66. Barbet, *Calvary*, p.119; Heller, *Report*; R. Bucklin, "A Pathologist Looks at the Shroud of Turin," in *La Sindone E La Scienza*, p. 117, and "Legal and Medical Aspects"; Bollone, "A Pathologist Observes"; personal communications with pathologist Dr. William Drake, 1985-86; M. Bocca, E. Messina and S. Salvi. "Comments on the Wounds of Anatomic-Function of a Nailed Hand, with Reference to the Sindon of Turin," *La Sindone E La Scienza*, 1983; and Lavoie, "Medical Aspects."

67. Barbet, *Calvary* [110 lbs]; Wilson, *The Shroud of Turin* [110 lbs.]; Edwards, Gabel, and Hosmer, "On the Physical Death," [75-125 lbs]; Bucklin, "Legal and Medical Aspects" [80 lbs]; and Humber, *The Sacred Shroud* [80 lbs].

68. Barbet, *Calvary*; Caselli, "Ascertainments"; Bucklin, "Legal and Medical Aspects"; Edwards, Gabel, and Hosmer, "On the Physical Death"; P. J. Smith, "Appendix II" in Barbet, *Calvary*, p. 119.

69. Bucklin and Gambescia, in *Update*; Bucklin, "Legal and Medical Aspects," p. 24; Caselli, "Ascertainments"; Smith, "Appendix II" in Barbet, *Calvary*, p. 119; C. T. Davis, "The Crucifixion of Jesus: The Passion of Christ from a Medical Point of View," *Arizona Medical* (1965): 183-87; Bloomquist, "A

Doctor Looks at Crucifixion," *Christian Herald* (March 1964): 46-48.

70. La Cava, "La Passione"; Rodante, "Coronation of Thorns." See also Zugibe, *The Cross and the Shroud*, pp. 24-27.

CHAPTER THREE

1. R. K. Wilcox, *Shroud* (New York: Bantam Books, 1979), p. 64. One physician, Dr. Gilbert Lavoie, duplicated some of the features of one Shroud blood mark on linen, the epsilon shaped clot seen on the man's forehead. While Dr. Lavoie is a staunch advocate of the authenticity of the Shroud, his best samples do not even duplicate all the features or characteristics of this one blood mark, let alone all the features on all the blood marks on the Shroud. As discussed below, Dr. Lavoie's blood mark lacks several critical features found on the Shroud's blood marks. Dr. Lavoie's blood mark lacks human whole blood serum around the edges. His blood mark has not stayed red, let alone remained red after the passage of many centuries. His best blood clots were not even transferred from a human body, let alone transferred in the same shape and form as when they flowed and coagulated on a human body. Nor does Dr. Lavoie's blood mark appear embedded into the cloth sample like the Shroud's blood marks. The Shroud's blood marks are visible on the outer or opposite side of the cloth in the same or similar shape and form as they appear on the inner side of the cloth. Dr. Lavoie's blood mark does not demonstrate these necessary qualities.

Unlike most of the blood marks on the Shroud, the epsilon blood mark produced on cloth by Dr. Lavoie lacks serum surrounding its border. This necessarily fails to duplicate an important feature of the Shroud's blood marks. Lavoie's best results were obtained when he first placed blood horizontally on saran (plastic) wrap for ½ hour. After these blood samples clotted, and while still on the Saran Wrap, they were placed in the vertical position. Linen pieces of cloth were then taped vertically to the blood clot samples opposite of the Saran Wrap at ½ hour intervals. All of the transfer occurs well within this time period according to Dr. Lavoie. (After four hours in the vertical position, the samples were then placed in the horizontal position with the Saran Wrap being removed about 16 to 20 hours later.) From the time the clots were placed in the vertical position (still adhered to the Saran Wrap), their serum continued to drip from the clots for 40 minutes. Dr. Lavoie's best results occurred when the serum dripped off his Saran Wrap samples. Serum has been identified throughout the Shroud's pristine blood marks, including those on the front and back of the head. The only time Dr. Lavoie's samples were placed on warm human skin (horizontally during his fourth procedure), the serum dried and little of it dripped away when later placed vertically and taped to cloth samples. Dr. Lavoie also moistened these clots, but these transfers appeared the least like the epsilon clot (G. R. Lavoie, B. B. Lavoie, V. J. Donovan, and J. S. Ballas, "Blood on the Shroud of Turin: Part 2," *Shroud Spectrum International* 8 (September 1983): 2-10).

The man in the Shroud would have initially received most or all of his approximately 130 wounds while he was in the vertical position. He would not have initially received most or all of them in the horizontal position, which is the initial point for all of Dr. Lavoie's samples. Furthermore, the linen Shroud cloth would never have been placed on the man while he was in the vertical position, but only when he was in the horizontal position, as long as 12 hours after he received his first wounds. None of Dr. Lavoie's procedures involved a small wound being inflicted (or blood even placed) on human skin while in the vertical position, with cloth then laid over the blood mark in the horizontal position. For these and other reasons, Dr. Lavoie necessarily failed to produce blood marks on cloth that were mirror images of clots that formed naturally on skin.

The man's blood marks are so embedded in the Shroud of Turin that they also appear on the outer (or opposite) side of the cloth in the same approximate shape and form as they appear on the inner side of the cloth. This was confirmed by photographs taken of the outer side of the Shroud in 2002, when its backing cloth from 1534 was removed (and later replaced with a new backing cloth). The man's head wounds on the front of his head, including the epsilon clot on his forehead, are quite visible on photographs of the outside of the Shroud cloth. Blood marks from the back of the man's head; his wrist; the flows on both of his arms; the large side wound; the flow across the small of his back; flows from both the front and back of his feet; as well as the scourge marks from his back and the back of his legs, all appear on the outer side of the Shroud of Turin. Dr. Lavoie's blood clots on cloth are not shown to be visible on the outer or opposite side of his cloth sample, in the same approximate shape as they appear on the inner side, which is another critical feature of the Shroud's blood marks that must be duplicated.

I have gone to a little length here to concentrate on the good Dr. Lavoie's results. Based on erroneous interpretations and misunderstandings of his very limited procedures, some in the Shroud community erroneously concluded the Shroud's blood marks could have naturally resulted from direct contact between the body and the cloth. Jelly-like coagulated blood has never been known to have transferred experimentally or naturally from a body to cloth, let alone become embedded in the cloth. Furthermore, these blood marks would have to be transferred without becoming broken or smeared at their edges, both when they were embedded into the cloth and when it was removed from the body. The Shroud's blood marks range from the faintest and smallest scourge marks that were incurred among the earliest of his wounds, to the heaviest and most recent postmortem blood flows, which also involved a considerable amount of watery fluid from the pleural cavity. Yet, all the various coagulated blood marks and their serum (and the watery fluid) appear to have transferred into the cloth in the same original condition as when they formed and flowed on the body. I have known Dr. Lavoie for decades; he is an excellent Shroud investigator and doctor. Yet, his procedures and results cannot be accurately interpreted to mean the Shroud's blood stains and blood marks were naturally transferred by direct contact with the body any more than can Dr. Barbet's, Dr. Vignon's or any other inferior results.

Physicist Arthur Lind recently conducted the most extensive experiments to date to encode blood marks in cloth by direct contact and other naturalistic and artistic attempts. Although his methodology was superb, he could not duplicate all the Shroud's extraordinary blood mark features on even one blood mark, let alone on all the blood marks that were encoded on this cloth in different shapes and intensities and in different time periods, after they first flowed from and onto the body. See G. R. Lavoie, B. B. Lavoie, V. J. Donovan, and J. S. Ballas, "Blood on the Shroud of Turin: Part 2," *Shroud Spectrum International* 8 (September 1983): 2-10; Lavoie, *Unlocking the Secrets of the Shroud* (Allen, TX: Thomas More, 1998); the preceding and this chapter; and A. Lind and M. Antonacci "Hypothesis that Explains the Shroud's Unique Blood Marks and Several Critical Events in the Gospels," *Shroud of Turin: The Controversial Intersection of Faith and Science*, International Conference, St. Louis, MO, October 9-12, 2014, http://www.shroud.com/pdfs/stllindpaper.pdf. I would also like to thank George de Kay of M. Evans and Co., Inc. for having allowed me over the years to use the descriptions of the man's blood marks and injuries from my earlier book.

2. P. Barbet, *A Doctor at Calvary*, (Garden City, New York: Doubleday Image Books, 1963), p. 27.

3. Neither John 19:34-37 nor Isaiah 53:5 would have indicated to a forger on which side Jesus was pierced.

4. Sister Damian of the Cross [E. L. Nitowski], *The Field and Laboratory Report of the Environmental Study of the Shroud in Jerusalem* (Salt Lake City: Carmelite Monastery, 1986).

5. J. P. Jackson, E. J. Jumper, and W. R. Ercoline, "Correlation of Image Intensity on the Turin Shroud with the 3-D Structure of a Human Body Shape," *Applied Optics* 23.14 (July 1984): 244-70; and J. P. Jackson, E. J. Jumper, and W. R. Ercoline, "Three-Dimensional Characteristic of the Shroud Image," *IEEE 1982 Proceedings of the International Conference on Cybernetics and Society* (October 1982); 559-575; E. J. Jumper, "Considerations of Molecular Diffusion and Radiation as an Image Formation Process on the Shroud," *Proceedings of the 1977 United States Conference of Research on the Shroud of Turin* (Albuquerque, N. M.: Holy Shroud Guild, March 1977), 182-89.

6. S. F. Pellicori, "Spectral Properties of the Shroud of Turin," *Applied Optics*, 19.12 (1980): 1913-1920; S. F. Pellicori and M. Evans, "The Shroud of Turin Through the Microscope," *Archaeology* (January/February 1981): 32.43. Several writers postulate that the large quantity of myrrh and aloes mentioned in John 19:40 were in the form of dry block or in powdered or granulated form. H. Daniel-Rops, *Daily Life in Palestine at the Time of Christ* (London: Weidenfeld, 1962); H. Daniel-Rops, *Daily Life in the Time of Jesus* (New York: Hawthorn Books, 1962); Wilson, *Shroud of Turin*, rev. ed. (Garden City, N.Y.: Doubleday Image Books, 1979), pp. 56-57; P. Barbet, *A Doctor at Calvary*, (Garden City, New York: Doubleday Image Books, 1963), pp. 163-

165; F. C. Tribbe, *Portrait of Jesus*? (New York: Stein and Day Publishers, 1983), pp. 89-92; J. A. T. Robinson, "The Shroud of Turin and the Grave-Clothes of the Gospels, in *Proceedings*, pp. 24-25; P. Vignon, "The Problem of the Holy Shroud," trans. E. A. Wuenschel, *Scientific American* 93.163 (1937): 162-164; T. Humber, *The Sacred Shroud* (New York: Pocket Books, 1978), p. 66; E. A. Wuenschel, *Self-Portrait of Christ* (Esopus, N.Y.: Holy Shroud Guild, 1957), p. 48; and J. Marino, *The Burial of Jesus and the Shroud of Turin* (Anaheim: Jeff Richards, 1999). The amount of spices mentioned in John would be about 50-75 lbs. in our weight today. Such a large amount would have been used to postpone purification temporarily until the provisional burial was completed. Such a large amount of myrrh and aloes in any of the above forms could also have been placed around the enshrouded body.

7. L. A. Schwalbe and R. N. Rogers, "Physics and Chemistry of the Shroud of Turin," *Analytica Chimica Acta* 135 (1982): 3-49. If this man was Jesus, or another crucifixion victim that received an individual burial, any bodily fluids such as perspiration, would likely have evaporated before burial. The evaporation process would have begun not long after death while he was still on the cross. Joseph of Arimathea or someone else would also had to have gone to Pilate's office and wait to secure an official order allowing him to take the body from the cross for an individual burial. (Crucifixion victims were normally left on the cross after death or thrown on a heap to be devoured by scavenging animals or birds of prey.) Joseph or another person would then have to stop and purchase a new linen cloth in the (crowded Passover/Sabbath) market place. He would then have to take the body down and transport it to the burial tomb.

8. Jackson, et al., "Correlation" and "Three Dimensional Characteristic."

9. Jackson, et al., "Correlation" and "Three Dimensional Characteristic"; Jumper, "Considerations of Molecular Diffusion."

10. P. Vignon, *The Shroud of Christ* 1902 (New Hyde Park, NY; University Books, 1970); Barbet, *Calvary*; Jumper, "Considerations of Molecular Diffusion."

11. Jackson, et al., "Correlation" and "Three Dimensional Characteristic."

12. Jackson, et al., "Correlation"; E. J. Jumper, A. D. Adler, J. P. Jackson, S. F. Pellicori, J. H. Heller, and J. R. Druzik, "A Comprehensive Examination of the Various Stains and Images on the Shroud of Turin," *ACS Advances in Chemistry No. 205 Archaeological Chemistry* (1984): 447-476.; S. F. Pellicori, "Spectral Properties of the Shroud of Turin," *Applied Optics* 19.12 (1980): 1913-1920; and E. Jumper, cited in C. Murphy, "Shreds of Evidence," *Harper's* 263.1578 (November 1981): 42-65.

13. J. H. Heller, *Report on the Shroud of Turin* (New York: Houghton Mifflin, 1983), p. 208.

14. Jackson et al., "Correlation" and "Three-Dimensional Characteristic." See also Jumper, "Considerations of Molecular Diffusion."

15. S. Rodante, "The Imprints of the Shroud Do Not Derive Only from Radiations of Various Wavelengths," *Shroud Spectrum International* 7 (June 1983): 21-24; S. Rodante, "Oily Migma and Shroud Marks. Exclusion of Apparent Death," in *La Sindone Scienz E Fede* (a cura di, Lamberto, Coppini e Francesco Svazzuti, 1978); Vignon, *The Shroud of Christ.*

16. J. Jackson, "Blood and Possible Images of Blood on the Shroud," *Shroud Spectrum International* 24 (Sept. 1987): 3-11. Jackson calls for further testing of the Shroud to confirm these observations and whether they exist on any of its other blood marks. Examining the Shroud at the molecular level with multi-spectral imaging could help resolve these questions as well.

17. G. F. Carter, "Formation of the Image on the Shroud of Turin by X Rays: A New Hypothesis," *ACS Advances in Chemistry No. 205 Archaeological Chemistry* III, J. B. Lambert, ed. American Chemical Society, 1984, pp. 425-446.

18. See for example, Dr. Alan Whanger, surgeon, in M. and A. Whanger, *The Shroud of Turin: An Adventure of Discovery* (Providence House Publishers, Franklin, Tenn., 1998), pp. 111-115; A. D. Accetta, "Nuclear Medicine and Its Relevance to the Shroud of Turin," "*Sindone 2000" Orvieto Worldwide Congress*, Italy (August 2000); A. D. Accetta, "Experiments with Radiation as an Image Formation Mechanism," *Shroud of Turin International Research Conference*, Richmond, Va., June 18-20, 1999; A. D. Accetta, Lecture at Dallas Shroud Meeting, Dallas, Tex., November 6-8, 1998; J. P. Jackson, "Is the Image on the Shroud Due to a Process Heretofore Unknown to Modern Science?" *Shroud Spectrum International* 34 (March 1990): 3-29, 18; J. P. Jackson, "An Unconventional Hypothesis to Explain All Image Characteristics Found on the Shroud Image," *History, Science, Theology and the Shroud*, A. Berard, ed. (St. Louis: Richard Nieman, 1991), pp. 325-344, 333-335; Michael Blunt, Professor of Anatomy, University of Sydney in I. Wilson, *The Blood and the Shroud* (New York: The Free Press, 1998), p. 29.

19. Dr. John Jackson, personal communications, 1991-1994; M. and A. Whanger, *The Shroud of Turin*, pp. 116-117; Carter, "Formation of the Image," p. 433; A. Accetta, K. Lyons, J. Heiserodt, J. Jackson, B. Farley, "Nuclear Radiation and the Shroud: Head Image," *Second International Dallas Conference*, October 25-28, 2001.

20. Accetta, "Nuclear Medicine and Its Relevance," "Experiments with Radiation" and Dallas lecture; Jackson, "An Unconventional Hypothesis," pp. 333-335; Jackson, personal communications, 1991-1994.

21. Carter, "Formation of the Image," pp. 433-434; M. and A. Whanger, *The Shroud of Turin*, pp. 117-118; Dr. Accetta, personal communication, August 9, 1999.

22. Carter, "Formation of the Image," p. 433.

23. M. and A. Whanger, *The Shroud of Turin*, p. 118.

CHAPTER FOUR

1. T. W. Case, *The Shroud of Turin and the C-14 Dating Fiasco* (Cincinnati: White Horse Press, 1996), p. 93.

2. T. Grieder and A. B. Mendoza, "La Galgada—Peru Before Potter," *Archaeology* (March/April 1981).

3. *Frontiers of Science*, May/June 1981, cited in F.C. Tribbe, *Portrait of Jesus?* (New York: Stein and Day, 1983), p. 154.

4. R. Gilbert, Jr., and M. M. Gilbert, "Ultraviolet-Visible Reflectance and Fluorescence Spectra of the Shroud of Turin," *Applied Optics* 19:12 (June 1980): 1930-1936.

5. G. Raes, "Examination of the 'Sindone,'" in *Report of the Turin Commission on the Holy Shroud*, trans., ed. M. Jepps, (London: Screenpro Films, 1976), pp. 79-83.

6. I. Wilson, *The Mysterious Shroud* (Garden City, N.Y.: Doubleday, 1986).

7. F. Testore, *Le Sainte Suiare, Examine et prelevements effectues le 21 April 1988,* Symposium, Paris, France, 7-8 September 1989, p. 5.

8. A. Braulik, "Altaegyptische Weberei," *Dinglers Polytechnisches Journal* 311 (1899): 13, Fig. 2, cited in E. A. Wuenschel, "The Truth About the Holy Shroud," *American Ecclesiastical Review* (1953): 10.

9. A. Braulik, "Altaegyptische Weberei," p. 31, Fig. 2, cited in Wuenschel, "Truth," p. 10.

10. A. Braulik, *Altaegyptische Gewebe* (Stuttgart: A. Bergstraesser, 1900), p. 15, Fig. 22, cited in Wuenschel, "Truth," p. 10.

11. A. Braulik, *Altaegyptische Gewebe*, Fig. 27.

12. L. Roth, *Ancient Egyptian and Greek Looms* (Bankfield Museum Notes, 2nd series, no. 2, 1913, pp. 24, 84-96, figs. 6, 7, 9, 10), cited in Wuenschel, "Truth," p. 10. See

also D. Fulbright, "Alkeldama repudiation of Turin Shroud omits evidence from the Judean Desert," *International Workshop on the Scientific Approach to the Archeiropoietos Images*, Frascati, May 4-6, 2010.

13. M. Flury-Lemberg, "Notiz zum Fischgratmuster auf dem Leinentuch," *Sindon* No. 19-20, pp. 121-124 (2003), as stated in Fulbright, "Akeldama repudiation."

14. According to P. Savio, in *Report of the Turin Commission* and "Alcuni documenti sopra in Santa Sindone," La Santa Sindone nella ricerche monderne, L.I.C.E., R. Berruti e C., p. 35 (1951), cited in Fulbright, "Alkeldama repudiation."

15. O. Petrosillo and E. Marinelli, *The Egnima of the Shroud: A Challenge to Science* (San Gwann, Malta: PEG, 1996), pp. 198-199.

16. I. Wilson, *The Shroud of Turin*, rev. ed. (Garden City, NY: Doubleday Image books, 1979) and Wilson, *Mysterious Shroud.* See also W. Bulst, *The Shroud of Turin* (Milwaukee: Bruce, 1957), p. 29.

17. Petrosillo and Marinelli, *Enigma of the Shroud,* p. 198.

18. M. Flury-Lemberg, "The Linen Cloth of the Turin Shroud: Some Observations of its Technical Aspects," *Sindon*, No. 16, December 2001, pp. 55-76.

19. Flury-Lemberg, "The Linen Cloth," p. 59.

20. Flury-Lemberg, "The Linen Cloth." I am not stating this is the only time that such a hem is found in history. I am repeating Dr. Flury-Lemberg's recognition that the Shroud's hem matches those from first century Masada to support her conclusion that the Shroud does not contain any weaving or sewing techniques which are inconsistent with its first century origin.

21. Flury-Lemberg, "The Linen Cloth," p. 60. PBS Interview, *Secrets of the Dead: Shroud of Christ?* Produced by Thirteen/WNET New York. © 2004 Educational Broadcasting Corporation. See also D. Edwards, "The Proof that this is the Face of Christ: Fresh Clue Shows Turin Shroud may be Genuine," *London Daily Mirror,* 3 April 2004.

22. Raes, "Examination of the 'Sindone,'" and Curto, "The Turin Shroud: Archaeological Observations Concerning the Material and the Image," in *Report of the Turin Commission.*

23. Wilson, *Shroud of Turin.*

24. *Mishnah*, trans. H. Danby (Oxford: Oxford UP, 1954), Division I Zera'im, Trac-

tate 4 Kila-im, 8:1 and 9:1; Division II Mo'ed, Tractate 7 Betzah, 1:10; Division IV Nizikin, Tractate 5 Makkoth, 3:8-9; Division VI Tohorôth, Tractate 5 Parah, 12:9, and Tractate 12 Uktzin, 2:6, cited in I. Wilson, *Shroud of Turin,* p. 303.

25. E. J. Jumper, A. D. Alder, J. P. Jackson, S. F. Pellicori, J. F. Heller, and J. R. Druzik, "A Comprehensive Examination of the Various Stains and Images on the Shroud of Turin," *Archaeological Chemistry III* 205 (1984): 447-476.

26. Jumper et al., "Comprehensive Examination."

27. *British Society for the Turin Shroud Newsletter,* 8, October 1984.

28. A. Guerreschi and M. Salcito, "Photographic and Computer Studies Concerning the Burn and Water Stains Visible on the Shroud and Their Historical Consequences," *IV Symposium Scientific International*, Paris, 25-26 April, 2002; A. Guerreschi and M. Salcito, "Further Studies on the Scorches and the Watermarks," *Third International Dallas Conference*, September 5-6, 2005. See also *Secrets of the Dead: Shroud of Christ?* Produced by Thirteen/WNET New York. © 2004 Educational Broadcasting Corporation.

29. I. Wilson, *Shroud of Turin*, p. 48.

30. Tribbe, *Portrait?*, p. 84; T. Humber, *The Sacred Shroud* (New York: Pocket Books, 1974), p. 51; G. R. Habermas and K. E. Stevenson, *Verdict on the Shroud* (Ann Arbor, MI: Servant Books, 1981), pp. 36, 118-119; and P. Barbet, *A Doctor at Calvary* (Garden City, N.Y.: Doubleday Image Books, 1963), p. 48. See also Acts 16:22-23, 37-40.

31. Bulst, *Shroud of Turin,* p. 68; A. Pauly, G. Wissowa, and W. Kroll, eds., *Rea Encyclopedia der Klassechen Altertumwissenschaft* (Stuttgart, 1893), entries under "Hasta," "Lancea," and "Pilum," cited by Wilson, *Shroud of Turin*, p. 48.

32. W. Meacham, "The Authentication of the Turin Shroud: An Issue in Archaeological Epistemology," *Current Anthropology* 24.3 (June 1983): p. 290.

33. G. M. Ricci, "Historical, Medical and Physical Study of the Holy Shroud," in *Proceedings of the 1977 United States Conference of Research on the Shroud of Turin,* March 23-24, 1977 (Albuquerque, N.M.: Holy Shroud Guild, 1977), p. 58; R. K. Wilcox, *Shroud* (New York: Bantam Books, 1979), pp. 41, 180; and Meacham, "Authentication," p. 292.

34. V. Tzaferis, "Crucifixion—The Archaeological Evidence," *Biblical Archaeology Review* II (January/February 1985): pp. 44-53; V. Tzaferis, "Jewish Tombs at and

Near Giv'at ha-Mivtar, Jerusalem," *Israel Exploration Journal* 20.1/2 (1970): pp. 18-32; N. Haas, "Anthropological Observations on the Skeletal Remains from Giv'at ha-Mivtar," *Israel Exploration Journal* 20.1/2 (1970): pp. 38-59.

35. Z. Greenhut, "Burial Cave of the Caiaphas Family," *Biblical Archaeology Review* (September/October 1992): 28-36, 76; R. Hachlili and A. Killebrew, "Was the Coin-on-Eye Custom a Jewish Burial Practice in the Second Temple Period?" *Biblical Archaeologist* 46 (Summer 1983): 147-153; W. Meacham, "On the Archaeological Evidence for a Coin-on-eye Jewish Burial Custom in the First Century A.D." *Biblical Archaeologist* 49 (March 1986): 56-59.

36. Meacham, "Archaeological Evidence"; R. Hachlili, "Ancient Burial Customs Preserved in Jericho Hills," *Biblical Archaeology Review* 5.4 (July 1979): 28-35. See also Hachlili and Killebrew, "Coin-on-Eye Custom."

37. F.L. Filas. *The Dating of the Shroud of Turin from Coins of Pontius Pilate,* 2nd ed. (Youngstown, Ariz.: Cogan Publications, 1982).

38. Filas, *Dating of the Shroud* F.C. Tribbe, *Portrait of Jesus*? (New York: Stein and Day, 1983); and A. Brame, "The Dating of the Shroud of Turin: Two Rare, Previously Unrecognized, Lituus Dipleta Issued A.D. 24-25 by Valerius Gratus and A.D. 29-30 by Pontius Pilate," *The Augustan* 22.94 (1984): 66-78. Brame has claimed that the lituus could also have been used on a coin minted by Pilate's predecessor, Valerius Gratus, but this coin would have been minted in A.D. 24-25, which would easily be in circulation at the time of Christ's death.

39. Filas, *Dating of the Shroud.*

40. Ibid.

41. Ibid.

42. Ibid., p. 5.

43. Ibid., pp. 11-12. These features showed up best on photographs taken of an enlargement of the Shroud face made from a sepia print based on the original 1931 photographic prints of Giuseppe Enrie. While these features are seen in varying degrees on numerous photographic negatives of the Shroud taken by different photographers, they are most clearly visible on an enlargement of the entire two thirds life-size photograph. The Enrie photographs were taken with film that emphasized contract, whereas subsequent photographs used improved film that tended to downplay contrasts. Also, subsequent photographers secured the Shroud to its frame with magnets, which produced tiny folds or draping effects rather than the stretched tautness of the Shroud cloth that was

obtained by Enrie, who is thought to have used metal tacks. Unfortunately this means that STURP's many photos do little to prove or disprove the existence of these coins. Further imaging of the Shroud should take Enrie's method into account so we may learn more about this theory. All according to Filas, *The Dating of the Shroud.*

44. C was an alternate form of Σ, and U or V alternates of Υ. Z. H. Klawans and K.E. Bressef, ed. *Handbook of Ancient Greek and Roman Coins* (Racine, Wis.: Western Publishing Company, 1995), pp. 21, 24.

45. Rev. Adam J. Otterbein, CSSR., personal communication on September 23, 1986. Rev. Otterbein was the president of the Holy Shroud Guild, to whom Father Filas left his Pilate coin collection on his death in 1985.

46. R. Reich, "Caiaphas' Name Inscribed on Bone Boxes," *Biblical Archaeology Review* (September/October, 1992): 38-44, 76, p. 38.

47. Reich, "Caiaphas' Name Inscribed"; Z. Greenhut, "Burial Cave of the Caiaphas Family," *Biblical Archaeology Review* (September/October 1992): 28-36, 76.

48. Reich, "Caiaphas' Name Inscribed," p. 41.

49. Josephus, *Antiquities* 18.35 and 18.95, as stated in Reich, "Caiaphas' Name Inscribed," pp. 41 and 76.

50. R. Hachlili, "Ancient Burial Customs" and Greenhut, "Burial Cave of the Caiaphas Family."

51. L.Y. Rahmani, "Jason's Tomb," *Israel Exploration Journal* 17 (1967): 61-113; as cited in Hachlili and Killebrew, "Coin-on-Eye Custom," pp. 151-152.

52. E. L. Sukenik, "A Jewish Tomb North-West of Jerusalem," *Tarbiz* I (1930): pp. 122-124 (in Hebrew); and A. Kloner, "The Necropolis of Jerusalem in the Second Temple Period" (doctoral thesis, Hebrew University of Jerusalem, 1980), as cited in Hachlili and Killebrew, "Coin-on-Eye Custom," pp. 151-52.

53. A. Kloner, "The Necropolis of Jerusalem," cited in Hachlili and Killebrew, "Coin-on-Eye Custom," p. 152.

54. Hachlili and Killebrew, "Coin-on-Eye Custom," p. 152 (oral communication with Y. Meshorer). The locations of the coins that this and the previous endnote refer to were not provided in Hachlili and Killebrew.

55. Hachlili and Killebrew, "Coin-on-Eye Custom," p. 152.

56. Hachlili, "Ancient Burial Customs," pp. 34-35. Interestingly, after Shroud of Turin researchers discussed the appearance of a coin over the eye of the man in the Shroud and drew connections with ancient Jewish burial customs, Hachlili changed her interpretation of her findings: She stated her new belief that the coins found at Jericho had been placed in the mouth, not on the eyes, as part of a pagan Greek custom of placing a coin(s) in the deceased's mouth as payment to Charon, the ferryman of Greek mythology who carried the spirits of the dead across the River Styx to Tartarus. (Hachlili and Killebrew, "Coin-on-Eye Custom," P. 150). However, according to analyses of an anatomist, physical anthropologist, and archaeologist, it is not possible for a coin to drop from the mouth into the skull when a body is in an ordinary, supine position because the foramen magnum—the only hole through which an object could pass — would be blocked by intact cervical vertebrae. (W. Meacham, "On the Archaeological Evidence for a Coin-on-Eye Jewish Burial Custom in the First Century A.D." *Biblical Archaeologist* 49 (March 1986): 56-59. Coins placed in the mouth would ultimately fall into the throat near the cervical vertebrae, or even the upper thorax, but not the skull. Even if the head were tilted 15 to 20 degrees, the possibility of falling into the skull is still only slight. Meacham reports that dozens of exhumations of recent burials have revealed loose teeth near the cervical vertebrae, the shoulders and even among the ribs, but that none had managed to fall inside the skull (Ibid. p. 57). Recent experiments by Mario Maroni have demonstrated that it is possible for coins to fall into the skull through the upper eye sockets only and not through the mouth. (M. Moroni, "Pontius Pilate's Coin on the Right Eye of the Man in the Holy Shroud, in the Light of the New Archaeological Findings," *History, Science, Theology and the Shroud,* ed. A. Berard, (St. Louis: Richard Nieman, 1991), pp. 276-301, figs. 6,7.)

It is also logical to conclude that a family so close to the Caiaphas family as to be buried with them would be following Jewish burial customs and not pagan customs, which called for placing a coin in the mouth. While little is positively known about the reason for laying coins over the eyes, William Meacham writes, "The ritual significance of closing the eyes of the deceased is noted in the Bible (Genesis 46:4) and in the first/second-century Mishnah…The use of the coins for this purpose may have had a special significance, for instance in rare types of death, or may have occurred more randomly…" ("Archaeological Evidence," p. 58). It may very well be that Jesus' death and burial (hurriedly before the Sabbath and sundown) fit in to such random circumstances, and that the use of coins by knowledgeable respected members of the council who buried Jesus (both Joseph of Arimathea and Nicodemus were members) would not violate, but uphold, Jewish laws and rituals. The use of coins at burials by families closely associated with the family of high priests is far more likely to have followed Jewish burial customs than to have followed pagan rituals.

57. M. Gichon, "Excavations at En Boqeq." *Qadmoniot* 12 (1970): 138-141 (in Hebrew), cited in Hachlili and Killebrew, "Coin-on-Eye Custom," p. 152; also discussed in Meacham, "Archaeological Evidence."

58. Meacham, "Archaeological Evidence."

59. Ibid., p. 58.

60. Greenhut, "Burial Cave of Caiaphas Family," p. 36.

61. Sister Damian of the Cross, OCD (Dr. Eugenia L. Nitowski), *The Field and Laboratory Report of the Environmental Study of the Shroud in Jerusalem,* (Salt Lake City: Carmelite Monastery, 1986).

62. Sister Damian of the Cross, *Field and Laboratory Report*; J. Kohlbeck and E. Nitowski, "New Evidence May Explain Image on Shroud of Turin," *Biblical Archaeology Review,* 12.4 (July/August 1986).

63. Sister Damian of the Cross, *Field and Laboratory Report*; Kohlbeck and Nitowski, "New Evidence."

64. Kohlbeck and Nitowski, "New Evidence."

65. R. Levi-Setti, G. Crow, Y. L. Wang, "Progress in High Resolution Scanning Ion Microscopy and Secondary Ion Mass Spectrometry Imaging Microanalysis," *Scanning Electron Microscopy* 2 (9185): 535-52.

66. Kohlbeck and Nitowski, "New Evidence."

67. M. Frei, "Nine Years of Palinological Studies on the Shroud," *Shroud Spectrum International,* 3 (June 1982): 3-7.

68. Ibid.

69. Ibid.

70. W. Bulst, "The Pollen Grains on the Shroud of Turin," *Shroud Spectrum International* 10 (March 1984): 20-28.

71. Ibid.

72. Ibid.

73. The majority of the Jerusalem pollens are insect pollinated and could not have landed on the Shroud as a result of winds. A. Danin, A. Whanger, W. Baruch, M. Whanger, *Flora of the Shroud of Turin* (St. Louis: Missouri Botanical Garden Press, 1999) pp. 14-15, 24.

74. M. Whanger and A. Whanger, *The Shroud of Turin: An Adventure of Discovery* (Franklin, Tenn.: Providence House Publishers, 1998), p. 78.

75. Whanger and Whanger, *The Shroud: An Adventure, p. 80; CBS Evening News*, interview of Dr. Danin, April 12, 1997.

76. Whanger and Whanger, *The Shroud; An Adventure*, p. 79.

77. *CBS Evening News*, interview of Dr. Danin, April 12, 1997; A. Danin, lecture at the Missouri Botanical Garden, St. Louis, MO, June 6, 1997; A. Whanger, "Flowers on the Shroud: Current Research," *CSST News* 1.1 (November 1997). Drs. Danin, Whanger, and Baruch are also able to conclude on the basis of extensive floral images and/or pollens on the Shroud of four narrowly distributed flowers that they could only derive from the vicinity of Jerusalem. A. Danin et al., *Flora of the Shroud*, p. 18.

78. *CBS Evening News*, interview of Dr. Danin; A. Danin, lecture at the Missouri Botanical Garden; Whanger and Whanger, *The Shroud: An Adventure*, p. 78.

79. Danin et al., *Flora of the Shroud*, p. 22.

80. Whanger and Whanger, *The Shroud: An Adventure*, pp. 74-75, 80.

81. These authors have advocated or discussed vertically collimated X-radiation, corona discharge and Kirlian photography.

82. A. Danin, Presentation at the International Workshop on the Scientific Approach to the Archeiropoietos Images, Frascati, Italy, May 4, 2010; Presentation at the Missouri Botanical Garden, St. Louis, August 18, 2008; "Botany of the Shroud of Turin," *Proceedings of the 2008 Columbus International Conference,* ed. G. Fanti, August 14-17, 2008.

83. Danin, lecture at the *Third International Congress*; Danin et al., *Flora of the Shroud*, p. 16.

84. Wilcox, *Shroud*, p. 133; see also T. D. Stewart, cited in Wilcox, who finds that the man is Caucasian.

85. Gressman, "Festschrifte for K. Budde," cited in Bulst, *Shroud of Turin*; H. Daniel-Rops, *Daily Life in Palestine at the Time of Christ* (London: Weidenfeld, 1962); H. Daniel-Rops, *Daily Life in the Time of Jesus* (New York: Hawthorn Books, 1962); and Wilson, *Shroud of Turin,* p. 47.

86. Bulst, *Shroud of Turin*, p. 104.

87. Daniel-Rops, *Palestine;* Daniel-Rops, *Time of Jesus;* H. Gressman, "Festschrifte for K. Budde," Appendix to *Zeitschrift fur die alttestamentliche Wissenschaft* 34 (1920); 60-68, cited in Bulst, *Shroud of Turin*; and Wilson, *Shroud of Turin*, p. 47. Also according to several Orthodox Jewish rabbis and scholars, as cited in K. E. Stevenson and G. R. Habermas, *Verdict on the Shroud* (Ann Arbor, Mich.: Servant Books, 1981), p. 36.

88. *Mishnah* (Shab. 23:5); John 11:44, 20:7, see also *The Jewish Encyclopedia*, vol. 3, pp. 434-436, 1925 edition, as cited in J. Iannone, *The Mystery of the Shroud of Turin* (New York: Alba House, 1998), p. 42.

89. J. P. Jackson, E. J. Jumper, B. Mottern, and K. E. Stevenson, "The Three-Dimensional Image on Jesus' Burial Cloth," *Proceedings of the 1977 United States Conference of Research on the Shroud of Turin* (Albuquerque, N.M.: Holy Shroud Guild, Mar. 1977), p. 91; and J. P. Jackson, in *Silent Witness* (Documentary, Released Nov. 8, 1978 (USA), Screenpro Films; David W. Rolfe, Producer). Scientific evidence for the chin band may be indicated in the vertical directionality study. J. Jackson, "The Vertical Alignment of the Frontal Image," *Shroud Spectrum International,* 32/33 (Sept/Dec 1989): 3-26, 12. Scientists are not certain what caused the impressions along the sides of the face, nor what looks like an unbound pigtail at the back of the head. Some think these impressions might be because, as with any ancient linen, the lots of the thread used in making the cloth vary. But such variations occur throughout the cloth, not just in these areas. In other places on the cloth where the lots of the weave vary, the image does not seem affected at all. That the image would be noticeably affected only in these particular locations, but nowhere else, would certainly be an amazing coincidence.

90. V. Barclay, *Catholic Herald,* March 18, 1960, cited in Wilcox, *Shroud*, and Tribbe, *Portrait?*; E. Wilson, *The Scrolls from the Dead Sea* (London: W. H. Allen, 1955), p. 60; and R. P. de Vaux, "Fouille au Khirbet Qumran," *Revue Biblique* 60 (1953): 102.

91. P. Maloney, "Modern Archaeology, History and Scientific Research on the Shroud of Turin," in *The Mystery of the Shroud of Turin: An Interdisciplinary Symposium,* video (Elizabethtown, Penn: Elizabethtown College, February 15, 1986), citation of letter from Dr. Frei.

92. Ian Dickinson, "Preliminary Details of New Evidence for the Authenticity of the Shroud: Measurement by the Cubit," *Shroud News* 58 (April 1990): 4-7.

93. E. A. Wuenschel, "The Shroud of Turin and the Burial of Christ/II—John's Account of the Burial," *Catholic Biblical Quarterly* 8 (1946): 161-66.

94. Meacham, "Authentication"; and Bulst, *Shroud of Turin*, pp. 90-91.

95. K. Little, "The Holy Shroud of Turin and the Miracle of the Resurrection," *Christian Order* (April 1994): 218-31, 223.

96. V. Tzaferis, "Crucifixion"; and V. Tazaferis, "Jewish Tombs at or near Giv' at ha-Mivtar, Jerusalem," *Israel Exploration Journal* 20/1.2 (1970): 18-32.

97. Tzaferis, "Crucifixion."

98. John 18:30-31.

99. Tzaferis, "Crucifixion."

100. Don Lynn, Image Processing Specialist for NASA and STURP scientist, personal communication on June 3, 1984.

101. E. M. Meyers , J. F. Strange, and C. L. Meyers, *Excavations at Ancient Meiron, Upper Galilee, Israel* (Durham, NC: American Schools of Oriental Research and Duke University, 1981), according to Dr. Michael Fuller, chairman, Sociology Department, St. Louis Community College at Florissant Valley.

102. S. Kraus, *Talmudische Archaologie* (Leipzig: 1910-1911), cited in Meacham, "Authentication."

103. S. Gansfield, *Code of Jewish Law,* rev. ed., trans. H. E. Goldin (New York: Hebrew Publishing, 1927), Ch. CXCVII, No. 9, pp. 99-100.

104. B. Lavoie, G. Lavoie, D. Klutstein, and J. Regan, "In Accordance with Jewish Burial Custom, the Body of Jesus Was Not Washed," *Shroud Spectrum International* 3 (June 1982): 8-17.

105. Lavoie et al., "In Accordance with Jewish Burial Custom," p. 15.

106. Wilson, *The Blood and the Shroud* (New York: The Free Press, 1998), p. 55; see also I. Wilson, *Mysterious Shroud,* p. 46. A few sindonologists attempt to claim that part or all of the body of the man in the Shroud was washed. Yet, this would have correspondingly altered some or all of the man's blood marks. These sindonologists claim these blood marks would then have re-bled and reformed in the tomb after the body was washed. Yet, the evidence indicates that all the approximately 130 blood marks formed and flowed in the various vertical positions that the man was in when he initially incurred the wounds and before he was laid in the tomb. It should also be noted that when a corpse is laid horizontally on its back, its internal blood will tend to accumulate by gravity on the dorsal side. Thus, it would not re-bleed in a manner similar to the various upright positions the man was in while he was alive.

107. John 19:14.

108. Deuteronomy 21:22, 23 and Jewish law prohibited a crucifixion victim from remaining on the cross at night. *Temple Scroll* 64:10-13; Philo, *De Spec. Leg.* 3.28 S. 151-52, *De Post.* 61, *De Som.* 2.31 S.213. See also the statement by Josephus: "The Jews are so careful about funeral rites that even those who are crucified because they were found guilty are taken down and buried before sunset." J.W. 4.5.2, S. 317. All above according to R. E. Brown, "The Burial of Jesus (Mark 15:42-47)," *Catholic Biblical Quarterly* 50 (1988).

109. Matt. 27:61; Mark 15:47; 16:2; Luke 23:55 – 24:1.

110. J.A.T. Robinson, "The Shroud of Turin and The Grave—Clothes of the Gospels," in *Proceedings*; and A. Feuillet, "The Identification and the Disposition of the Funerary Linens of Jesus' Burial According to the Data of the Fourth Gospel," *Shroud Spectrum International* 4 (September 1982) 13-23, reprint from *La Sindone E La Scienze,* II Congresso Internationale di Sindonologia, Turin, Italy, 1978.

111. Daniel-Rops, *Palestine*; Daniel-Rops, *Time of Jesus*; Wilson, *Shroud of Turin*, pp. 56-57; Barbet, *Calvary*, pp. 163-165; Tribbe, *Portrait?* Pp. 89-92; Robinson, "Grave Clothes," pp. 24-25; P. Vignon, "The Problem of the Holy Shroud," trans. E. A. Wuenschel, *Scientific American* 93.163 (1937): 162-64; Humber, *Sacred Shroud*, p. 66; E. A. Wuenschel, *Self-Portrait of Christ* (Esopus, N.Y.: Holy Shroud Guild, 1957), p. 48; and J. Marino, *The Burial of Jesus and the Shroud of Turin* (Anaheim: Jeff Richards, 1999). (The amount of spices mentioned in John would be about 75-100 lbs. in our weight today.).

112. Daniel-Rops, *Palestine*; Daniel-Rops, *Time of Jesus;* Wilson, *Shroud of Turin*, pp. 56-57; Barbet, *Calvary*, pp. 163-165; Tribbe, *Portrait?* pp. 89-92; Robinson, "Grave-Clothes," pp. 24-25; P. Vignon, "The Problem of the Holy Shroud," Humber, *Sacred Shroud*, p. 66; Wuenschel, *Self-Portrait*, p. 48; and Marino, *The Burial of Jesus.*

113. John not only notes the Sabbath was fast approaching, but this particular Sabbath fell on Passover (John 19:14, 42), in which Jesus' death parallels that of the sacrificial lamb. Perhaps, John notes the presence of myrrh after Jesus' death in parallel with its presence after Jesus' birth.

114. Personal communication from Robinson to STURP scientist Don Lynn in 1979.

115. *Mishnah*, Division II Mo'ed, *Shabbath* 23:5, p. 120. See also Robinson, "Grave-Clothes," p. 27; and Tribbe, *Portrait?* P. 89.

116. J. A. T. Robinson, "Grave-Clothes," p. 23.

117. Ibid.

118. J. P. Jackson, et al., "Three Dimensional Image," in *Proceedings*, p. 91; and J. P. Jackson, in *Silent Witness.*

119. Mark 15:43; Luke 23:50; John 3:1 and 7:50.

120. John 19:1; Matt. 27:26; Mark 15:15.

121. Mark 14:65, 15:19; Matt. 26:67, 27:30; Luke 22:63; John 18: 22 and 19:3.

122. Mark 15:16-20; Matt. 27:27-31; John 19:2,3.

123. John 19:17.

124. Matt. 27:32; Mark 15:21; Luke 23:26. Another school of thought thinks that the man in the Shroud caried the full cross instead of the cross beam. However, these same injuries could also have occurred from carrying a full cross.

125. Luke 24:39, 40; John 20:20, 25, 47.

126. John 19:32, 33.

127. John 19:34, 35.

128. Matt. 27:59-62 and 28:1; Mark 15:42, 46-16:3; Luke 23:52-24:2; John 19:12, 41-20:1.

CHAPTER FIVE

1. L. A. Schwalbe and R. N. Rogers, "Physics and Chemistry of the Shroud of Turin," *Analytica Chemica Acta* 135 (1982): 3-49; R. H. Dinegar, "The 1978 Scientific Study of the Shroud of Turin," *Shroud Spectrum International* 4 (September 1982: 3-12; E. J. Jumper, A. D. Adler, J. P. Jackson, S. F. Pellicori, J. H. Heller, and J. R. Druzik, "A Comprehensive Examination of the Various Stains and Images on the Shroud of Turin," *ACS Advances in Chemistry No. 205 Archaeological Chemistry III,* J. B. Lambert, ed. American Chemical Society (1984), pp. 447-476;

2. R. Gilbert, Jr., and M. M. Gilbert, "Ultraviolet-Visible Reflectance and Fluorescence Spectra of the Shroud of Turin," *Applied Optics* 19:12 (June 1980): 1930-1936; Schwalbe and Rogers, "Physics and Chemistry."

3. Gilbert, Jr., and Gilbert, "Ultraviolet-Visible Reflectance and Fluorescence";

Schwalbe and Rogers, "Physics and Chemistry", V. D. Miller and S. F. Pellicori, "Ultraviolet Fluorescence Photography of the Shroud of Turin," *Journal of Biological Photography* 49.3 (July 1981: 71-85.

4. Schwalbe and Rogers, "Physics and Chemistry"; Dinegar, "The 1978 Scientific Study."

5. Schwalbe and Rogers, "Physics and Chemistry"; Dinegar, "The 1978 Scientific Study"; Jumper et al., "A Comprehensive Examination."

6. Jumper et al., "A Comprehensive Examination"; J. H. Heller and A. D. Adler, "A Chemical Investigation of the Shroud of Turin," *Can. Soc. Forens. Sci.* J. 14.3 (1981): 81-103; L. Gonella, "Scientific Investigation of the Shroud of Turin – Problems, Results and Methodological Lessons," in *Turin Shroud – Image of Christ?* (Hong Kong: Cosmos Printing Press Ltd. 1987), pp. 29-40.

7. Schwalbe and Rogers, "Physics and Chemistry"; Dinegar, "The 1978 Scientific Study"; J. H. Heller and A. D. Adler, "A Chemical Investigation."

8. Jumper et al., "A Comprehensive Examination," p. 456.

9. S. F. Pellicori and M. Evans, "The Shroud of Turin Through the Microscope," *Archaeology* (January/February 1981): 32.43; S. F. Pellicori, "Spectral Properties of the Shroud of Turin," *Applied Optics* (June 15, 1980); 1913-1920, G. G. Gray, "Determination and Significance of Activation Energy in Permanence Tests," in *Preservation of Paper and Textiles of Historic and Artistic Value,* Advances in Chemistry series 164 (Washington, DC: American Chemical Society, 1977), as cited in Pellicori "Spectral Properties"; S. Pellicori and R. A. Chandos, "Portable Unit Permits UV/vis Study of 'Shroud'," *Industrial Research & Development*, February 1981, 23: 186-189, 187; J. Rinaudo, "Protonic Model of Image Formation on the Shroud of Turin," *Third International Congress on the Shroud of Turin,* Turin, Italy, June 5-7, 1998; J. Rinaudo, "A Sign of Our Time," *Shroud Sources Newsletter,* May/June 1996, pp. 2-4; J. Jackson, E. Arthurs, L. Schwalbe, R. Sega, D. Windisch, W. Long, E. Stappaerts, "Infrared Laser Heating for Studies of Cellulose Degradation" *Applied Optics,* 15 Sept., 1988, 27:3937-3943.

10. Pellicori, "Spectral Properties"; Rinaudo, "Protonic Model," and "A Sign"; Jumper et al., "A Comprehensive Examination."

11. L. Gonella, "Scientific Investigation," p. 31

12. Ibid.

13. J. P. Jackson, E. J. Jumper, and W. R. Ercoline, "Correlation of Image Intensity on the Turin Shroud with the 3-D Structure of a Human Body Shape," *Applied Optics* 23.14 (July 1984): 2244-70.

14. T. Phillips, "Shroud Irradiated with Neutrons?" *Nature* 337 (1989): 594.

15. K. Little, "The Holy Shroud and the Miracle of the Resurrection," *Christian Order* (April 1994): 218-231, 221.

16. K. Little, "The Formation of the Shroud's Body Image," *British Society for the Turin Shroud Newsletter*, No. 46, November/December 1997, pp. 19-26, 20.

17. Little , "The Formation" and "The Holy Shroud."

18. Ibid.

19. Little, "The Formation."

20. Rinaudo, "Protonic Model" and "A Sign"; J. Rinaudo, in *British Society for the Turin Shroud Newsletter,* No. 38, Aug/Sep: 13-16, 1994; J. Rinaudo, "A New Stage," *II est Vivant*, No. 89, March/April, 1992.

21. Rinaudo, "Protonic Model" and "A Sign."

22. Rinaudo, "Protonic Model."

23. J. P. Jackson, "Is the Image on the Shroud Due to a Process Heretofore Unknown to Modern Science?" *Shroud Spectrum International* No. 34, (March 1990): 3-29; J. P. Jackson, "An Unconventional Hypothesis to Explain all Image Characteristics Found on the Shroud Image," *History, Science, Theology and the Shroud*, A. Bernard, ed. (St. Louis: Richard Nieman, 1991), pp. 325-344; Rinaudo, "Protonic Model"; G. F. Carter, "Formation of the Image on the Shroud of Turin by X-Rays: A New Hypothesis," *ACS Advances in Chemistry No. 205 Archaeological Chemistry III,* J. B. Lambert, ed. American Chemical Society, 1984, pp. 425-446; G. Carter Interview, May 23, 1999, http://earthfiles.com/earth025.html.

24. J. P. Jackson, E. J. Jumper and W. R. Ercoline, "Correlation of Image Intensity on the Turin Shroud with the 3-D Structure of a Human Body Shape," *Applied Optics* 23.14 (July 1984): 2244-2270; J. P. Jackson, E. J. Jumper, and W. R. Ercoline, "Three Dimensional Characteristics of the Shroud Image," *IEEE 1982 Proceedings of the International Conference on Cybernetics and Society* October 1982): 559-575; J. P. Jackson, E. J. Jumper, B. Mottern, and K. E. Stevenson, "The Three-dimensional Image on Jesus Burial Cloth," *Proceedings of the 1977 United States Conference of Research on the*

Shroud of Turin (Albuquerque, N.M.: Holy Shroud Guild, Mar. 1977), 74-94; C. Avis, D. Lynn, J. Lorre, S. Lavoie, J. Clark, E. Armstrong, and J. Addington, "Image Processing of the Shroud of Turin," *IEEE 1982 Proceedings of the International Conference on Cybernetics and Society (*October 1982): 554-558; G. Tamburelli, "Some Results in the Processing of the Holy Shroud of Turin," *IEEE Transactions on Pattern Analysis and Machine Intelligence* PAMI-3.6 (November 1981): 670-676.

25. J. P. Jackson, "The Vertical Alignment of the Frontal Image," *Shroud Spectrum International* 32/33 (1989): 3-26; Jackson, et al., "The Three-Dimensional Image"; W. R. Ercoline, R. C. Downs, Jr., J. P. Jackson (1982), "Examination of the Turin Shroud for Image Distortions," *IEEE 1982 Proceedings of the International Conference on Cybernetics and Society* (October): 576-579.

26. E. J. Jumper, "Considerations of Molecular Diffusion and Radiation as an Image Formation Process on the Shroud," *Proceedings of the 1977 United States Conference of Research on the Shroud of Turin* (Albuquerque, NM: Holy Shroud Guild, March 1977), pp. 182-189; J. P. Jackson, "A Problem of Resolution posed by the Existence of a Three-Dimensional Image on the Shroud of Turin," *Proceedings of the 1977 United States Conference of Research on the Shroud of Turin* (Albuquerque, NM: Holy Shroud Guild, March, 1977): 223-233; Jackson et al., "The Three-Dimensional Image."

27. P. Di Lazzaro and D. Murra, "Shroud like coloration of linen, conservation measures and perception of patterns onto the Shroud of Turin," *2014 Workshop on Advances in the Turin Shroud Investigation* (ATSI 2014), Bari, Italy, Sep 4-5, http://www.shsconferences.org/articles/shsconf/pdf/2015/02/shsconf_atsi2014_00004.pdf; P. Di Lazzaro, D. Murra, A. Santoni, and G. Baldacchini, "Sub-micrometer coloration depth of linens by vacuum ultraviolet radiation," *Proceedings of the International Workshop on the Scientific Approach to the Acheiropoietos Images*, Turin, Italy, May 4-6, 2010, pp. 3-18; G. Baldacchini, P. Di Lazzaro, D. Murra, and G. Fanti, "Coloring linens with excimer lasers to simulate the body image of the Turin Shroud," *Applied Optics*, Vol. 47, No. 9, 20 March 2008, pp. 1278-1285; Jackson, "Is the Image" and "An Unconventional Hypothesis."

28. Di Lazzaro et al, "Sub-micrometer coloration;" Baldacchini et al, "Coloring linens with excimer lasers;" Jackson, "Is the Image" and "An Unconventional Hypothesis.

29. Jumper et al., "A Comprehensive Examination"; Heller and Adler, "A Chemical Investigation"; Schwalbe and Rogers, "Physics and Chemistry"; Jackson et al., "Vertical Alignment," "Correlation," "Three Dimensional Characteristics," and "Three Dimensional Image"; C. Avis, D. Lynn, J. Lorre, S. Lavoie, J. Clark, E. Armstrong, and J. Addington, "Image Processing of the Shroud of Turin," *IEEE 1982 Proceedings of the International Conference on Cybernetics and Society* (October 1982): 554-558; G. Tamburelli, "Some Results in the Processing of the Holy Shroud of Turin," *IEEE Transac-*

tions on Pattern Analysis and Machine Intelligence PAMI-3.6 (November 1981): 670-676; G. Tamburelli, Reading the Shroud, Called the Fifth Gospel, with the Aid of the Computer," *Shroud Spectrum International* 2 (March 1982): 3-11; G. Tamburelli, "An Image Resurrection of the Man of the Shroud," *Shroud Spectrum International* 15 (June 1985): 3-6; Jackson, "Is the Image"; W. R. Ercoline, R. C. Downs, Jr., and J. P. Jackson, "Examination of the Turin Shroud for Image Distortions," *IEEE 1982 Proceedings of the International Conference on Cybernetics and Society* (October 1982): 576-579; Di Lazzaro et al., "Sub-micrometer coloration,"; J. H. Heller, *Report on the Shroud of Turin* (New York: Houghton Mifflin, 1983); G. Fanti, J. A. Botella, P. di Lazaro, T. Heimburger, R. Schneider, N. Svensson, "Microscopic and Macroscopic Characteristics of the Shroud of Turin Image Superficiality," *Journal of Imaging Science and Technology* (Jul.-Aug. 2010): 040201-1 – 040201-8; G. Fanti, J. A. Botella, F. Cresilla, F. Lattarullo, N. Svensson, R. Schneider, A. Whanger, "List of evidences of the Turin Shroud," *Proceedings of the International Workshop on the Scientific Approach to the Acheiropoietos Images,* Turin, Italy, May 4-6, 2010, pp. 67-75.

30. STURP scientist John Heller, as quoted in W. McDonald, "Science and the Shroud," *The World and I,* (Oct. 1986): 420-428; L. Gonella, "Scientific Investigation," p. 31.

31. Little, "The Holy Shroud," p. 225; Jackson, "Is the Image" and "An Unconventional Hypothesis."

CHAPTER SIX

1. Robert Rucker, "MCNP Analysis of Neutrons Released From Jesus' Body In The Resurrection," *Shroud of Turin: The Controversial Intersection of Faith and Science, International Conference,* St. Louis, MO, October 9–12, 2014, http://www.shroud.com/pdfs/stlruckerppt.pdf.

2. Thermal energy is reached when the speed of the neutron is in equilibrium with the surrounding atoms at room temperature. As a neutron passes through matter, it will ricochet (scatter) off the nuclei of the various atoms in the Shroud and the tomb, gradually losing energy and slowing down. Even if the neutrons are emitted in the body at thermal energy, the average number of ricochets that a neutron will experience before being captured is approximately 158 and the captures (or conversions) will occur within a fraction of a second, all according to nuclear engineer Robert Rucker.

3. Specific types of atoms containing the same number of protons within their nuclei (or with the same atomic number) are called elements, such as chlorine or calcium. When the name or abbreviation of a chemical element is given along with its *total* number of protons and neutrons (or its atomic mass), it is called an isotope, such as Cl-35, Cl-36, Ca-40 or Ca-41.

4. Calcium-40 (Ca-40) comprises 97% of all calcium atoms or isotopes while chlorine-35 (Cl-35) comprises 75.78% of all chlorine atoms or isotopes.

5. M. Flury-Lemberg, *Sindone 2002 Preservation*, (Torino, Editrice ODPF, 2003). Although the Shroud's patches were also removed in 2002, I have continued to use the full-length images taken by STURP photographer Vernon Miller in 1978, and not the very professional photos taken after the patches' removal. Vernon Miller's and Barrie Schwortz's 1978 photographs still seem to provide the best contrast and context. Part of this reason may be due to the presence of the patches, which seem to provide more of a framework for the torso and the upper part of the arms.

6. Nitrogen-14 (N-14) comprises 99.63% of all nitrogen.

7. When this conversion occurs one of the electrons in the atom also disappears.

8. Some highly-energetic particles from outer space also produce cosmic rays in the atmosphere.

9. 77.8% of the earth's atmosphere or air is comprised of N-14.

10. These are the general principles upon which radiocarbon dating operates. It does not include modifications to the calibration curve for fluctuations in the radioactive level of the atmosphere or for different levels of atmospheric carbon dioxide in the oceans, neither of which directly concerns the radiocarbon dating of the Shroud.

11. Nitrogen (N-14) would have been acquired by the flax plant from rain water or the soil in which it grew. Dr. Robert Hedges, who participated in the Shroud's radiocarbon dating at the Oxford Laboratory, estimated linen contained 1000 ppm. R. Hedges, "Hedges Replies," *Nature* 337 (1989): 594. Modern unbleached flax linen utilized in experiments by Dr. Arthur Lind contained approximately 566-732 ppm. A. Lind, M. Antonacci, D. Elmore, G. Fanti, and J. Guthrie, "Production of Radiocarbon by Neutron Radiation on Linen," *International Workshop on the Scientific Approach to the Archeiropoietos Images*, Frascati, Italy, May 4-6, 2010, pp. 255-262.

12. P. Jennings, "Still Shrouded in Mystery," *30 Days in the Church and in the World* 1.7 (1988): 70-71, 71.

13. Jennings, "Still Shrouded," p. 71.

14. W. Meacham, "Radiocarbon Measurements and the Age of the Turin Shroud; Possibilities and Uncertainties," in *Turin Shroud Image of Christ?* (Hong Kong: Cosmos Printing Press Ltd., 1987), pp. 41-56: O. Petrosillo, and E. Marinelli, *The Enigma of the Shroud,* translated from Italian (San Swann, Malta: PEG 1996); I. Wilson, "Is

This the News We have Been Waiting For?" *British Society for the Turin Shroud Newsletter* 14, September 1986; 3-4; D. Sox, *The Shroud Unmasked* (Basingstroke, Hampshire: Lamp, 1988); N. Rufford, "Vatican Steels Itself for 'Fake' Result," *London Times,* August 7, 1988, pp. 1-3; D. Nelson, R. Morlan, J. Vogel, J. Southen, and D. Having, "New Dates on Northern Yukon Artifacts: Holocene Not Upper Pleistocene," *Science* 232 (1986): 749-751.

15. Meachem, "Radiocarbon Measurements"; Rufford, "Vatican Steels Itself"; Nelson et al., "New Dates on Northern Yukon Artifacts."

16. A. Goude, *Environmental Change* (Oxford: Clarendon, 1988), p. 10, as cited in Meacham, "Radiocarbon Measurements," p. 43.

17. Lind et al.

18. Lind et al., "Production of Radiocarbon."

19. Confirmed by personal communications with physicist Arthur C. Lind in 2010.

20. Lind et al., "Production of Radiocarbon."

CHAPTER SEVEN

1. Physicist Arthur Lind, personal communication, March 12, 2015.

2. Robert Rucker, "MCNP Analysis of Neutrons Released From Jesus' Body In The Resurrection," *Shroud of Turin: The Controversial Intersection of Faith and Science, International Conference*, St. Louis, MO, October 9–12, 2014, http://www.shroud.com/pdfs/stlruckerppt.pdf; Nuclear engineer Robert Rucker, personal communications, 2014 and 2015.

3. Ibid.

4. This would be less true if the sides of the cloth were tucked under the legs of the man in the Shroud at burial. For further discussion, see endnote 19 of Chapter Twelve.

5. What Did Jesus' Tomb Look Like? An Interview with Leen Ritmeyer (Part 2), July 24, 2008, http://thegospelcoalition.org/blogs/justintaylor/2008/07/24/what-did-jesus-tomb-look-like-interview/.

6. Matt 27:61; Luke 23:55; John 20:5.

7. Rucker, "MCNP Analysis of Neutrons Released."

8. The organic or indigenous amounts of chlorine, calcium and nitrogen in dried human blood would be the same for all humans. The amount of N-14 in a linen sample can be acquired by blasting several small representative parts of individual threads (aliquot) with an enormous, but known amount of neutrons, whose C-14 to C-12 date would translate to more than 100,000 years into the future. This C-14 to C-12 ratio would reveal the amount of N-14 within the sample that produced this large number of C-14 atoms. Since X-ray fluorescence measurements of organic chlorine and calcium for small linen samples are now possible, they should be measured even though these amounts would be revealed in the Cl-36 to Cl-35 and the Ca-41 to Ca-40 ratios.

The organic amounts of chlorine within linen refer to chlorine that is bound within it as a result of the growth of the flax plant or manufacturing process of linen. Inorganic amounts of chlorine are those acquired from handling, breathing, water or other sources. Because the amounts of chlorine, calcium and nitrogen will vary somewhat from point to point on a sample, representative parts of each individual thread should be used for Cl-36 (or Ca-41) measurements, along with C-14 and N-14 measurements (except only a tiny fraction would be used for N-14 measurements). These organic amounts would have been present at the time of the hypothetical neuron radiating event. In case any inorganic chlorine or calcium was also present at the time of the neutron radiating event, the inorganic material removed from the samples should also be tested for any above-background presence of Cl-36 and Ca-41.

9. While Cl-36 to Cl-35 and Ca-41 to Ca-40 ratios are routinely measured on solid objects such as rocks at AMS laboratories, chemical techniques would have to be adapted and perfected to capture all of the trace amounts of chlorine and calcium from milligram size samples such as neutron irradiated linen and blood before submitting them for AMS analysis. These techniques would have to be performed at laboratories with certain facilities. (See Appendix A.)

10. Since Cl-36 and Ca-41 are created by neutron radiation at very precise rates, scientists would only have to deduct the infinitesimally natural levels of Cl-36 and Ca-41 from the Cl-36 to Cl-35 and Ca-41 to Ca-40 ratios. Whenever we discuss the amounts or levels of Cl-36 and Ca-41 created by neutron radiation within any cloth or blood samples, we are referring to those amounts above the natural infinitesimal levels of Cl-36 and Ca-41.

11. The amount of neutron radiation would be expressed in terms of neutrons per centimeter squared. For example, the amount of neutron radiation that the 1988 radiocarbon site received is estimated at $8.3 \times 10^{13}\ n \cdot cm^{-2}$. However, the part of the cloth that lay under the middle of the man's back would have received approximately ten times this amount of neutron radiation.

12. The number of new C-14 atoms created within a neutron irradiated sample can

be calculated based upon the areal density of the cloth, its indigenous N-14 content and the amount of neutrons (or neutron fluence) it received.

13. These age calculations would naturally include a percentage of certainty or a plus or minus (+/-) range, as is found in radiocarbon dating. The Shroud of Turin was carbon dated with 95% certainty from 1260–1390.

14. K. Little, "The Formation of the Shroud's Body Image," *British Society for the Turin Shroud* Newsletter, No. 46, November/December 1977, pp. 19-26.

15. Ibid.

16. This would be even more true if the Shroud was older than the first century. Additionally, if the Shroud was older than the first century, then the modern neutron "irradiator" would necessarily be the *second* undocumented forger to have performed an impossible task with this burial cloth. Under this scenario, the first forger would had to have encoded the Shroud's unique body images and blood marks some time after the first century occurrence of the various events depicted on the Shroud and before its medieval arrival in Europe. (If the Shroud is the burial garment of the historical Jesus Christ, then it would naturally derive from the first century.)

17. If the Shroud was neutron irradiated by a forger, he could only have attempted this in modern times. However, as we have shown, a modern forger could not have calculated, let alone deposited the correct amounts of Cl-36, Ca-41 and C-14 atoms that still remain in various amounts and locations throughout the Shroud's linen and its blood marks. There are additional problems that a modern forger would have encountered in attempting to leave the correct amounts of all radioactive atoms in all parts of the Shroud.

If a forger deposited the amounts of particle radiation found throughout the Shroud, he would have had to irradiate the cloth before 1988. Yet, it was only *after* the cloth's medieval dating that any scientist or anyone else proposed that neutron particles irradiated the Shroud. This was first offered in 1989 as a scientific explanation for the cloth's aberrant radiocarbon dating by Harvard physicist Thomas Phillips, and expounded upon in the 1990s by nuclear physicist Kitty Little, biophysicist Jean-Baptiste Rinaudo, and then in my 2000 book. Since then, the Test the Shroud Foundation has also sponsored scientific research that further advances these positions.

A forger would not have known the location from which the radiocarbon sample would have been removed. Two scientific protocols prior to 1988 that were agreed upon by the scientists that were to test the Shroud called for several samples to be removed from a number of possible locations. A forger would had to have uniformly neutron irradiated the entire Shroud before 1988 to be sure the removed sample(s) were contaminated to medieval times. He would not have heard of the body image forming hypotheses in which neutrons, protons, alpha particles, deuterons, electrons (and, perhaps, gamma rays and other electromagnetic radiation) emanated from the length, width and depth of the corpse wrapped in the Shroud until 2000. How would a forger know to encode the entire cloth and all the blood marks with varying amounts of neutron or particle radiation? Shroud scientists and investigators never heard of this until nuclear engineer Robert Rucker proposed it to an international conference on the Shroud in 2014 in St. Louis.

A forger wouldn't have known the indigenous or organic amounts of Cl-35, Ca-40 and N-14 that were at each location on the Shroud. He wouldn't know how many Cl-36, Ca-41 or C-14 atoms to leave at each location. These amounts have never been measured from the Shroud and the techniques to measure the ratios have not been developed. If he somehow managed to get the Cl-36 and Ca-41 amounts correct, he would still leave too many C-14 atoms.

Nor would a forger have had access to the sophisticated computer codes that can follow the paths of trillions upon trillions of neutron particles that nuclear engineer Robert Rucker utilized in 2014. Access to the cruder codes available before 1988 was extremely limited. He would had to have acquired such codes from an institution such as the Los Alamos National Laboratory and would likely have needed a security clearance. Assuming he could overcome these hurdles, he would have needed access then to a supercomputer on which to learn and run the code, which would also have cost tens of thousands of dollars for just a few weeks. Assuming he could make extremely sophisticated and accurate calculations, he would also need exclusive possession of the Shroud for six months while at a nuclear generator to intimately encode every fractional part of the 14'3" x 3'7" cloth and blood marks with the correct amounts of neutron radiation.

All of the above access to computer codes, super-computers, financial resources, security clearances and the Shroud itself could never have occurred anonymously without leaving some sort of a trail or a record. An anonymous modern forger who irradiated the Shroud is as impossible as the anonymous medieval forger that Bishop D'Arcis claimed painted the Shroud, or that some think invisibly rewove the Shroud at the 1988 radiocarbon site. Remember, a forger would not only have to match the Cl-36 and Ca-41 atoms contained throughout the Shroud, but also leave the correct amount of C-14 atoms at every location. Yet, even assuming that a medieval "reweaver" and a modern neutron "irradiator" both did the impossible, they would still need for another impossible forgery to have occurred, which encoded this cloth's unparalleled body images and blood marks.

18. Rucker, "MCNP Analysis of Neutrons Released"; Nuclear engineer, Robert Rucker, personal communication, March 13, 2015.

19. Nuclear engineer, Robert Rucker, personal communications, August 6, 2014 and April 20, 2015.

20. Physicists Arthur Lind and David Elmore, nuclear engineer, Robert Rucker, personal communications, 1999-2014.

21. Yahya, History, L. Kratchkovsky and A. Vasiliev, eds, *Histoire de Yahya-ibn-Sa 'id d'Antioche,* Fasc. II, *Patrologia Orientalis* 23.3, [up to 1013; Arabic text with French translation] (Paris ,1932); Yahya, *History,* L. Cheikho, B. Carra de Vaux and H. Zayyat, eds. *Annales Yahia Ibn Sa'id Antiochensis,* CSCO 51 (= Scriptores arabici, 3rd ser. 7), 207-73, [the only available version for 1013-1027/34: Arabic text only] (Beirut and Paris, 1909); L. Vincent and F. Abel, *Jerusalem: Recherches de topographie, d'archeologie et d'historire,* ii, *Jerusalem nouvell,* (fasc. I and II, Paris, 1914), p. 249; M. Gil, *A History of Palestine,* (Cambridge, 1992), 634-1099, all as stated in M. Biddle, *The Tomb of Christ,* (Glouchestershire: Sutton Publishing Limited, 1999), pps. 72-73, 113-115.

22. M. Biddle, *The Tomb of Christ,* pps. 73, 103, 115. Prof. Biddle also thinks the possible remaining bench located to the right of the entrance to the burial tomb could have been recessed into the wall.

23. M. Biddle, *The Tomb of Christ*, p. 88.

CHAPTER EIGHT

1. http://www.chem.agilent.com/en-US/products-services/Instruments-Systems/Molecular-Spectroscopy/Cary-620-FTIR-Microscopes/pages/default.aspx

2. G. Fanti and R. Maggiolo, "The double superficiality of the frontal image of the Turin Shroud," *J. Opt. A, Pure Appl. Opt.* 6, 491-503 (2004).

3. J. S. Chickos and J. Uang, "Chemical Modification of Cellulose, The Possible Effects of Chemical Cleaning on Fatty Acids Incorporated in Old Textiles," *Approfondimento Sindone,* 2001, 241-4.

4. J. H. Heller and A. D. Adler, "A Chemical Investigation of the Shroud of Turin," *Can. Soc. Forens. Sci. J.* 14.3 (1981): 81-103; J. H. Heller, *Report on the Shroud of Turin* (New York: Houghton Mifflin, (1983); Dr. Alan Adler, personal communication, January 1998.

5. R. A. Freer–Waters and A. J. T. Jull, "Investigating a Dated Piece of the Shroud of Turin," *Radiocarbon,* Vol. 52, Nr 4, 2010, pps. 1521-1527.

6. J. Heller and A. Adler, as stated in T. W. Case, *The Shroud of Turin and the C-14 Dating Fiasco* (Cincinnati: White Horse Press, 1996), p. 76; W. Meacham in "Turin Shroud Dated to A.D. 200-1000," press release, October 14, 1988. Starch has been identified on other Shroud samples according to A.A.M. Van der Hoeven, "Internal Selvedge in starched and dye temple mantle – No invisible repair in Turin Shroud – No Maillard reaction," pps. 7-10, May 9, 2012, www.JesusKing.info.

7. Heller and Adler, as stated in Case, *The Shroud and the C-14 Fiasco.* According to archaeologist W. Meacham, STURP scientists also informed him of this, as stated in "Turin Shroud Dated to 200-1000 A.D." press release; STURP scientist Thomas D'-Muhula also stated this at a lecture on May 11, 1995, according to J. Kerlin at http://childrensermons.com/shroud/present.1.htm. Several STURP scientists have also confirmed these datings to me and other researchers after this information was first published.

8. The observations or identifications of cotton by Drs. Giulio Fanti, Thibault Heimburger, Robert Villareal, Eugenia Nitowski, Rachel Freer-Waters and Timothy Jull, and chemist, Raymond Rogers are collectively discussed in A.A.M. Van der Hoeven, "Internal selvedge," pps. 3-7.

9. G. Raes, "Examination of the 'Sindone'" in *Report of the Turin Commission on the Holy Shroud,* trans., ed. M. Jepps, (London: Screenpro Films, 1976), pp. 79-83.

10. M. Flury-Lemberg, "The Linen Cloth of the Turin Shroud: Some Observations on its Technical Aspects," *Sindon* New series, No. 16, December 2001, 55-61, 59.

11. Flury-Lemberg, "The Linen Cloth of the Turin Shroud," p. 59.

12. Ibid., p. 60.

13. G. Fanti, P. Baraldi, R. Basso, AS. Tinti, "Non-destructive dating of ancient flax textiles by means of vibrational spectroscopy. *Vibrational Spectroscopy* (2013), http://dx.doi.org/10.1016/j.vibspec.2013.04.001; "The Turin Shroud," *Secrets of the Bible,* American Heroes Channel, aired January 4, 2015.

14. Ibid.

15. G. Fanti and P. Malfi, "A New Cyclic-Loads Machine for the Measurement of Micro-Mechanical Properties of Single Flax Fibers Coming From the Turin Shroud," *XXI Congresso Associone Italiana di meceanica Teorica c applicate, AIMETA*, Torino, September 17-20, 2013; "The Turin Shroud," *Secrets of the Bible*, American Heroes Channel, January 4, 2015.

16. These were in the form of press releases from the International Center of Sindonology in Turin and Mons. Cesare Nosiglia, Archbishop of Turin, Custodian of the Shroud.

17. A. Ivanov and D. Kouznetsov, "Biophysical correction to the old textile radiocarbon dating results," *L'Identification Scientifique de l'homme du Linneul Jesus de Nazareth:* Actes Du Symposium Scientifique International, Rome, 1993 (Paris: Francois-Xavier Guibert, 1995), pp. 229-236; D. A. Kouznetsov, A. A. Ivanov, and P. R. Veletsky, "A re-evaluation of the Radiocarbon Date of the Shroud of Turin Based on

Biofractionation of Carbon Isotope and a Fire-Simulating Model," *Archaeological Chemistry: Organic, Inorganic, and Biochemical Analyses*, Mary Virginia Orno, ed., (American Chemical Society, 1996): 229-247; D. A. Kouznetsov, A. A. Ivanov, and P R. Veletsky, "Effects of fires and biofractionation of carbon isotopes on results of radiocarbon dating of old textiles: the Shroud of Turin," *Journal of Archaeological Science* 23 (1996): 109-121.

18. L. A. Garza-Valdes, *The DNA of God?* (London: Hodder & Stoughton, 1998); *The Mysterious Man of the Shroud,* directed by Terry Landau, CBS documentary aired on April 1, 1997; L. A. Garza-Valdes and F. Cervantes-Ibarrola; "Biogenic Varnish and the Shroud of Turin," *L'Identification Scientifique de l'homme du Linneul Jesus de Nazareth: Actes Du Symposium Scientifique International,* Rome, 1993 (Paris: Francois-Xavier Guibert, 1995). Dr. Garza also stated this at a public conference on the Shroud of Turin held at the University of Southern Indiana on February 12, 1994 and has repeated this claim a number of times to other Shroud researchers.

19. The inorganic amounts of chlorine, calcium and nitrogen will first need to be thoroughly rinsed from the linen control samples. This should be done even though the organic amounts of chlorine and calcium will be revealed in the Cl-36 to Cl-35 and Ca-41 to Ca-40 ratios taken from the control samples. The organic amounts of chlorine within linen refer to chlorine that is bound within it as a result of the growth of the flax plant or manufacturing process of linen. (Inorganic amounts of chlorine are those acquired from handling, breathing, water or other sources.) These organic amounts would have been present when the control samples were neutron irradiated. In case any inorganic chlorine or calcium was also present when they were irradiated, the inorganic material rinsed or removed from the control samples should also be tested for any above-background presence of Cl-36 and Ca-41. This would be excellent practice for the same procedure with inorganic material from the Shroud.

20. This is consistent with the 2014 positions of STURP physicist, John Jackson, who moderated an open and hearty discussion on future testing of the Shroud of Turin at an international conference held in St. Louis in October of 2014, and of Professor Bruno Barberis of the International Center of Sindonology in Turin who attended as a special guest. They took the position that until an explanation for the Shroud's aberrant radiocarbon dating can first be provided or demonstrated, then another C-14 or radiocarbon dating would do more harm than good.

21. The amount of N-14 in a linen sample can be acquired by blasting several very small representative parts of *individual* warp and weft threads (aliquot) with an enormous, but known amount of neutrons, whose C-14 to C-12 date would translate to more than 100,000 years into the future. This C-14 to C-12 ratio would reveal the amount of N-14 within the sample that produced this large number of C-14 atoms.

22. Since washing or rinsing inorganic chlorine and nitrogen from blood is very difficult, the inorganic amounts that accumulated in the blood can be calculated and subtracted from both numbers in the Cl-36 to Cl-35 ratios in blood.

23. Varying amounts of chlorine and nitrogen will appear in linen woven from flax plants from various fields. Since the material used in woven linen comes from *within* the long thin flax stalks, the lengths of the individual threads would reflect more consistent amounts of chlorine and nitrogen than would small cross sections of the cloth.

24. A side-by-side cloth and blood sample from the middle of the man's back should be removed and tested, however, in order to help determine whether the blood marks momentarily disappeared along with the body during the neutron radiating event. See Chapter Twelve.

25. *The Mysterious Man of the Shroud,* directed by Terry Landeau, CBS documentary, aired April 1, 1997; L. A. Garza-Valdes, *The DNA of God?* (London: Hodder & Stoughton, 1998), pp. 41, 42.

26. P. Barbet, *A Doctor at Calvary* (Garden City, New York: Doubleday Image Books, 1963; P. Barbet, Le Cinque Piaghe di Cristo "The Five Wounds of Christ," SEI, Turin. 1940. In: C Goldoni, "The Shroud of Turin and the bilirubin blood stains," *Proceedings of the 2008 Columbus International Conference,* Shroud Science Internet Group, August 14-17, edited by G. Fanti. http://www.ohioshroudconference.com/papers/p04.pdf.

27. C. Goldoni, "The Shroud of Turin and the bilirubin blood stains."

28. A. D. Adler, "Chemical Investigation on the Shroud of Turin" in *The Mystery of the Shroud of Turin Interdisciplinary Symposium* video, Elizabethtown, Penn.: Elizabethtown College, February 15, 1986; A. D. Adler, "The Origin and Nature of Blood on the Shroud of Turin," excerpts from lecture of the Dept. of Anatomy, Univ. of Hong Kong, in *Turin Shroud – Image of Christ?* (Hong Kong: Cosmos Printing Press Ltd., 1987); E. J. Jumper, A. D. Adler, J. P. Jackson, S. F. Pellicori, J. H. Heller and J. R. Druzik, "A Comprehensive Examination of the Various Stains and Images on the Shroud of Turin," *ACS Advances in Chemistry No. 205 Archaeological Chemistry III,* J. B. Lambert, ed. American Chemical Society (1984), PP. 447-476.

29. B. Barberis, "The Future of Research on the Shroud," *Shroud of Turin: The Controversial Intersection of Faith and Science, International Conference,* St. Louis, MO, October 9-12, 2014, http://shroud.com/pdfs/stlbarberispaper2.pdf; B. Barberis, "Perspectives for the Future Study of the Shroud," *1st International Congress on the Holy Shroud in Spain – Valencia,* Valencia, Spain, April 28-30, 2012, http://shroud.com/pdfs/barberisv.pdf; P. Di Lazzaro, A. Danielis, M. Guarneri, M. Missori, D. Murra, V. Piraccini, V. Spizzichino and S. Bollanti, "Multidisciplinary

Study of the Shroud of Arquata, 'extractum ab originali'," *Workshop on Advances in the Turin Shroud Investigation,* Bari, Italy, September 4-5, 2014, http://www.frascati.enea.it/fis/lac/excimer/sindone/report%20arquata.pdf.

CHAPTER NINE

1. R. N. Rogers, "Studies on the radiocarbon sample from the Shroud of Turin," *Thermochimica Acta,* 425 (2005) pps. 189-194, 189.

2. Ibid., p. 191.

3. Ibid., p. 192.

4. Ibid.

5. Ibid., p. 193.

6. Ibid.

7. Ibid.

8. J. Marino and M. S. Benford, "Evidence for the Skewing of the C-14 Dating of the Shroud of Turin Due to Repairs," *"Sindone 2000" Orvieto Worldwide Congress,* Orvieto, Italy, August 27-29, 2000, http://www.shroud.com/pdfs/marben.pdf.

9. M. S. Benford and J. Marino, "Textile Evidence Supports Skewed Radiocarbon Date of Shroud of Turin," (August 2002), http://www.shroud.com/pdfs/textevid.pdf; R. N. Rogers and A. Arnoldi, "Scientific Method Applied to the Shroud of Turin," (2002), http://www.shroud.com/pdfs/rogers2.pdf; R. Rogers, "Supportive comments on the Benford-Marino 16th century repairs hypothesis," *British Society for the Turin Shroud Newsletter*, No. 54, November 2001, pp. 28-33; J. Marino and S. Benford, "Could the Shroud's Radiocarbon Date have been Skewed due to 16th Century Repairs?," *British Society for the Turin Shroud Newsletter*, No. 54, November 2001, pp. 18-27; M. S. Benford and J. Marino, "Historical Support of a 16th Century Restoration in the Shroud C-14 Sample Area," (August 2002), http://www.shroud.com/pdfs/histsupt.pdf; M. S. Benford and J. Marino, "New Historical Evidence Explaining the 'Invisible Patch' in the 1988 C-14 Sample Area of the Turin Shroud," (September 2005) *Third International Dallas Conference*, http://www.shroud.com/pdfs/benfordmarino.pdf; M. S. Benford and J. Marino, "Discrepancies in the radiocarbon dating area of the Turin shroud," *Chemistry Today*, vol 26 n 4, (July-August 2008),http://www.shroud.com/pdfs/benfordmarino 2008.pdf.

10. Benford and Marino, "Textile Evidence."

11. Rogers and Arnoldi, "Scientific Method," p. 21.

12. R. N. Rogers, "Shroud No Hoax, Not Miracle," Letter to the Editor, *Skeptical Inquirer*, July/August 2004.

13. The Shroud was carbon dated from 1260-1390. The water stains at the radiocarbon site could not have been encoded later than 1532.

14. M. Flury-Lemberg, *Sindone 2002* (Torino: Editrice ODPF, 2003), p.60. English translation: Rosamund Bandi and Susie Clavarino Phillips.

15. M. Flury-Lemberg, "The Invisible Mending of the Shroud in Theory and Reality." *British Society for the Turin Shroud Newsletter*, No. 65, June 2007, pp. 10-27.

16. Flury-Lemberg, "The Invisible Mending of the Shroud in Theory and Reality."

17. Ibid., p. 15.

18. Flury-Lemberg, "The Invisible Mending of the Shroud in Theory and Reality."

19. In addition, unlike other repairs of the Shroud, there is no historical record for this particular repair, which would have been the finest of all. There would certainly be no need to keep this a secret from the public or from any private records among the owners, or any number of church officials or members, or by the masterful restorer(s). To repair the Shroud is certainly nothing to be ashamed of. On the contrary, it would be something to be proud of. Assuming the repair was so excellent that no person or scientific instrument (to this day) could even see where it occurred, would clearly be something to be extremely proud of.

20. Flury-Lemberg, "The Invisible Mending of the Shroud in Theory and Reality," p. 26.

21. L. A. Schwalbe and R. N. Rogers, "Physics and Chemistry of the Shroud of Turin," *Analytica Chimica Acta* 135 (1982); 3-49, note 6, p. 47.

22. John Jackson's comments on Antonacci's response to Rogers. E-mail to Shroud Science Group on May 31, 2005.

23. Ibid.

24. The radiographs were reported in R. A Morris, L. A. Schwalbe and J. E. London,

X-ray Spectrom, 9 (1980) 40, according to Schwalbe and Rogers, "Physics and Chemistry of the Shroud of Turin." The radiograph of the entire lower left portion of the Shroud of Turin can be found in the latter publication by Schwalbe and Rogers. At the St. Louis Shroud Conference held in 2014, Jackson stated in an open forum on future testing that the density bands attributed to the Shroud could belong to its backing cloth.

25. John Jackson's comments on Antonacci's response to Rogers. E-mail to Shroud Science Group on May 31, 2005.

26. M. Antonacci, *The Resurrection of the Shroud* (New York: M. Evans and Company, Inc., 2000) pps. 168 & 304; See also J. Marino, The Shroud of Turin and the Carbon 14 Controversy," *Fidelity,* (February 1989): 35-47.

27. A. Adler, "Updating Recent Studies on the Shroud of Turin," *Archaeological Chemistry: Organic, Inorganic, and Biochemical Analyses*, Mary Virginia Orna, ed. American Chemical Society (1996): 223-228; A. D. Adler, A. Whanger, and M. Whanger, "Concerning the Side Strip on the Shroud of Turin," http://www.shroud.com/adler2.htm,(October 27, 1997); Dr. Alan Adler, personal communications, June 1998 and February 1999.

28. Rogers, "Studies on the radiocarbon sample," p. 189.

29. Ibid., p. 192.

30. Adler, "Updating Recent Studies," p. 225.

31. A. Geurreschi and M. Salcito, "Photographic and Computer Studies Concerning the Burn and Water Stains Visible on the Shroud and Their Historical Consequences." *IV Symposium Scientifique International,* Paris, April 25-26, 2002; A. Guerreschi and M. Salcito, "Further Studies on the scorches and the Water stains," *Third International Dallas Conference*, September 5-6, 2005.

32. Ibid.

33. R. A. Freer-Waters and A. J. T. Jull, "Investigating a Dated Piece of the Shroud of Turin," *Radiocarbon,* Vol. 52, Nr 4, 2010, 1521-1527, 1524.

34. Ibid., p. 1526.

35. Flury-Lemberg, "The Invisible Mending of the Shroud in Theory and Reality."

36. Rogers and Arnoldi, "Scientific Method," p. 21.

37. Rogers, "Shroud Not Hoax, Not Miracle," Letter to the Editor.

38. R. Villarreal, B. Schwortz and M. S. Benford, "Analytical Results on Threads Taken from the Raes Sampling Area (Corner) of the Shroud," (2008), *Proceedings of the 2008 Columbus International Conference*, Shroud Science Internet Group, edited by G. Fanti, 319-336, 322.

39. Ibid.; Villarreal, video of presentation, at 19:23 "looks very much like cotton," 12:30 "both regions, region 1 and region 2, are cotton", and 14:24 slide "unexpected silicon", and 33:53 "Silicon all through the tread", according to A. A. M. van der Hoeven, "Internal selvedge in starched and dyed temple mantle – No invisible repair in Turin Shroud – No Maillard reaction," http://www.jesusking.info, May 9, 2012.

40. van der Hoeven, "Internal Selvedge", p. 26.

41. "If the Shroud had been stored at a constant 25° C, it would have taken about 1319 years to lose a conservative 95% of its vanillin. At 23° C, it would have taken about 1845 years. At 20° C, it would take about 3095 years. If the Shroud had been produced between A.D. 1260 and 1390, as indicated by the radiocarbon analysis, lignin should be easy to detect. A linen produced in A.D. 1260 would have retained about 37% of its vanillin in 1978. The Raes threads, the Holland cloth, and all other medieval linens gave the test for vanillin wherever lignin could be observed on growth nodes. The disappearance of all traces of vanillin from the lignin in the Shroud indicates a much older age than the radiocarbon laboratories reported." Rogers, "Studies on the radiocarbon sample," p. 191.

42. Rogers, "Studies on the radiocarbon sample," pps. 190 and 191.

43. Rogers, "Studies on the radiocarbon sample," p. 192.

44. E-mail from John Jackson and Keith Propp to Shroud Science group on February 9, 2005.

45. Benford and Marino, "Textile Evidence," p. 9.

46. "Home" page of "The French Re-Weavers," http://www.thefrenchreweavers.com; "Alterations-Repairs-Reweaving" page of "Arrow Fabricare Services," http://www.arrowcare.com/alterations-repairs--reweaving.html; See also video, http://www.withoutatrace.com/reweaving.html. This is the video that I was referred to by Joe Marino.

Exercise no. 19 of the Fenway System talks about reweaving linen cloth that has been woven in a basket weave. This weave is approximately six under and two over for both the weft (horizontal) and the warp (vertical). However, a 3:1 herringbone twill means that the weft thread passes under three warp threads and then over one, with each successive weft thread beginning at an ascending point one warp thread earlier, *and then*, in series, at a descending point to form a diagonal herringbone pattern. As you can see, a herringbone twill is much more complicated and could not have been rewoven or blended invisibly into surrounding linen in the 1500s or earlier. I doubt if it could be invisibly blended or rewoven to match all of the surrounding area with state of the art microscopes today.

Even if you could somehow blend herringbone twill 360° around the damaged area, the overlapping threads would be denser and more numerous. The exercise even states, "The fact that you periodically must pass over and under several threads requires you to begin weaving over intact portions of the damaged threads at a greater distance from the damage than you have to do for simple weaves in order to get sufficient anchorage." This would only make the additional, blended threads more detectable, not only to the naked eye, but also to all the technology that was applied to the Shroud. http://shrouduniversity.com/frenchreweavinginstructionbook.pdf.

47. According to a translation and e-mail provided by Antonio Lombatti to Shroud-Science@yahoogroups.com on February 8, 2005.

CHAPTER TEN

1. Greek text in A. Heisenberg, *Nicholas Mesarites-Die Palas-revolution des Johannes Comnenos* (Wurzburg, 1907), p. 30, according to I. Wilson, *The Shroud of Turin*, (Garden City, N.Y.: Doubleday, 1979) pp. 167-68.

2. D. Crispino, "1204: Deadlock or Springboard?" *Shroud Spectrum International* 4 (September 1982): 24-30; P. Dembowski, "Sindone in the Old French Chronicle of Robert de Clari," *Shroud Spectrum International* 2 (March 1982): 13-27; I. Wilson, *The Shroud of Turin*, p. 169; I. Wilson, *The Mysterious Shroud,* (Garden City, NY: Doubleday, 1986), p. 104; M. Green, "Enshrouded in Silence: In Search of the First Millennium of the Holy Shroud," *Ampleforth Journal* 74 (Autumn 1969): 321-345; See also R. Andes, as stated in F. C. Tribbe, *Portrait of Jesus?* (New York: Stein and Day Publishers, 1983), p. 56.

3. *Story of the Image of Edessa*, Appendix C, in Wilson, *The Shroud of Turin; Evagrius' Ecclesiastical History,* From 431 to 594 A. D. (London: Samuel Bagster and Sons, 1846); *St. John of Damascus on the Divine Images*, trans. By D. Anderson, (Crestwood, N.Y.: St. Vladimir's Seminary Press, 1980); *Acts of the Holy Apostle Thaddaeus,* trans. in A. Roberts and J. Donaldson, eds., *The Ante-Nicene Fathers*, Vol. 8 (New York: Charles Scribner's Sons, 1899; rpt. Grand Rapids, Mich.: Eerdmans, 1951); *The Doctrine of Addai,* as stated in J. B. Segal, *Edessa The Blessed City* (Oxford, 1970).

4. Wilson, *The Mysterious Shroud,* p. 110.

5. P. Vignon, *The Shroud of Christ* (Westminster, 1902; trans. New Hyde Park, N.Y.: University Books, 1970); E. A. Wuenschel, *Self-Portrait of Christ* (Esopus, N.Y.: Holy Shroud Guild, 1957); Green, "Enshrouded in Silence."

6. A. Whanger and M. Whanger, "Polarized image overlay technique: a new image comparison method and its applications," *Applied Optics* 24.6 (March 15, 1985): 766-772; Wilson, *The Mysterious Shroud,* color plates 23-27; Tribbe, *Portrait of Jesus?*, p. 241.

7. M. Whanger and A. Whanger, *The Shroud of Turin: An Adventure of Discovery* (Franklin, Tenn.: Providence House Publishers, 1998); Alan Whanger, personal communications, February 10 and April 16, 2000. See also Tribbe, *Portrait of Jesus?,* p. 241.

8. I am indebted to Dr. Alan Adler, who first pointed out this line of reasoning to me.

9. D. Mercieri, "Ancient coin portrays shroud-like Jesus," Image (Turin Shroud Center of Colorado Newsletter) 3.1 (Spring 1995): 6-7.

10. W. Bulst, "The Pollen Grains on the Shroud of Turin," *Shroud Spectrum International* 10 (March 1984): 20-28; M. Frei, "Nine Years of Palinological Studies on the Shroud," *Shroud Spectrum International* 3 (June 1982): 3-7.

11. M. Symeon, "De Const. Porph. Et Romano Lecapeno," sec. 50, p. 491 of Ms. and 748 of *Corpus scriptorum historiae byzantinae* (Bonn, 1978) as stated in Wilson, *The Shroud of Turin*, p. 116.

12. Wilson, *The Shroud of Turin,* p. 115.

13. This term is first used by Evagrius in his *Ecclesiastical History* written in the sixth century and continues to be used to describe the Image of Edessa and Mandylion until its disappearance in the thirteenth century.

14. Wilson, *The Shroud of Turin,* p. 112.

15. J. Heller, KMOX radio interview, St. Louis, MO, on December 29, 1983.

16. Wilson, *The Shroud of Turin*; Tribbe, *Portrait of Jesus*?; R. Drews, *In Search of the Shroud of Turin* (Totowa, NJ: Roman & Allanheld, 1984).

17. *Story of the Image of Edessa*, appendix C, para. 15, in Wilson, *The Shroud of Turin*, p. 280.

18. A. Grabar, "LaSainte Face de Laon et le Mandylion dans l'art orthodoxe," *Sem-*

inarium Knodakovianum (Prague, 1935), p. 16, as stated in Wilson, *The Shroud of Turin,* p. 114.

19. Wilson, *The Shroud of Turin*, pp. 114-115.

20. *Vita Alexilus*, Monday, August 5, Ulr.III, Massiman, 176 T. according to I. Wilson in "The Shroud and the Mandylion: A reply to Professor Averil Cameron," *Turin Shroud – Image of Christ?* (Hong Kong: Cosmos Printing Press Ltd., 1987), pp. 19-28; See also W. Bulst, *The Shroud of Turin* (Milwaukee: Bruce, 1957), pp. 42, 125.

21. Matthew 27:59; Mark 15:46; Luke 23:53.

22. E. Von Dobschutz, *Christusbilder* (Leipzig 1899), Beilage III, pp. 130-135, as stated in R. Drews, *In Search of the Shroud,* pp. 39, 46-48; See also Wilson in "The Shroud and the Mandylion," pp. 23-24.

23. Von Dobschutz, *Christusbilder*, Document 30b (Kap. V), p. 189, according to Drews, *In Search of the Shroud*; John of Damascus *De.Fid.Orth*. IV, 16, in Migne, *Patrologia Graeca* 94, 1173 according to Wilson in "The Shroud and the Mandylion."

24. E. Von Dobschutz, *Christusbilder*, Document 71 (Kap V), p. 217, according to Wilson, "The Shroud and the Mandylion," and Drews, *In Search of the Shroud.*

25. Wilson, *The Shroud of Turin*, pp. 119-120.

26. *Acta Thaddaei 3* (from *Acta apostolorum apocrypha,* ed. R. A. Lipsius (Leipzig, 1891), I, p. 274; trans. In A. Roberts and J. Donaldson, eds., *The Ante-Nicene Fathers* (Grand Rapids, MI: Eerdmans, 1951), Vol. VIII, pp. 558-59, as stated in I. Wilson, "The Shroud and the Mandylion;" see also Wilson, *The Shroud of Turin*, pp. 120, 307.

27. "Liturgical Tractate" according to E. Von Dobschutz, *Christusbilder*, Beilage II, C pp. 110-114 as stated in Wilson, "The Shroud and the Mandylion," pp. 21, 27; "Monthly Lection," as stated in Drews, *In Search of the Shroud of Turin*, pp. 39-40, 116.

28. Paul Maloney states 88 pollen gains were counted from a 2 cm^2 location from the dorsal side strip and 163 on a tape from the same size area on the left arm, but that about 300 were counted from a comparable size area near the face. "Is the Shroud of Turin Really Medieval?" *Newsletter of the Association of Scientists and Scholars International or the Shroud of Turin*, Ltd. (ASSIST), I (I): 5-7.

29. J. P. Jackson, "Foldmarks as a Historical Record of the Turin Shroud," *Shroud Spectrum International* II (June 1984): 6-29.

30. Wilson, *The Mysterious Shroud*, p. 120.

31. *Story of the Image of Edessa; Evagrius' Ecclesiastical History; St. John of Damascus on the Divine Image Acts of the Holy Apostle Thaddeus; The Ante-Nicene Fathers; The Doctrine of Addai*, as stated in Segal, *Edessa The Blessed City*.

32. The principal Syriac texts are: "The Doctrine of Addai," published by Dr. W. Cureton in *Ancient Syriac Documents Relative to the Earliest Establishment of Christianity in Edessa,* 1864, from two manuscripts of the fifth and sixth centuries from the Nitrian collection; *The Doctrine of Addai the Apostle*, translated by G. Phillips and Wright, 1876, from a manuscript then in the Imperial Library of St. Petersburg, according to I. Wilson, *The Shroud of Turin*.

33. Wilson, *The Shroud of Turin*, p. 128.

34. *Story of the Image of Edessa.*

35. "The Teaching of Thaddaeus the Apostle," trans., in A. Roberts and J. Donalson, eds., *The Ante-Nicene Fathers*, Vol. 8, p. 665, as stated in Wilson, *The Shroud of Turin,* p. 131.

36. J. Wilkinson, *Egeria's Travels*, rev., ed. (Jerusalem: Ariel Publishing House, Warminister, England: Aris & Phillips, 1981).

37. S. Runciman, "Some Remarks on the Image of Edessa," *Cambridge Historical Journal* III (1929-31).

38. *Evagrius' Ecclesiastical History*, From A.D. 431 to 594.

39. Evagrius, "Ecclesiastical History," original text in Migne, *Patrologia graeca*, 86.2: 2748-2749, translation from Bohn's Ecclesiastical Library (1854), as stated in Wilson, *The Shroud of Turin*, p. 137.

40. Green, "Enshrouded in Silence;" Wilson, *The Shroud of Turin*; Wilson, *The Mysterious Shroud*; Humber, *The Sacred Shroud* (New York: Pocket Books, 1974); R. Wilcox, *Shroud* (New York: Bantam Books, 1979); Tribbe, *Portrait?*

41. As stated in Wilson, *The Shroud of Turin*; D. Scavone, "The History of the Shroud to the 14th Century," *History, Science, Theology and the Shroud*, ed. A. Berard, (St. Louis: Richard Nieman, 1991), pp. 171-204; Tribbe, *Portrait?*

42. The one account that mentions any image of Jesus prior to the sixth century is found in the "Doctrine of Addai" written in about A.D. 400. This account describes

the image as an ordinary painting. It does not even describe it as having been painted on cloth. Unlike the Image of Edessa, no miracles are associated with this painting. Neither is it described as "not made by human hands" nor as an image of Christ miraculously imprinted onto cloth.

43. E. J. Jumper, A. D. Adler, J. P. Jackson, S. F. Pellicori, J. H. Heller, and J. R. Druzik, "A Comprehensive Examination of the Various Stains and Images on the Shroud of Turin," *ACS Advances in Chemistry No. 205 Archaeological Chemistry III, ed. J. B. Lambert, American Chemical Society* (1984), pp. 447-76, p. 453.

44. Jumper et al., "Comprehensive Examination," p. 453.

45. Attorney Jack Markwardt points out that if the Shroud is folded once widthwise and once lengthwise (as you would a towel for example), that several small circular burn marks on the dorsal side line up over each other and match. These burn marks are separate from the burn and scorch marks left by the fire of 1532. He speculates that the small round burn marks resulted from pitch-soaked firebrand being administered to the cloth folded in this manner while the Edessans were trying to start the fire that burned Chosroe's timber mound in 544. One could speculate that small burning wood embers, or sparks, could have cause them as well. Markwardt further speculates the Shroud, at that point, could have been placed in its long-standing doubled in four fold configuration within a frame to conceal the damning evidence of their treatment of the revered cloth.

46. W. Meacham, "The Authentication of the Turin Shroud: An Issue in Archaeological Epistemology," *Current Anthropology* 24.3 (June 1983), pp. 283-311.

47. W. H. Carroll, "The Dispersion of the Apostles: Jude and the Shroud," *Faith and Reason* (Fall 1981): 235-243.

48. J. B. Chabot, "Anonymi auctoris Chronicon ad annum Christi 1234 pertinens," *Carpus scriptorium christianorom orientalium*, 81-82, Scr. Syri 36-37, 1953, quoted and translated in Segal, *Edessa The Blessed City*.

49. Translation from Green, "*Enshrouded in Silence*," p. 333.

50. Vatican Library Coder No. 5696, fol. 35, published I P. Savio *Ricerche storiche sulla Santa Sindone* (Turin, 1957), footnote 31, p. 340; translation by Green, as stated in Wilson, *The Shroud of Turin,* pp. 158, 312.

51. Ordericus Vitalis, *Historia ecclesiastica*, part III, bk. IX, 8, "De Gestis Bolduini Edessae principatum obtinet," according to Wilson, *The Shroud of Turin*, pp. 158-312.

52. Wilson, *The Shroud of Turin*, Chapter XVIII.

53. Ibid.

54. Ibid.

55. J. Jackson and R. Jackson, "New Evidence that the Shroud of Turin Pre-Dates the Radiocarbon Date by Centuries," *Third International Congress on the Shroud of Turin,* Turin, Italy, June 5-7, 1998.

56. Ibid.

57. H. Evans and W. Wixon, eds., *The Glory of Byzantium: Art and Culture of the Middle Byzantine Era*, A.D. 843-1261 (New York: The Metropolitan Museum of Art, 1997); Jackson and Jackson, "New Evidence."

58. Jackson and Jackson, "New Evidence."

59. Bulst, *Das Turiner grabtuch und das Christusbild.*

60. Translated by and reported in the *British Society for the Turin Shroud Newsletter* 18 (January 1988): pp. 7-8.

61. Ibid.

62. Translation from P. Johnstone, *The Byzantine Tradition in Church Embroidery*, (London, 1967), p. 54, as stated in Wilson, *The Shroud of Turin*, pp. 157-312.

63. Wilson, *The Shroud of Turin,* pp. 145-146 and 156-157.

64. *St. John of Damascus on the Divine Images*, trans. By D. Anderson, in A. Roberts and J. Donaldson, eds., *The Ante-Nicene Fathers;* Theodore of Studium, in Migne, *Patrologia graeca* 177.64, p. 1288, as stated in Wilson, *The Shroud of Turin*, p. 148; Segal, *Edessa The Blessed City*, p. 215; See also S. Runciman, "Some Remarks on the Image of Edessa," *Cambridge Historical Journal* 111.3 (1931): 238-52.

65. This theory was originated by Ian Wilson in 1978 and is presented in great detail in his book *The Shroud of Turin*. See in particular chapter XIX.

66. Tribbe, *Portrait?* p. 57.

67. Wilson, *The Shroud of Turin*, Chapter XIX.

68. Wilson, *The Shroud of Turin*, p. 183.

69. de Puy, *History of the Military Order of the Templars Paris,* 1713), as cited in Tribbe, *Portrait?* p. 57.

70. Wilson, *The Shroud of Turin*, p. 183.

71. All such descriptions here and elsewhere of the Templar image or idol can be found in Wilson, *The Shroud of Turin*, Chapter XIX.

72. *Chronicles of St. Denis*, art. III, quoted in de Puy, *Histoire de l' Ordre Militaire des Templiers* (1713), p. 25, as stated in Wilson, *The Shroud of Turin*, p. 189.

73. Wilson, *The Shroud of Turin.*

74. See Crispino, "1204: Deadlock or Springboard?"

75. P. Savio, *Ricerche storiche sulla Santa Sindone* (Turin, 1957), according to Prof. D. Scavone, personal communication, August 18, 1987. A summarized version of Savio's theory appears in Tribbe, *Portrait?*

76. As stated in Wilson, *The Mysterious Shroud*, and in N. Currer-Briggs, *The Holy Grail and the Shroud of Christ* (Middlesex: England, ARA Publications, 1984).

77. Wilson, *The Mysterious Shroud.*

78. As stated in Currer-Briggs, *The Holy Grail and the Shroud of Christ.*

79. J. Walsh, *The Shroud* (New York and Toronto: Random House, 1963), pp. 44-45; Vignon, *The Shroud of Christ*, pp. 56-57; Tribbe, *Portrait?*, pp. 62-63; J. Jannone, *The Mystery of the Shroud of Turin* (New York: Alba House, 1998); and W. K. Muller, as stated in the *British Society for the Turin Shroud Newsletter* 13 (April 1986).

80. Wuenschel, *Self-Portrait of Christ*; L. Fossati, "A Critical Study of the Lirey Documents," *Shroud Spectrum International* 41 (December 1992), p. 4.

81. Translation by Herbert Thurston found in Appendix B in Wilson, *The Shroud of Turin.*

82. After this account, the Shroud was thereafter referred to by most people not as the true Shroud of Christ, but as a painting or even a likeness or representation, as did the Avignon Pope Clement VII (although its exhibitors and subsequent popes cer-

tainly did not). The Avignon Pope, however, not only allowed the exhibit to continue, but imposed perpetual silence upon Bishop d'Arcis about this matter. For an additional account on the background and motives attributed to the Avignon Pope and to the de Charny family, see Wilson, *The Shroud of Turin*, Chapter XX, particularly p. 208.

83. Scavone's views can be found in his review of a work by Robert Babinet published in the *British Society for the Turin Shroud Newsletter,* 46 (November/December 12997): 36-37.

84. The discovery of the Chevalier manipulation was made by Hilda Leynen of Antwerp and the total explanation is contained in D. Crispino "Literary Legerdemain," *Shroud Spectrum International*, "Spicilegium" (1996): pp. 63-66.

85. Crispino, "Literary Legerdemain," p. 64.

86. L. Fossati, "The Lirey Controversy," *Shroud Spectrum International* 8, (September 1983): 24-34, Crispino, "Literary Legerdemain," p. 65.

87. Crispino, "Literary Legerdemain," p. 66.

88. Fossati, "The Lirey Controversy," pp. 28-30.

89. Bishop de Poitiers would subsequently also allow his niece to marry Geoffrey II de Charny, who arranged for the Shroud to be exhibited in 1389. It is questionable that such permission would have been granted if scandal had followed the de Charny family name.

90. *Promptuarium Sacrarum Antiquitatum Tricassinae Dioecesis*, 1610, as stated in E. A. Wuenschel, "The Holy Shroud of Turin: Eloquent Record of the Passion," *The Ecclesiastical Review*, 93 (November 1935), p. 444; Humber, *The Sacred Shroud*.

91. Frei, "Nine Years of Palinological Studies;" Bulst, "The Pollen Grains."

92. S. Shafersman, in letter to Walter McCrone in the *Microscope*, 30 (1982): 344-352.

93. P. Maloney, "Modern Archaeology, History and Scientific Research on the Shroud of Turin," in *Mystery of the Shroud of Turin: An Interdisciplinary Symposium*, video, Elizabethtown, PA: Elizabethtown College, (February 15, 1986).

94. A Danin, A. Whanger, U. Baruch, M. Whanger, *Flora of the Shroud of Turin* (St. Louis: Missouri Botanical Garden Press, 1999) p. 24; Maloney, "Modern Archaeology."

CHAPTER ELEVEN

1. J. P. Jackson, "Is the Image on the Shroud Due to a Process Heretofore Unknown to Modern Science?," *Shroud Spectrum International* 34 (March 1990): 3-29, 9.

2. J. P. Jackson, "An Unconventional Hypothesis to Explain All Image Characteristics Found on the Shroud Image," *History, Science, Theology and the Shroud*, A. Berard, ed. (St. Louis: Richard Nieman, 1991), pp. 325-344; J. P. Jackson, E. J. Jumper and W. R. Ercoline, "Correlation of Image Intensity on the Turin Shroud with the 3-D Structure of a Human Body Shape," *Applied Optics* 23.14 (July 1984): 2244-2270; and Jackson, "Is the Image."

3. *The Elements*, 3rd ed., Clarendon Press, Oxford, 1998.

4. J. P. Jackson, "The Vertical Alignment of the Frontal Image," *Shroud Spectrum International* 32/33 (1989): 3-26; J. P. Jackson, E. J. Jumper, B. Mottern, and K. E. Stevenson, "The Three-Dimensional Image on Jesus' Burial Cloth, "*Proceedings of the 1977 United States Conference of Research on the Shroud of Turin* (Albuquerque, N.M.: Holy Shroud Guild, Mar. 1977), 74-94, p. 83; and W. R. Ercoline, R. C. Downs, Jr., and J. P. Jackson, "Examination of the Turin Shroud for Image Distortions," *IEEE 1982 Proceedings of the International Conference on Cybernetics and Society* (October 1982): 576-579.

5. Physicists John Jackson and Arthur Lind, personal communications.

6. All of the dorsal body image features could also be encoded if the body vanished or disappeared vertically in the same direction in which the cloth collapsed.

7. Nuclear engineer, Robert Rucker, personal communication, January 25, 2015.

8. Ercoline et al., "Examination of the Turin Shroud."

9. I'm not aware of this image distortion being mentioned or reported until STURP made a brief reference to it without an explanation in Ercoline, et al., "Examination of the Turin Shroud."

10. G. Fanti, R. Maggiolo, "The double superficiality of the frontal image of the Turin Shroud," J. Opt. A, Pure Appl. Opt. 6 (2004): 491-503.

11. The Corona Discharge Hypothesis by Dr. Fanti also accounts for the outer side imaging observed at these locations.

12. E. J. Jumper, A. D. Adler, J. P. Jackson, S. F. Pellicori, J. H. Heller, and J. R.

Druzik, "A Comprehensive Examination of the Various Stains and Images on the Shroud of Turin," *ACS Advances in Chemistry No. 205 Archaeological Chemistry III,* J. B. Lambert, ed. American Chemical Society (1984), pp. 447-476, 456.

13. S. F. Pellicori and M. Evans, "The Shroud Through the Microscope," *Archaeology* (January/February 1981): 32-43; S. F. Pellicori, "Spectral Properties of the Shroud of Turin," *Applied Optics* (June 15, 1980); 1913-1920, G. G. Gray, "Determination and Significance of Activation Energy in Permanence Tests," in *Preservation of Paper and Textiles of Historic and Artistic Value,* Advances in Chemistry series 164 (Washington, DC: American Chemical Society, 1977), as cited in Pellicori "Spectral Properties"; S. Pellicori and R. A. Chandos, "Portable Unit Permits UV/vis Study of 'Shroud'," *Industrial Research & Development,* February 1981, 23: 186-189, 187; J. Rinaudo, "Protonic Model Image Formation on the Shroud of Turin," *Third International Congress on the Shroud of Turin,* Turin, Italy, June 5-7, 1998; J. Rinaudo, "A Sign of Our Time," *Shroud Sources Newsletter,* May/June 1996, pp. 2-4; J. Jackson, E. Arthurs, L. Schwalbe, R. Sega, D. Windisch, W. Long, E. Stappaerts, "Infrared Laser Heating for Studies of Cellulose Degradation" *Applied Optics,* 15 Sept., 1988, 27:3937-3943.

14. Rinaudo, "Protonic Model" and "A Sign."

15. Rinaudo, "Protonic Model," p. 4.

16. Rinaudo, "Protonic Model" and personal communication.

17. These gamma rays would radiate at low energy, and when they hit the electrons in the heavier elements of the coin and flowers, their atoms could fluoresce long-wave X-rays from the objects' surfaces, or fluoresce as short-wave X-rays, as ultraviolet rays or even as visible light.

18. Rinaudo, "Protonic Model" and "A Sign."

19. Ibid.

20. Some scientists have claimed that the radiation in Rinaudo's Protonic Model is essentially too strong. They claim that some of the linen fibers that were encoded at the levels found within his model are more than 2-3 fibers deep, and that some of their coloring extends beyond the primary cell wall (or that the centers of the fibers are colored). (G. Fanti, "Hypothesis Regarding the Formation of the Body Image on the Turin Shroud. A Critical Compendium," *Journal of Imaging Science and Technology* 55(6): 060507-1-060507-14, 2011.)

The initial energy source of Rinaudo's particle radiation derives from gamma rays with an ideal initial energy of 4.5 MeV, which split deuterium nuclei at the surface of the body. This results in protons and neutrons with energies of at least 1.135 MeV.

Whereas the energy range for Rinaudo's proton irradiations were between 1.135 and 1.4 MeV, the energy range of protons and neutrons in the Historically Consistent Hypothesis could be much less. Rinaudo's Protonic Model inherently cannot contain protons (or neutrons) at lesser energies than stated above. Robert Rucker's MCNP calculations determined that the neutrons left behind by the body at thermal energy (0.0253 eV) is one possible explanation why the Shroud dated to medieval times. In addition, the energy required to break chemical bonds is only two to ten eV. (https://en.wikipedia.org/wiki/Bond-dissociation energy). In light of these points, the energy range of the protons and neutrons within the Historically Consistent Hypothesis could be considerably less than the minimum 1.135 MeV contained within Dr. Rinaudo's landmark Protonic Model.

Some sindonologists also claim that some Shroud photomicrographs show striations or abrupt interruptions of color on its threads, which proton or other radiation may not be able to encode. (G. Fanti, et al., "Microscopic and Macroscopic Characteristics of the Shroud of Turin Image Superficiality." *Journal of Imaging Science and Technology* 54(4): 040201-1-040201-8, 2010. Yet, of the hundreds or more of photomicrographs that were taken of the Shroud only two have been published that possibly indicated this effect. During his Moderator's Remarks at the lengthy Open Dialogue for Future Testing at the St. Louis Shroud Conference on October 11, 2014, physicist John Jackson stated this perceived effect could merely be due to lighting.

These same sindonologists also note, with only one of the above supporting Shroud photomicrographs, that straw-yellow coloring can be seen on fibers between the crevices of crossing threads. Yet, this sole photomicrograph indicates these colored fibers are located among other colored fibers. When linen is irradiated by protons, other highly charged particles, UV or certain other forms of radiation, superficial straw-yellow coloring will result on the fibers if they were in the field of radiation. These photomicrographs are from the parts of the cloth that were located at or near the "foot," as well as "the heel." Yet, these indistinct locations on both the draped frontal cloth, or the bottom dorsal side, could be quite different in their geometric relation to the foot or the heel; as they could be for many parts of the cloth and the body as a whole. The amounts of radiation that each part of the cloth receives will vary depending upon its original distance from and its original position on, over, under or at an angle to the body, as well as its movement(s) during the encoding process.

It should also be noted that the abrupt changes in coloring on the superficial fibers on the Shroud's threads and in the crevices of the threads also seemingly appear on a photomicrograph taken of "clear cloth." I think the striations and crevices considerations by some sindonologists coincide with the appearance of previously unpublished photomicrographs from Mark Evans, who unfortunately passed away over a decade ago. I think these observations may be due to lighting as stated by Jackson.

21. P. Barbet. Le cinque piaghe di Cristo (The five wounds of Christ), SEI, Turin (1940) In: C. Goldoni, "The Shroud of Turin and the bilirubin blood stains," Proceedings of the 2008 Columbus International Conference, Shroud Science Internet

Group, August 14-17, edited by G. Fanti, http://www.ohioshroudconference.com/papers/po4.pdf. The personal testimonies of several witnesses are also contained in Goldoni's above article. They have seen a noticeable difference in the reddish color of the Shroud's blood stains when they are exposed to artificial lighting or to sunlight (UV) and when they are not. See also P. Barbet, *A Doctor at Calvary*, (Garden City, New York: Doubleday Image Books, 1963).

22. An initial exposure to ultraviolet light would be expected to produce similar results.

CHAPTER TWELVE

1. P. Barbet, *A Doctor at Calvary* (Garden City, New York: Doubleday Image Books, 1963), p. 27.

2. J. H. Heller and A. D. Adler, "A Chemical Investigation of the Shroud of Turin," *Can. Soc. Forens. Sci. J.* 14.3 (1981): 81-103; V. D. Miller and S. F. Pellicori, "Ultraviolet Fluorescence Photography of the Shroud of Turin," *Journal of Biological Photography* 49.3 (July 1981): 71-85; Dr. P. Scotti, as cited by Dr. Alan Adler in "The Turin Shroud Lecture," Department of Chemistry, Queen Mary College, London, July 20, 1984; Adler, "Chemical Investigation on the Shroud of Turin" in *The Mystery of the Shroud of Turin Interdisciplinary Symposium* video, Elizabethtown, Penn.: Elizabethtown College, February 15, 1986; J. H. Heller, *Report on the Shroud of Turin* (New York: Houghton Mifflin, 1983); A. Adler, as stated in T. W. Case, *The Shroud of Turin and the C-14 Dating Fiasco* (Cincinnati: White Horse Press, 1996), p. 76; E. Jumper, as cited in Adler, "The Turin Shroud Lecture"; L. Gonella, "Scientific Investigation of the Shroud of Turin: Problems, Results and Methodological Lessons," *Turin Shroud—The Image of Christ?* (Hong Kong: Cosmos Printing Press Ltd., 1987).

3. A. Lind and M. Antonacci "Hypothesis that Explains the Shroud's Unique Blood Marks and Several Critical Events in the Gospels," *Shroud of Turin: The Controversial Intersection of Faith and Science, International Conference*, St. Louis, MO, October 9-12, 2014, www.shroud.com/pdfs/stlindpaper.pdf.

4. Even hypothetical uses of the side strip to loosely bind the Shroud around the man cannot account for the complete and intimate contact by all the blood marks on both sides of his body. Nor can such binding explain the blood marks being embedded in the Shroud in the same shape as when they formed and coagulated on the body.

5. While speaking of his resurrection in the days before he died, the Gospels record Jesus referring to his body and his shed blood as two vitally similar, yet distinct things. According to the Gospels, when the women and the apostles went to Jesus' tomb on Easter morning they merely looked at his burial cloths and were naturally astounded at Jesus' disappearance. In the case of the women, they had prepared spices and oint-

ments and were returning so that they could anoint his body (Mark 16:1; Luke 23:56 – 24:1). They were overwhelmed when they arrived at the tomb and realized the body was missing and were told by an angel that he has risen. They ran to tell the apostles. When Peter and John ran to the tomb and saw the undisturbed burial cloths they began to believe and realized that according to scripture, Jesus must rise from the dead. Yet they returned home. Neither they nor the women unfolded or spread out the undisturbed burial cloths. Whenever this did occur with this burial shroud, the bloodstains by themselves without any context or relation to even a negative body image, would not have made a visible impression on the viewer.

6. S. Rodante, "The Coronation of Thorns in the Light of the Shroud," *Shroud Spectrum International* I (December 1981): 4-24, Translated and reprinted from *Sindon* 24 (Oct. 1976).

7. J. P. Jackson, "Blood and Possible Images on the Shroud," *Shroud Spectrum International* 24 (Sept. 1987): 3-11, 3.

8. As one of the volunteers in Dr. Lind's experiments, I can verify that the dried, coagulated blood flows (that occurred within minutes on our arms) were only removed after pouring water on them and rubbing vigorously. According to John 20, Mary Magdalene was the first person to see Jesus on Easter morning. She saw him so soon after his reappearance that Jesus would not allow her to hug him (John 20:17). Even though Mary Magdalene initially confused him with the gardener, she did not observe any of Jesus' numerous blood marks on his body, nor did anyone else on Easter Sunday.

9. Nuclear engineer Robert Rucker and physicist Arthur C. Lind, 2014-2015.

10. R. A. Rucker, "MCNP Analysis of Neutrons Released from Jesus' Body in the Resurrection," *Shroud of Turin: The Controversial Intersection of Faith and Science, International Conference*, St. Louis, MO, Oct. 9-12, 2014, www.shroud.com/pdfs/stl-ruckerppt.pdf.

11. M. Antonacci, "Particle radiation from the body could explain the Shroud's images and its carbon dating," *Scientific Research and Essays* Vol. 7(29), pp. 2613-2623, (2012); See also A. Lind, M. Antonacci, D. Elmore, G. Fanti, and J. Guthrie, "Production of Radiocarbon by Neutron Radiation on Linen," *International Workshop on the Scientific Approach to the Archeiropoietos Images*, Frascati, Italy, May 4-6, 2010, pp. 255-262.

12. Nuclear engineer, Robert Rucker, personal communications, December 12, 2014 and January 20, 2015. Rucker calculates that even if the number of neutrons are all released at 10.0 MeV, the total kinetic energy of the neutrons would only be enough

to heat 100 pounds of water about 46° F. While the number of neutrons (or energy) that are required to be released if the cloth draped completely naturally over the body would be 4.2×10^{18}, this would not significantly affect the amount of energy released. Nor would a similar percentage of neutrons being released from the disappearing blood or the neutrons being released from a body that remained in rigor mortis in a very cool cave or tomb.

13. Little, "The Holy Shroud," p. 227. Matthew is the only Gospel that discusses earthquakes and it discusses two. The first one, described in 27:51-52 occurs after Jesus' death on the cross. It says that the earth shook, rocks were split and tombs were opened. While the second event is described in 28:2 as a great earthquake, which appears to have occurred on Easter morning (after the women went to see the tomb and just before an angel descended), it has a limited and localized effect that does not damage Jesus' burial cloth or the tomb. See also R. Rucker, "The Disappearance of Jesus' Body," Appendix H.

14. Nuclear engineer, Robert Rucker first explained this to me. Physicists Arthur Lind and John Jackson also think this is a plausible scientific explanation. Personal communications, 2014 and 2015.

15. R. Rucker, "The Disappearance of Jesus' Body," Appendix H. In this section of the Appendix, Rucker concludes "But there is no evidence for an earthquake occurring at any of the ten post-resurrection appearances, which argues against the concept that a transition into an alternate dimensionality would necessarily cause an earthquake." However, the original transition into an alternate dimensionality would be more complex than other post-resurrection transitions. See endnote 20 below.

16. R. Rucker, "The Disappearance of Jesus' Body." Appendix H.

17. Matt. 20:18-19, 16:21, 17:23, 27:40, 63, 16:4, 12:39-40, Mark 8:31, 9:9, Luke 9:22, 24:6-7, John 2:19-22.

18. Regarding the power to raise Jesus from the dead, see also John 10:17-18 (RSV), in which Jesus states ". . . I lay down my life that I may take it again. No one takes it from me, but I lay it down of my own accord. I have power to lay it down, and I have power to take it again; this charge I have received from the Father."

19. Matt. 28:9-10, 16-20, Luke 24:13-32, 36-49, John 20:11-21, 26-30, 1 Corin. 15:5-8.

20. R. Rucker, "The Disappearance of Jesus' Body," Appendix H and personal communication, March 21, 2015. Rucker also acknowledges that Jesus' post-resurrection transitions were less complex than the original transition in which Jesus' dead body was not only resurrected back to life, but transformed to a glorified or resurrection

body that could make such transitions on its own accord. In the original transition, Jesus' soul would also have been reunited with his body to make it alive again.

21. Nuclear engineer, Robert Rucker, personal communications, December 12, 2014.

22. This other possible explanation as to what happened to the historical Jesus Christ, or the man in the Shroud, when his body (and/or blood) disappeared was first introduced in a highly respected scientific journal in 1935 by Albert Einstein and Nathan Rosen, "The Particle Problem in the General Theory of Relativity," *Physical Review* 48 (1935): 73-77. They first devised the concept of a shortcut in space-time travel based on Einstein's theory of general relativity that allows a person or object to pass through a bridge or "wormhole" in space and time. According to modern physicists, mathematical theories of space-time travel are not only possible under Einstein's theory of general relativity, but these bridges or wormholes are completely consistent with tested theories of gravity and would allow travel between two points in different universes or two points within the same universe. This form of travel could circumvent the speed of light barrier and may even permit travel to past or future times. The famous British physicist Steven Hawking has published and lectured on bridges or wormholes, and his bestselling book, *A Brief History of Time*, devotes whole chapters to this subject. At this time the science of wormholes is not only mature, but in the words of physicist Matt Visser in *Lortenzian Wormholes: From Einstein to Hawking*, "...the theoretical analysis of Lortenzian wormholes is 'merely' an extension of known physics – no new physical principles of fundamentally new physical theories are involved." (New York: American Institute of Physics, 1996), p. 369.

As the disappearing body enters the wormhole, some elementary particles such as protons, neutrons, electrons and, perhaps, gamma rays could be left behind. Keep in mind that only a tiny fraction of the atoms are needed to disintegrate to encode the Shroud's superficial images. More than 99.999999% of the body would travel through the bridge. A key element of this hypothesis states that as matter passes through the wormhole or bridge, the entrance and exit mouths of the hole gain and lose mass, which can be acquired from and returned anew to the matter itself. When the body entered the wormhole under this hypothesis, the Shroud cloth would have collapsed or risen to the point of the body's departure. The Shroud itself would have been right at the mouth of the wormhole entrance and may have received some of the small mass left behind in the form of the basic building blocks of matter — protons, neutrons, electrons and alpha particles. The large gravitational force of the wormhole could be so strong as to separate the disintegrating body into its elemental particles. Hypothetically, this force does not have to operate on other objects that are not in the local wormhole, such as the Shroud. The analogous example of nearby objects getting sucked into Black Holes as large as those in outer space does not necessarily apply to objects surrounding a very localized wormhole. Please also keep in mind that this localized wormhole, like the image-encoding event, would last only a fraction of a second; whereas, analogous Black Holes in outer space are continuous.

This momentary gravitational force could also explain or compliment the vacuum, wherein the dorsal and even frontal sides of the cloth are briefly drawn into the body region at its sudden disappearance or dematerialization.

The above abbreviated wormhole with its specific application to, and travel directions for, the man in the Shroud and his blood is not postulated to exist naturally, of course. Even if its hypothetical existence did not violate the laws of science, it would seemingly have to be the result of an intentional act. Only a power like that of God could ostensibly cause such an event. Space-Time travel would not involve an explosion and could also be said to be another *possible* means for Jesus to have reportedly traveled between heaven and earth, as well as disappeared and reappeared on earth. Hypothetically, an object that passed even through a local wormhole could then have a different appearance. When Jesus was first seen after his resurrection, his close acquaintances and friends did not initially recognize him. *John* 20:14-17; *Luke* 24:15-16; *Mark* 16:12. See also *John* 21.4. Of course, God could have used some method or utilized a principle or concept that science has not realized yet or discovered.

23. Among the conditions are whether the cloth draped normally over a reclined body as seen in the three figures a, b and c below.

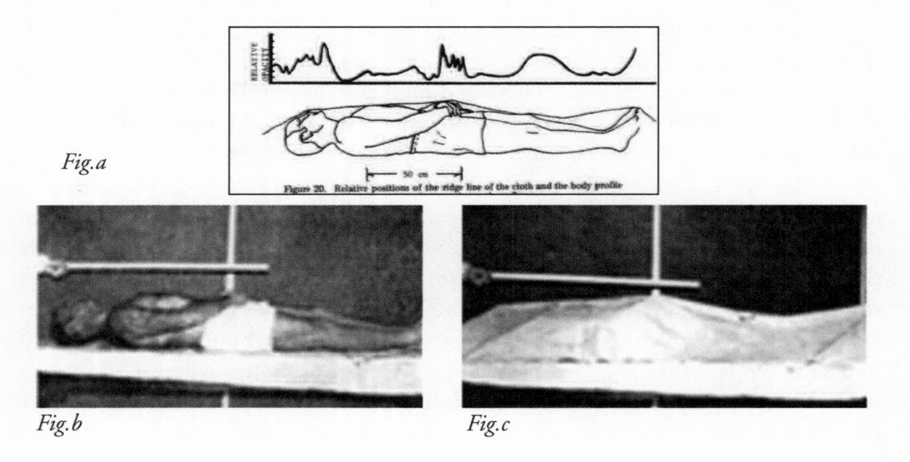

Figure 20. Relative positions of the ridge line of the cloth and the body profile

Fig.a

Fig.b

Fig.c

Other possible conditions that could be modeled would be if the right and left ends of the enveloped cloth were folded onto the shins or folded under the feet. The 1988 radiocarbon sampling site is near the left end. These ends could have been folded at roughly 45° angles, landing near or over the man's shins. This would have increased the amount of neutrons that the 1988 radiocarbon site received. Even though most of the rest of the cloth would not have received an additional amount of radiation, the two ends of the burial cloth could have. Similarly, the right and left ends of the cloth could have been folded sideways toward and tucked under the man's feet, thus, exposing them to more neutron radiation.

If the cloth draped naturally over a reclined body, as seen above, approximately 4.2 x 10^{18} neutrons would need to be released uniformly throughout the length and width and depth of the body for samples from the 1988 radiocarbon site to date 1200 years to the future. If the two ends of the enveloped cloth were folded sideways toward the body, and the 1988 radiocarbon site was folded under and landed in between the back of the feet, 3.0 x 10^{18}neutrons would have to be released uniformly throughout the length, width and depth of the body for this location to date 1200 years to the future. If the 1988 site was folded at an approximately 45° angle and landed directly over the man's shins, then even fewer neutrons would need to be released uniformly throughout the length, width and depth of the body for this site to date 1200 years to the future. This amount would comprise even less than 0.000000015% of the total neutrons in the body.

The MCNP code could be run with the 1988 radiocarbon site being in the above or other configurations with the body giving off the above or other corresponding amounts of neutron (and similar proton) radiation uniformly throughout its length, width and depth. (Although neutrons will comprise about 45% of the particles released from the body, and protons will comprise about 55%, the total infinitesimal percentage of both particles released from the body should be roughly the same.) The amounts of neutron radiation that were received by cloth, charred material or blood at every location on this burial cloth could be calculated in various models by the MCNP code. The calculations should assume that the blood was on the Shroud throughout the radiating event. Remember that if you first non-destructively measure the indigenous amounts of Cl-35, Ca-40 and N-14 in the above Shroud samples, their subsequent Cl-36 to Cl-35, Ca-41 to Ca-40 and C-14 to C-12 ratios will tell us the amount of neutrons radiation that every sample received. These amounts can be compared to the Cl-36 to Cl-35, Ca-41 to Ca-40 and C-14 to C-12 ratios calculated for each above model to help ascertain the cloth's configuration and the total amount of neutron radiation that uniformly or otherwise emanated from the body of the man in the Shroud. They can also be helpful in determining whether the blood marks received any, some or all of the neutron radiation calculated by the MCNP models discussed above.

24. One possible source of insight could come from sampling small blood samples from two "off-image" areas with two small blood samples from nearby body image locations. The first of these samples could be taken from the back of the man's right foot along with the off-image blood just to the side of the back of the right foot. The second set of these samples could come from the blood from the man's right forearm and the off-image blood mark off his right elbow. This sampling could possibly add insight as to whether only the blood on the body disappeared, or whether all the man's shed blood (on or off the body) disappeared and reappeared.

Testing these two off-image areas are further complicated by the real possibilities that one or both of these off-image blood marks may have been on the man's body at the time the body images and blood marks were encoded or embedded in the Shroud. If the ends of the cloth were tucked under the feet at burial as discussed in endnote 20, or if the cloth was once tucked around the man's arm as seen below, both blood

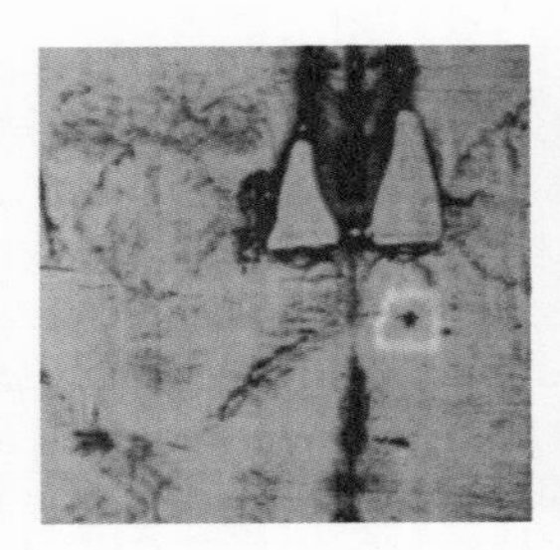

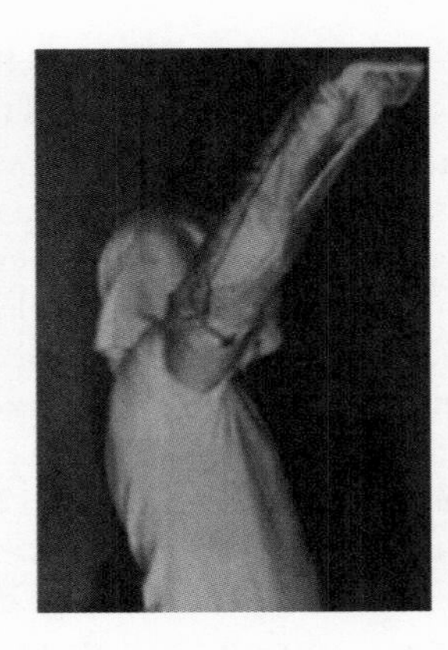

marks could have hypothetically disappeared and reappeared along with the Shroud's other blood marks. Cl-36 to Cl-35 or Ca-41 to Ca-40 ratios and C-14 to C-12 ratios of these two pairs of on-image and off-image blood marks, along with MCNP calculations under various fold and non-fold configurations, could provide key insight into the questions of whether and how much neutron radiation the blood marks on the man in the Shroud received, and if they appeared and disappeared at the time of the neutron radiating event.

25. J. Rinaudo, "Protonic Model of Image Formation on the Shroud of Turin," *Third International Congress on the Shroud of Turin,* Turin, Italy, June 5-7, 1998; J. Rinaudo, "A Sign for Our Time," *Shroud Sources Newsletter,* May/June 1996, pp. 2-4; J. Rinaudo, in *BSTS Newsletter*, No. 38, August/September 1994, pp. 13-16; J. Rinaudo, "A New Stage," *Il est Vivant,* No. 89, March/April, 1992.

26. J. P. Jackson, "Is the Image on the Shroud Due to a Process Heretofore Unknown to Modern Science?" *Shroud Spectrum International* 34 (March 1990): 3-29, 18; J. P. Jackson, "An Unconventional Hypothesis to Explain All Image Characteristics Found on the Shroud Image," *History, Science, Theology and the Shroud*, A. Berard, ed. (St. Louis: Richard Nieman, 1991), pp. 325-344.

27. J. Rinaudo, "Protonic Model" and personal communications in 2006.

CHAPTER THIRTEEN

1. While this summary is similar in many respects to my Summation in 2000, it contains important medical and scientific developments and analyses from the last 15 years.

2. Although Professor Conrad is enjoying his retirement and did not wish to upgrade his decades old survey, I would think, if anything, that the numbers of Gospel and New Testament papyri or parchment writings that have been discovered have only increased.

3. According to Professor Conrad, without some qualification the figures given above could be misleading. Note that for some authors over two hundred manuscripts are indicated, while twenty or fewer are listed for others. This reflects figures available to the researcher; the figures were drawn fundamentally from the most recent scholarly editions of the Greek or Latin texts in question and the information presented in each instance by the editor, who indicates the manuscripts he or she has employed to construct the text there printed. Of concern to such editors is not the total number of manuscripts available, but the major older manuscripts offering significant variant readings of the text. In the case of most authors, numerous manuscripts survive from centuries following the earliest extant manuscripts, but these can easily be grouped into families and shown to have been copied from extant earlier manuscripts, so that they do not enter into account in the establishment of the most authoritative text of an author.

These manuscripts were kept on either parchment or papyri. Parchment is a much higher grade of paper, made from carefully processed sheepskin or other animal skin. Papyri, which is where the modern term *paper* comes from, is probably the oldest form of paper we have. It was made from reed grown in marshes. In time, it generally rotted everywhere except in arid climates such as Egypt. The papyri are always fragmentary, as is the case also with the earliest surviving papyrus fragment of the New Testament, a fragment containing only the Gospels of John 18:31-33, 37-38; as the Gospel of John is supposed to have been composed about A.D. 90, this papyrus fragment is only about twenty years later than the text of which it is a copy.

The terms *majuscule and minuscule* refer to styles of handwriting. Majuscule used larger, capital or uncial letters throughout. Minuscule was a script of smaller letters, and eventually superseded the former style.

In addition to manuscripts and papyri of the Greek and Latin texts of ancient authors and of the New Testament, there are manuscript anthologies containing short citations of ancient authors, lectionaries containing short biblical passages, and early translations of the New Testament into Latin, Coptic, Syriac, and other languages used by Christians in the later ancient world. These have not been counted in the table listing.

4. F. F. Bruce, *The Books and the Parchments,* rev. ed., (London: Pickering and Inglis Ltd., 1963), p. 178.

5. J. W. Montgomery, *History and Christianity* (Downers Grove, IL: Inter-Varsity Press, 1971), p. 29.

6. I Corinthians 15:3-15 is recognized by all historians and scholars as having been written no later than the A.D. mid-50s. Furthermore, many scholars think that the historical accounts contained in the Acts of the Apostles and the Synoptic Gospels were written prior to or no later than the A. D. mid to late 60s. Part of their reasons is that the book of Acts closes with Paul awaiting trial under house arrest in A.D. 59,

after having been in custody for two years in Rome, yet we are not told of the outcome. These historical sources make no mention at all of the very critical facts of Paul's death, Peter's death, or the complete destruction of Jerusalem. Most historians agree that these events occurred in the mid to late 60s.

The destruction of Jerusalem, the centuries-old capital and center of Jewish culture, would have been as traumatic to Jews as a civil war. Several of the books of the New Testament center on Paul's travels around the ancient world. Peter was the leader of the apostles and the rock upon which Jesus built his church. Similarly, the death of James, the brother of Jesus himself, is not mentioned. James did not believe in the divinity of Jesus (John 7:5; cf. Mark 3:21), but after he saw the resurrected Christ (1 Corinthians 15:7), became a prominent leader in the Jerusalem Church. Josephus (Ant. 20.9.1) and Hegesippus (in Eusebius *Hist. Eccl.* 2:23) both tell of the martyrdom of James in Jerusalem ca. A.D. 62 or 66. The absence of the deaths of all three of these important leaders is all the more glaring when we read of the deaths of other lesser-known individuals and martyrs, such as Stephen, in ca. A.D. 36 and James, the brother of John, in ca. A.D. 41-44, clearly mentioned in Acts 7:54-8:1 and 12:1, 2. J. Finegan *The Archaeology of the New Testament* (London: Croon Held Ltd., Boulder, Col.: Westview Press, Inc., 1981).

7. 1 Corinthians 15:3-8; Matthew 28:9, 10, 28:16-20; Mark 16:9, 16:14-20; Luke 24:13-52; John 20:14-29, 21:1-25; Acts 1:1-9, 2:22-24, 32, 3:15, 4:33, 5:30-32, 13:30, 31; 1 Peter 1:3.

8. Matthew 17:2, 5; Mark 9:2, 3, 7; Luke 9:28, 29, 34, 35. In two other eyewitness accounts in Acts, Jesus appears to Paul and his companions in a light from heaven. Although Jesus temporarily blinded Paul, the light did not harm his body or clothing, or those of his companions (Acts 9:3-8; 26:13-16).

CHAPTER FOURTEEN

1. D. H. Sox, *The Shroud Unmasked* (Basingstoke, Hampshire: The Lamp Press, 1988), p. 95; E. T. Hall, "The Turin Shroud: An Editorial Postscript," *Archaeometry* 31.1 (1989): 92-95; See H. E. Gove, *Relic, Icon or Hoax?* (Bristol and Philadelphia: Institute of Physics Publishing, 1996), p. 104 wherein it is reported that Dr. Garman Harbottle of the Brookhaven Laboratory stated that Gove was "famous for grabbing the ball and running with it."; Gove, *Relic?*, p. 106.

2. R. Dinegar and L. Schwalbe, "Isotope Measurements and Provenance Studies of the Shroud of Turin," *Archaeological Chemistry* IV; Ralph O., Allen, ed., *Advances in Chemistry Series 220* (American Chemical Society: Washington, D.C., 1989), pp. 409-417.

3. Ibid.

4. R. Burleigh, M. Leese, and M. Tite, "An Intercomparison of Som e AMS and Small Gas Counter Laboratories," *Radiocarbon,* Vol. 28, No. 2A (1986): 571-577.

5. Burleigh et al., "An Intercomparison," p. 571.

6. Ibid.

7. Ibid., p. 577.

8. Ibid., p. 577.

9. Ibid., p. 576.

10. Ibid.

11. See comments of Garman Harbottle, Brookhaven National Laboratory, quoted in J. Marino, "The Shroud of Turin and the Carbon 14 Controversy," *Fidelity,* (February 1989): 35-47, 39.

12. Dinegar and Schwalbe, "Isotopic Measurements and Provenance Studies," p. 412.

13. Sox, *The Shroud Unmasked*, p. 96; O. Petrosillo, and E. Marinelli, *The Enigma of the Shroud*, translated from Italian (San Gwann, Malta: PEG 1996).

14. H. E. Gove, "Turin Workshop on Radiocarbon Dating the Turin Shroud"; *Nuclear Instruments and Methods in Physics Research*, B29 (1987) 193-195, 194. See also Sox, *The Shroud Unmasked,* pp. 108-09; Petrosillo and Marinelli, *The Enigma of the Shroud*, pp. 34-35; and Marino, "The Shroud of Turin and the Carbon 14 Controversy," p. 41.

15. Michael Tite, letter to *Nature,* Vol. 332, April 7, 1988.

16. H. E. Gove, *Relic, Icon or Hoax*? (Bristol and Philadelphia: Institute of Physics Publishing, 1996).

17. Gove, *Relic?*, p. 48. See also M. Antonacci, *The Resurrection of the Shroud* (M. Evans and Company, Inc., 2000) p. 193.

18. Gove, *Relic?*, p. 57. See also Antonacci, *Resurrection*, p. 193.

19. Gove, *Relic?*, p. 87. See also Antonacci, *Resurrection*, p. 193.

20. Gove, *Relic?*, pp. 191 and 192. See also Antonacci, *Resurrection*, p. 200.

21. Gove, *Relic?*, p. 192 and 193. See also Antonacci, *Resurrection*, p. 200.

22. Gove, *Relic?*, p. 113. See also Antonacci, *Resurrection*, p. 197.

23. Antonacci, *Resurrection*, p. 200. See also Gove, *Relic?*, p. 214.

24. Gove, *Relic?*, p. 192. See also Antonacci, *Resurrection*, p. 200.

25. Gove, *Relic?*, p. 216. See also Antonacci, *Resurrection*, p. 200.

26. *Il Giornale*, May 12, 1989, as quoted in Petrosillo and Marinelli, *The Enigma of the Shroud*, (San Gwann, Malta: PEG, 1996), pp. 119-120.

27. Gove, *Relic?*, pp. 155-56.

28. R. Van Haelst, "Analyzing Radiocarbon data using Burr statistics," http://shroud.com/pdfs/vanhaelst8.pdf 2011; R. Van Haelst, "A critical review of the radiocarbon dating of the Shroud of Turin. ANOVA – a useful method to evaluate sets of high precision AMS radiocarbon measurements." *International Workshop on the Scientific Approach to the Archeiropoietos Images,* Frascati 2010; R. Van Haelst, "Radiocarbon Dating the Shroud of Turin: The Nature Report," (June 1999), http://www.shroud.com/vanhels5.pdf. If any statements regarding the raw data are incorrect, I invite the Arizona laboratory, the other radiocarbon labs and the British Museum to supply all the raw data from their initial report and their final report, including the locations of the subsamples.

29. P. E. Damon, et al. "Radiocarbon Dating of the Shroud of Turin," *Nature* 337 (1989): 611-615; Van Haelst, "Radiocarbon Dating," "A critical review" and "Analyzing Radiocarbon data."

30. Van Haelst, "A critical review," "Radiocarbon Dating" and "Analyzing Radiocarbon data."

31. Damon, et al. "Radiocarbon Dating of the Shroud of Turin"; Van Haelst, "A critical review, " "Radiocarbon Dating" and "Analyzing Radiocarbon data."

32. Ibid.

33. Ibid

34. If combining these dates statistically worked against the acquisition of an accept-

able degree of certainty in my description or in some formulistic or statistical manner, I'm confident this was more than offset by the benefits, and that the combinations discussed was certainly of a net benefit to acquiring a perceived acceptable range of certainty.

35. Van Haelst, "A critical review," and "Radiocarbon Dating."

36. B. Walsh, "The 1988 Shroud of Turin Radiocarbon Tests Reconsidered," *Shroud of Turin International Research Conference*, Richmond 1999.

37. M. Riani, A. Atkinson, F. Crosilla, G. Fanti, "A robust statistical analysis of the 1988 Turin Shroud radiocarbon dating results," *International Workshop on the Scientific Approach to the Acheiropoietos Images,* Frascati 2010. An expanded version of this paper titled "Carbon Dating of the Shroud of Turin: Partially Labeled Regressors and the Design of Experiments," can be found at http://www.lse.ac.uk/collections/statistics/research/researchreports_2010.htm; Walsh, "The 1988 Shroud of Turin Radiocarbon Tests Reconsidered," pp. 340-341; Van Haelst, "Radiocarbon Dating" and "A critical review."

38. Riani et al., "Carbon Dating of the Shroud of Turin,"; personal communication with one of the co-authors.

39. Walsh, "The 1988 Shroud of Turin Radiocarbon Tests Reconsidered," p. 326, 333.

40. Walsh, "The 1988 Shroud of Turin Radiocarbon Tests Reconsidered," p. 339. See also, Riani et al., "Carbon Dating of the Shroud of Turin."

41. Walsh, "The 1988 Shroud of Turin Radiocarbon Tests Reconsidered," p. 339.

42. T. Phuillips, "Shroud irradiated with neutrons?" *Nature* 337 (1989): 594.

43. R. Hedges, "Hedges Replies," *Nature* 337 (1989): 594.

44. Hedges, "Hedges Replies."

45. W. Meachem, "Thoughts on the Shroud ^{14}C Debate," *Proceedings of the International Scientific Symposium*, 2-5 March 2000, Turin, Italy, pp. 441-454, 443.

46. See Antonacci, *Resurrection*, p. 183.

47. P. Jennings, "Still Shrouded in Mystery," *30 Days in the Church and in the World* 1.7 (1988): 70-71.

48. Meachem, "Thoughts on the Shroud ^{14}C Debate."

49. Gove, *Relic?*, p. 165; Antonacci, *Resurrection*, pp. 199.

50. Meacham, "Thoughts on the Shroud ^{14}C Debate," p. 449.

51. Gove, *Relic?*, p. 155, Antonacci, *Resurrection*, p. 199.

52. Meacham, "Thoughts on the Shroud ^{14}C Debate," p. 447.

53. Meacham, "Thoughts on the Shroud ^{14}C Debate," p. 444.

CHAPTER FIFTEEN

1. P. Damon et al. "Radiocarbon Dating of the Shroud of Turin." *Nature* (1989): 611-615.

2. T. Phillips, "Shroud Irradiated with Neutrons?" *Nature* 337 (1989): 564.

3. R. Hedges, "Hedges Replies," *Nature* 337 (1989): 594.

4. Ibid.

5. P. Jennings, "Still Shrouded in Mystery," *30 Days in the Church and on the World* 1.7 (1988): 70-71, 71.

6. "HEDGES REPLIES — The processes suggested by Phillips were considered by the participating laboratories. However, for the reasons given below, the likelihood that they influenced the date in the way proposed is in my view *so exceedingly remote that it beggars scientific credulity.*" (italics added) Hedges, "Hedges Replies," p. 594.

7. Hedges, "Hedges Replies," p. 594.

8. Ecology.com estimates that in 2011 births per year were 131.4 million and deaths per year were 55.3 million.

9. Several of the previous pages were first discussed in my earlier paper "Scientists and Semantics" presented at the *Shroud of Turin: The Controversial Intersection of Faith and Science, International Conference*, St. Louis, Missouri, October of 2014. This paper can be read at http://shroud.com/pdfs/stlantonaccipaper.pdf.

10. The quotations from John 3:16, 17 and 10:17, 18 are from the Revised Standard

Version. The quotation from John 11:25, 26 is from the New Revised Standard Version.

11. If scientific evidence indicated that particle radiation emanated from the length, width and depth of Jesus' crucified corpse, then even if scientists did somehow duplicate all the unique features on both body images, 130 blood marks and all off-image features, by using advanced technology thousands of years later, this still would not mean that a miraculous event that occurred to Jesus' body did not cause the original mutually exclusive features.

APPENDIX A

1. The laboratory in which the scientists are working will need a license to receive radioactive linen, blood, limestone or other neutron irradiated material.

2. If this did not work, iodine could be considered as a carrier. X. Hou, W. Zhou, N. Chen, L. Zhang, M. Luo, Y. Fan, W. Liang, Y. Fu, "Determination of ultralow level 1291/1271 in natural samples by separation of microgram carrier free iodine and accelerator mass spectrometry detection" *Anal Chem.* 2010 Sep 15; 82 (18): 7713-21. doi: 10.1021/ac101558k.

APPENDIX B

1. R. Rucker, "MCNP Analysis of Neutrons Released From Jesus' Body In The Resurrection," *Shroud of Turin: The Controversial Intersection of Faith and Science, International Conference*, St. Louis, MO, October 9–12, 2014, http://www.shroud.com/pdfs/stlruckerppt.pdf.

APPENDIX C

1. R. N. Rogers and A. Arnoldi, "Scientific Method Applied to the Shroud of Turin," (2002), http://www.shroud.com/pdfs/rogers2.pdf; T. de Wesselow, *The Sign*, (New York, Penguin Group, 2012); H. Felzmann, *Resurrected or Revived?*, 1. Edition 2012; G. Fanti, "Hypotheses Regarding the Formation of the Body Image on the Turin Shroud. A Critical Compendium," *Journal of Imaging Science and Technology*, 55(6): 060507-1-060507-14, 2011.

2. G. Fanti, "Hypotheses Regarding the Formation."

3. G. Fanti, "Hypotheses Regarding the Formation."

4. G. Fanti, "Hypotheses Regarding the Formation."

5. This analysis of Mr. Rogers' image-forming hypothesis is similar to part of the "Combined Review of: 'The Sign' by Thomas de Wesselow and 'Resurrected or Revived?' by Helmut Felzman," which Patrick Byrne and I wrote in 2012.

6. R. N. Rogers, "The Shroud of Turin: Radiation Effects, Aging and Image Formation," p. 6, http://www.shroud.com/pdfs/rogers8.pdf.

7. Ibid., p. 1.

8. Ibid., p. 9. This inconsistency was first pointed out to me by Robert Rucker in his "Review of 'The Shroud of Turin: Radiation Effects, Aging and Image Formation' by Ray Rogers" in publication. Rucker's article points out other inconsistencies and unfounded conclusions in Rogers' 2005 radiation paper.

9. R. N. Rogers and A. Arnoldi, "Scientific Method Applied to the Shroud of Turin: A Review," https://www.shroud.com/pdfs/rogers2.pdf.

APPENDIX D

1. L. Garlaschelli, "Life-Size Reproduction of the Shroud of Turin and Its Image," *Journal of Imaging Science and Technology* 54(4): 04031-04031-14, 2010.

2. Garlaschelli, "Life-Size Reproduction," P. 040301-10.

3. Ibid., p. 040301-8.

4. Ibid., p. 040301-11.

5. T. Heimburger, "Comments About the Recent Experiment of Professor Luigi Garlaschelli," p. 1, November 2009, http://www.shroud.com/pdfs/thibault-lg.pdf.

6. Heimburger, "Comments About the Recent Experiment," p.2.

7. Ibid.

8. Dr. Paolo D. Lazzaro, personal communication to Russ Breault and Shroud Science Group, July 9, 2010.

9. Garlaschelli, "Life-Size Reproduction," pps. 040301-8 and 9.

10. Garlaschelli's images are found in Garlaschelli, "Life-Size Reproduction."

11. Garlaschelli, "Life-Size Reproduction," p. 040301-8.

12. Ibid.

13. Ibid.

APPENDIX E

1. A. Loth, *Le portrait de N.-S. Jésus-Christ d'aprés le Saint-Suaire de Turin* (Librairie Religieuse H. Oudin, Paris, France, 1900), p. 53; O. Scheuermann, *Turiner Tuschbold Aufgestrahlt?* (VDM Verlag Dr. Müller, Saarbrucken, Germany, 2007), p. 51; A Whanger and M. Whanger, *The Shroud of Turin, an Adventure of Discovery* (Providence House, Franklin, TN, 1998); G. Fanti, F. Lattarulo, and O. Scheuermann, "Body image formation hypotheses based on corona discharge," *III Dallas International Conference Shroud of Turin,* Dallas, Texas, 2005; www.dim.unipd.it/fanti/corona.pdf, accessed August 2011; G. B. Judica Cordiglia, "La Sindone immagine elettrostatica," *Proc. III Congr. Naz. di Studi sulla Sindone, Trani 1984* (Edizioni Paoline, Cinisello Balsamo, MI, Italy, 1986), pp. 313-327; F. Lattarulo, "L'immagine sindonica spiegata attraverso un processo sismoelettrico," *III Congresso internazionale di studi sulla Sindone, Torino, 1998, Unpublished Proceedings on CD*, pp. 334-346.

2. G. Fanti, F. Lattarulo and G. Pesavento, "Experimental results using corona discharge to attempt to reproduce the Turin Shroud Image," *SHS Web of Conferences* 15, 00003 (2015), http://www.shs-conferences.org or http://dx.doi.org/10.1051/shsconf/20151500003; G. Fanti, "Can corona discharge explain the body image formation of the Turin Shroud?," *J. Imaging Sci. Technol.* 54, 020508 (2010); R. Basso and G. Fanti, "Optics research applied to the Turin Shroud: Past, present and future," *Optics Research Trends,* edited by P. V. Gallico (Nova Science Publisher, Inc., New York, 2007), p. 4; G. Fanti, *La Sindone, una sfida alla scienza moderna* (Aracne Ed., Roma, Italy, 2008), p. 201.

3. G. Fanti, "Hypotheses Regarding the Formation of the Body Image on the Turin Shroud, A Critical Compendium," *Journal of Imaging Science and Technology* 55(6): 060507-1 — 060507-14, 2011.

4. Lind explained that the electric field used in Fanti's method quickly moves charged air molecules or atoms to and from the linen many times (probably more than 60 times per second). This creates more and more of them, which collide with uncharged air molecules so that they become sufficient in number to create an effect on linen. He explains, however, that the intensity of this resulting image will not have the same relationship as exists with the distance between the body and the cloth. So far, I have not observed three dimensional body image information or focused resolution on body images produced by the Corona Discharge method similar to those found on the Shroud's body image.

5. Fanti, "Hypotheses Regarding the Formation of the Body Image," p. 060507-11.

6. Dr. Giulio Fanti, personal communication, July 1, 2015

7. Ibid., p. 060507-1.

8. Ibid., p. 060507-2.

APPENDIX F

1. A. Carpinteri, G. Lacidogna and O. Borla, "Is the Shroud of Turin in relation to the old Jerusalem historical earthquake?" *Meccanica*, published online, February 11, 2014.

2. Ibid., Abstract. The original quote has a small "6" under the upraised "14," however this was difficult to set in print and is confusing to a non-scientist reader, so the subscript "6" was left out of the quote.

3. Nuclear engineer, Robert Rucker, personal communications, January 30, 2015, May 31, 2015, June 1, 2015.

4. Carpinteri, et al., "Is the Shroud of Turin," Section 4.

5. Ibid., Abstract and introduction.

6. Ibid., Reference 8.

APPENDIX G

1. Revised Standard Version. Some translations state that the napkin "covered his face."

2. A. Hermosilla, "Commonalities between the Shroud of Turin and the Sudarium of Oviedo," *2014 Workshop on Advances in the Turin Shroud Investigation (ATSI 2014)*, Bari, Italy, September 4–5, 2014, http://www.shs-conferences.org; C. Barta, R. Alvarez, A. Ordonez, A. Sanchez, J. Garcia, "New Discoveries on the Sudarium of Oviedo," *Shroud of Turin: The Controversial Intersection of Faith and Science, International Conference*, St. Louis, MO, October 9–12, 2014, http://www.shroud.com/pdfs/stlbartapaper.pdf; J. Bennett, *Sacred Blood, Sacred Image* (Littleton, CO, Publications About Spain, 2011); M. Guscin, *The Oviedo Cloth* (Cambridge: The Lutterworth Press, 1998).

3. Guscin, *The Oviedo Cloth,* pg. 56; A. D. Adler, "The Shroud of Turin — Blood

Tests," interview by Linda Moulton Howe, May 23, 1999, http://earthfiles.com/earth026.html; A. Danin, A. Whanger, U. Barach, M. Whanger, *Flora of the Shroud of Turin* (St. Louis: Missouri Botanical Garden Press, 1999) pp. 23-24.

4. Danin, et al., *Flora of the Shroud.*

5. Guscin, *The Oviedo Cloth*, p. 22.

6. Barta, et al., "New Discoveries on the Sudarium"; Hermosilla,; "Commonalities between the Shroud and the Sudarium."

7. I. Wilson, "Controversy Over Oviedo Cloth Radiocarbon Dating," *British Society for the Turin Shroud Newsletter* 50 (November 1999): 12. See also Guscin, *The Oviedo Cloth*, pp. 76-84.

8. Robert Rucker, "MCNP Analysis of Neutrons Released From Jesus' Body In The Resurrection," *Shroud of Turin: The Controversial Intersection of Faith and Science, International Conference*, St. Louis, MO, October 9–12, 2014, http://www.shroud.com/pdfs/stlruckerppt.pdf.

APPENDIX H

1. "The Gospel According to John" by D. A. Carson, 1991, Eerdmans, see page 473, On the issue of who is "the disciple whom Jesus loved," after consideration of John 13:23, 19:26-27, 20:2-9, 21:1, and 21:20-25, D. A. Carson concludes that "If we compare the four canonical Gospels, by a process of elimination we arrive at John the son of Zebedee as the most likely identity of the disciple whom Jesus loved."

2. "John" by Edwin A. Blum in "The Bible Knowledge Commentary, An Exposition of the Scriptures by Dallas Seminary Faculty," Vol. 2 on the New Testament, Edited by John F. Walvoord and Roy B. Zuck, 1983, SP Publications, Comments on John 20:8 on page 342.

3. "The Gospel of John" by Merrill C. Tenney in "The Expositors Bible Commentary," Volume 9, Frank E. Gaebelein general editor, 1981, Zondervan, Comments on John 20:8 on page 188

4. "The Bible Exposition Commentary, New Testament, Volume 1" by Warren W. Wiersbe, 2001, Victor, Comments on Matthew 28:2 on page 105.

5. "Exploring the Gospel of John, An Expository Commentary" in "The John Phillips Commentary Series" by John Phillips, 2001, Kregel, page 375

6. "The Gospel of John, Volume 2", Revised Edition, by William Barclay in "The Daily Study Bible Series," 1975, Westminster Press

7. "Systematic Theology" by Lewis Sperry Chafer, 1948, Vol. 2, page 150

8. "Systematic Theology, An Introduction to Biblical Doctrine" by Wayne Grudem, 1994, Inter-Varsity Press, pages 831-832

9. "Lectures in Systematic Theology" by Henry C. Thiessen, revised by Vernon D. Doerksen, 1979, Eerdmans, Chapter 44, Section 2B, page 383

10. "Christian Theology" by Millard J. Erickson, Third Edition, 2013, Baker, page 1073

11. "Integrative Theology, Volume 3, Spirit-Given Life: God's People Present and Future" by Gordan R. Lewis and Bruce A. Demarest, 1994, Zondervan, page 471

12. "Basic Theology" by Charles C. Ryrie, 1986, Victor Books, page 269

13. "Understanding Christian Theology" edited by Charles R. Swindoll and Roy B. Zuck, 2003, Thomas Nelson, page 1274

14. "Systematic Theology, Biblical and Historical" by Robert Duncan Culver, 2001, Christian Focus Publications

15. "Introduction to Philosophy, A Christian Perspective" by Norman L. Geisler and Paul D. Feinberg, 1987, Baker

16. "Philosophical Foundations for a Christian World View" by J. P Moreland and William Lance Craig, 2003, InterVarsity Press

17. "Lifeviews, Understanding the ideas that shape society today" by R. C. Sproul, 1995, Revell

18. "The Universe Next Door, A Basic Worldviw Catalog" Third Edition by James W. Sire, 2009, InterVarsity Press

19. "Understanding the Times, The Story of the Biblical Christian, Marxist/Leninist, and Secular Humanist Worldviews" by David A. Noebel, 2006, Summit Press

20. "Reasonable Faith, Christian Truth and Apologetics" by William Lane Craig, 2008

21. "Why I am a Christian" edited by Norman L. Geisler and Paul K. Hoffman, 2006, Baker

22. “Scaling the Secular City, A Defense of Christianity” by J. P. Moreland, 1987, Baker

23. “The New Evidence that Demands a Verdict” by Josh McDowell, 1999

24. “Baker Encyclopedia of Christian Apologetics” by Norman L. Geisler, 1999, Baker

25. “The Case for the Resurrection of Jesus” by Gary R. Habermas and Michael R. Licona, 2004, Kregel

26. “Beyond Death: Exploring the Evidence for Immortality” by Gary R. Habermas and J. P. Morland, 2004, Wipf & Stock

27. “Christianity and the Nature of Science, A Philosophical Investigation” by J. P Moreland, 1999, Baker

28. “The Soul of Science, Christian Faith and Natural Philosophy” by Nancy R. Pearcey and Charles B. Thaxton, 1994, Crossway

29. “In Defense of Miracles, A Comprehensive Case for God’s Action in History” edited by R. Douglas Geivett and Gary R. Habermas, 1997, InterVarsity Press

30. “The Elements” by John Emsley, 3rd ed., 1998, Clarendon Press, Oxford,

31. “MCNP Analysis of Neutrons Released from Jesus’ Body in the Resurrection” by Robert A. Rucker, PowerPoint presentation at the international conference on the Shroud of Turin titled “Shroud of Turin: The Controversial Intersection of Faith and Science,” October 9-12, 2014, in St. Louis, Missouri.

32. “Image Formation on the Shroud of Turin Explained by a Protonic Model Affecting Radiocarbon Dating” by J. B. Rinaudo, Third International Congress on the Shroud of Turin, Turin, Italy, June 5-7, 1998

33. “The Physics of Christianity” by Frank J. Tipler, 2007, Doubleday, page 199

34. “The NIV Exhaustive Concordance” by Edward W. Goodrick and John R. Kohlenberger III, 1990, Zondervan, pages 1695 to 1696

35. “The Elegant Universe - Superstrings, Hidden Dimensions, and the Quest for the Ultimate Theory” by Brian Greene, second edition, 2010, W. W. Norton & Company

36. "The Road to Reality, A Complete Guide to the Laws of the Universe" by Roger Penrose, 2004, Vintage

37. Personal communication from Dr. K. N. Schwinkendorf, April 13, 2015

ACKNOWLEDGEMENTS

Art Lind has a doctorate in physics; however, the best way to describe him is that he is an extraordinary problem solver. He's now a retired Fellow from McDonnell Douglas Corporation and The Boeing Company, but when he worked there he was frequently temporarily assigned from his usual position in the McDonnell Douglas Research Laboratory to different projects when they had difficult problems to solve. He's a pure scientist who lets the evidence guide his research and speak for itself. His main attraction to the Shroud was to solve or explain its unique features and its unprecedented images. With his all-around brilliance in science and as the long-time lead researcher and supervising scientist for our Foundation, he may be close to understanding the causes of the Shroud's countless, unique features.

Art is even better at sharing and explaining scientific matters to lay people and the public. I've never had a conversation with him in which I didn't learn something. He has not only repeatedly discussed every Shroud matter under the sun, but every scientific matter that our research has touched on ranging from quantum mechanics to the cosmos and beyond. He usually has time for all your questions, but when he doesn't, he'll still send you an e-mail with his advice, even if it's sent at midnight. He's an even more extraordinary person than he is a scientist, and neither I nor the world could ever begin to repay him for his more than 20 years of patient analysis, brilliant research and advice.

I had the pleasure of talking to Bob Rucker for the first time a few years ago. While he has had a long time interest in the Shroud, he wasn't able to work at it nearly as much as he would have liked. He's more than making up for lost time. In fact, his timing is perfect. His work as a nuclear engineer in modeling the various radioactive effects caused by neutrons emanating uniformly from a body are ground breaking. His introductory presentation at the St. Louis International Shroud Conference in 2014 altered the thinking of everyone in sindonology. His initial computer modeling not only provides critical information for selecting Shroud samples in future testing, but clearly illustrates to everyone the miraculous nature of such a distribution and its source.

He is able to get startling results with an understated scientific manner. He understands that science and naturalism are not identical and that science has far more to learn about our universe than it can currently explain. Yet, his conscientiousness is what stands out the most. While most scientists up until now have avoided any involvement with Shroud research, it's hard to keep up with the scientific projects that Bob wants to conduct. He not only recognizes the need and utility for such projects, but that science has a duty to pursue and investigate the tangible evidence on this burial cloth. He has spent large amounts of his time making sure my understanding and explanations of neutrons, protons, particle radiation and their effects are correct. He also originated and explained how the body's transition to an alternate dimensionality is the best hypothetical explanation for the body's obvious disappearance. I struggle to keep up with his tireless and invaluable help, which is the nicest problem that I have ever had in this profession. I can't say enough about his help in the most difficult and important scientific matters of all.

Joe Marino was the first person I ever met in the field of sindonology. I couldn't have met a better guy. The reason he can help in more areas than anyone in the field is because he has read and acquired more material than anyone else. He was like this before the age of computers and the internet. Regardless what academic studies, professions, residences or personal challenges he has experienced, his quest for ultimate truth has always been the quiet, yet constant guide to his life. He instinctively knew decades ago that this subject may contain a demonstrable answer. Yet, even more impressive has been his willingness to share this information with anyone in any way.

Gary Habermas and Ken Stevenson wrote *The Verdict on the Shroud* in 1981, whose release that fall changed my entire perspective as a person and an attorney. Their analysis and approach to surprisingly new scientific evidence that was being acquired from a new area of investigation was the best introduction and foundation that a young lawyer could have had. I've had the pleasure of reading some of Gary's other works and making his acquaintance at conferences and in correspondence. His character and kindness are truly equal to his works, which is rarely the case for the rest of us writers.

The vast majority of the outside financial support for the Test the Shroud Foundation has come from Paul Ernst, Dick Nieman and Pat Byrne (and their wives). They have conscientiously studied the qualities of and the evidence for the authenticity of the Shroud for many years. Despite the past non-involvement of many scientists that we have approached, these men endured and continued to support the Foundation's efforts. Interestingly, all of these men have persevered in their support while struggling in their own ways with humanity's universal problem of death. Their character and friendship is even more consistent and impressive than their support.

I would like to thank John Schulte, Chuck Neff, Keith Plein and Laura Clark whose hard work and cooperation in a variety of ways over a number of years has helped many to understand the importance of this growing body of evidence. I'm sure we will continue to work in even more ways in the future.

Bill Mathis superbly photographically captured the blood marks from the Shroud for this book. He also helped me to start re-publicizing my works on the Shroud earlier than I planned, which turned out to be of great value for this book.

I want to thank Dick Schmidt, my honest, hard working accountant, who goes the extra mile for all of his clients and acquaintances. I'm one of many people who doesn't deserve all the valuable help that he provides, but so greatly appreciates it.

I'd also like to acknowledge the benefit and friendship I received from Monsignor Molloy, Mike Miller and Jim Rygelski, whose integrity and professionalism will not be forgotten.

I want to thank Julia Eberlin for editing all the chapters, helping promote an international Shroud conference and keeping me posted on a number of Shroud matters while I have been completing this book. She does many things well.

I want to thank Betty Gravlin for her devoted work on our website, promoting the St. Louis Shroud Conference and other ways of publicizing Shroud information. She has also helped our office on various technical matters that I don't have a clue about, but she understands well.

I sincerely want to thank Donna Herbst for all of her office skills and her strong and dedicated interest in promoting the Shroud conference and in completing this book. She not only works very hard, but she has come in on many Saturdays and Sundays. She stood out by doing everything she could to improve and complete this book.

I would like to thank Chris Hardgrave for performing a wide variety of legal and Shroud-related tasks with superb skill over the last nine years. Among her best skills was serving as a sounding board for a range of Shroud ideas. Donna and Chris worked with me every day, so their hardest job was putting up with me.

Cindy Sheltmire first volunteered to help at the St. Louis Shroud Conference in the fall of 2014. I was crazy not to enlist her help until June of this year because in such a short time she has devoted many long hours to finding many insightful edits to the text and ideas for the book cover that neither I nor anyone else had thought of. She is full of ideas and abilities that can advance the worldwide public's awareness of the Shroud's unique evidence and the testing that will take these results and their implications to a whole new level.

Ellie Jones' all around skills as an editor, designer, advisor and publisher result from her natural writing talents and outgoing personality, but it's her conscience that is the most impressive.

I want to thank my 2000 manuscript typist Carol Esmar for her and Jack's continued interest in the Shroud and their long-time friendship.

I want to thank my wife for enduring 23 of the 33 years I have worked on the Shroud. All the research, papers, conferences and manuscripts have always taken much longer than planned. The accompanying frustrations, late nights, working through weekends, missed trips and vacations, lost income and heavy expenses have all been worse than expected. She's not an expert on the Shroud, so the complex scientific, medical, archaeological and historical evidence would not have carried the day for her. In most people's eyes, the logic and utility of the project, along with an individual's patience, would naturally have expired after more than two decades of obvious social, personal and financial losses. Fortunately, her talent and personality can overcome any situation.

PHOTO CREDITS

The full-length frontal positive and negative images on the book cover were photographed by Vernon Miller and Barrie M. Schwortz, respectively (Copyright 1978, Vernon Miller and Copyright 1978 Barrie M. Schwortz Collection, STERA, Inc. All Rights Reserved). The facial and full length body images seen before the first chapter were photographed and copyrighted respectively by Giuseppe Enrie in 1931 and Vernon Miller in 1978.

Fig. 1	Courtesy Brooks Institute
Fig. 2	Shaam News Network
Fig. 3	Copyright 1978, Vernon Miller
Fig. 4	Copyright 1978, Vernon Miller
Fig. 5	Copyright 1978, Vernon Miller
Fig. 6	Copyright 1978, Vernon Miller
Fig. 7	Courtesy Brooks Institute
Fig. 8	Courtesy Jean Lorre, Jet Propulsion Laboratory
Fig. 9	Courtesy Jessica Dodson and Eric Stillwell
Fig. 10	Copyright 1978-81, Ernest Brooks II, Vernon Miller, and Barrie Schwortz
Fig. 11	Courtesy Arcidiocesi di Torino
Fig. 12	Copyright 1978, Vernon Miller
Fig. 13	Copyright 1978 Barrie M. Schwortz Collection, STERA, Inc. All Rights Reserved
Fig. 14	Copyright 1978, Vernon Miller
Fig. 15	Courtesy Brooks Institute
Fig. 16	Copyright 1978, Vernon Miller
Fig. 17	Vernon Miller
Fig. 18	Tipolitografia F. Ili Scaravaglio e C.
Fig. 19	Vernon Miller
Fig. 20	Copyright 1978, Vernon Miller
Fig. 21	Courtesy Dr. Sebastiano Rodante
Fig. 22	Copyright 1978, Vernon Miller
Fig. 23	Courtesy Dr. Sebastiano Rodante
Fig. 24	Courtesy Dr. Sebastiano Rodante
Fig. 25	Welcome Institute for the History of Medicine
Fig. 26	Copyright 1978, Vernon Miller
Fig. 27	Courtesy Ian Wilson
Fig. 28	Dr. Robert Bruce-Chwatt
Fig. 29	Tipolitografia F. Ili Scaravaglio e C.
Fig. 30	Courtesy Pauline Books & Media
Fig. 31	Courtesy Pauline Books & Media

Fig. 32 Copyright 1978, Vernon Miller
Fig. 33 Dr. Robert Bruce-Chwatt
Fig. 34 Copyright 1978, Vernon Miller
Fig. 35 Copyright 1978, Vernon Miller
Fig. 36 Courtesy Holy Shroud Guild
Fig. 37 Copyright 1978 Barrie M. Schwortz Collection, STERA, Inc. All Rights Reserved
Fig. 38 Copyright 1978, Vernon Miller
Fig. 39 Courtesy Dr. Eugenia Nitowski
Fig. 40 Courtesy Dr. Eugenia Nitowski
Fig. 41 Courtesy Dr. John Jackson and Optical Society of America
Fig. 42 Courtesy Dr. John Jackson and Optical Society of America
Fig. 43 Copyright 1978, Vernon Miller
Fig. 44 Copyright 1978, Vernon Miller
Fig. 45 Copyright 2003 ODPF
Fig. 46 Copyright 2003 ODPF
Fig. 47 Copyright 2003 ODPF
Fig. 48 Copyright 2003 ODPF
Fig. 49 Copyright 2003 ODPF
Fig. 50 Copyright 2003 ODPF
Fig. 51 Copyright 1978 Barrie M. Schwortz Collection, STERA, Inc. All Rights Reserved
Fig. 52 Copyright 1978, Vernon Miller
Fig. 53 Copyright 1978, Vernon Miller
Fig. 54 Copyright 1978, Vernon Miller
Fig. 55 Arcidiocesi di Torino
Fig. 56 Copyright 1978, Vernon Miller
Fig. 57 Copyright 2003 ODPF
Fig. 58 Courtesy Mechthild Flury-Lemberg
Fig. 59 Courtesy Mechthild Flury-Lemberg
Fig. 60 Courtesy Mechthild Flury-Lemberg
Fig. 61 Courtesy Mechthild Flury-Lemberg
Fig. 62 Courtesy Aldo Guerreschi
Fig. 63 Courtesy Aldo Guerreschi
Fig. 64 Courtesy Aldo Guerreschi
Fig. 65 Courtesy Aldo Guerreschi
Fig. 66 Courtesy Pauline Books & Media
Fig. 67 Fr. Francis Filas
Fig. 68 Fr. Francis Filas
Fig. 69 Fr. Francis Filas
Fig. 70 Dr. Max Frei and Shroud Spectrum International
Fig. 71 Courtesy Dr. Avinoam Danin and Dr. Alan Whanger
Fig. 72 Israel Exploration Journal

Fig. 73	Sculpture by Thomas Goyne featured in BSTS #10
Fig. 74	Copyright 1978 Barrie M. Schwortz Collection, STERA, Inc. All Rights Reserved
Fig. 75	Copyright 1978, Vernon Miller
Fig. 76	Copyright 1978, Vernon Miller
Fig. 77	Arcidiocesi di Torino
Fig. 78	Copyright 1978, Mark Evans
Fig. 79	Copyright 1978, Vernon Miller
Fig. 80	Reproduced by kind permission of the NDA
Fig. 81	Courtesy Jean-Baptiste Rinaudo
Fig. 82	Copyright 1978, Vernon Miller
Fig. 83	Courtesy Brooks Institute
Fig. 84	Courtesy Jessica Dodson and Eric Stillwell
Fig. 85	Copyright Nobel Media
Fig. 86	ANSTO Institute for Environmental Research
Fig. 87	Copyright 2003 ODPF
Fig. 88	Copyright 2003 ODPF
Fig. 89	Courtesy Arthur C. Lind
Fig. 90	Courtesy Arthur C. Lind
Fig. 91	Copyright 2014 ARC
Fig. 92	Arcidiocesi di Torino
Fig. 93	Reconstruction drawing by Dr. Leen Ritmeyer
Fig. 94	Copyright 1978, Vernon Miller
Fig. 95	Copyright 1978, Vernon Miller
Fig. 96	Arcidiocesi di Torino
Fig. 97	Copyright 1978, Vernon Miller
Fig. 98	Courtesy ITN
Fig. 99	Copyright 1978, Vernon Miller
Fig. 100	Copyright 1978, Vernon Miller
Fig. 101	Courtesy Robert Rucker
Fig. 102	Courtesy Arthur C. Lind
Fig. 103	Copyright St. Paul's, 1998
Fig. 104	Copyright 2003 ODPF
Fig. 105	Copyright 1978, Vernon Miller
Fig. 106	Copyright 1978, Vernon Miller
Fig. 107	Copyright 2015 Agilent Technologies
Fig. 108	Courtesy Aurelio Ghio
Fig. 109	Courtesy Mechthild Flury-Lemberg
Fig. 110	Courtesy Mechthild Flury-Lemberg
Fig. 111	Courtesy Mechthild Flury-Lemberg
Fig. 112	Courtesy Mechthild Flury-Lemberg
Fig. 113	Courtesy Christine Hardgrave
Fig. 114	Courtesy Fr. Francis Filas

Fig. 115 Courtesy Dr. Avinoam Danin and Dr. Alan Whanger
Fig. 116 Photograph by Melanie Lovell-Smith
Te Ara, the Encyclopedia of New Zealand http://www.teara.govt.nz/
© Crown Copyright 2005-2013 Manatū Taonga, The Ministry for Culture and Heritage, New Zealand
Fig. 117 Copyright 2003 ODPF
Fig. 118 Copyright 1978 Barrie M. Schwortz Collection, STERA, Inc. All Rights Reserved
Fig. 119 Courtesy Robert Rucker
Fig. 120 Courtesy Aurelio Ghio
Fig. 121 Copyright 1978, Vernon Miller
Fig. 122 Copyright 1980 Ernest Brooks
Fig. 123 Courtesy Mechthild Flury-Lemberg
Fig. 124 Courtesy Mechthild Flury-Lemberg
Fig. 125 Copyright 2003 ODPF
Fig. 126 Courtesy Aldo Guerreschi
Fig. 127 Courtesy Aldo Guerreschi
Fig. 128 Courtesy Aldo Guerreschi
Fig. 129 Courtesy Aldo Guerreschi
Fig. 130 Courtesy Aldo Guerreschi
Fig. 131 Copyright 2002, Raymond Rogers
Fig. 132 Copyright St. Paul's, 1998
Fig. 133 Copyright 1978 Barrie M. Schwortz Collection, STERA, Inc. All Rights Reserved
Fig. 134 Copyright 1978 Barrie M. Schwortz Collection, STERA, Inc. All Rights Reserved
Fig. 135 Copyright 1978 Barrie M. Schwortz Collection, STERA, Inc. All Rights Reserved
Fig. 136 Copyright 1978 Barrie M. Schwortz Collection, STERA, Inc. All Rights Reserved
Fig. 137 British Museum
Fig. 138 Werner Bultst, S.J.
Fig. 139 Shroud Spectrum International
Fig. 140 Courtesy Dorothy Crispino
Fig. 142 Courtesy Ian Wilson and Frank Tribbe
Fig. 148 Courtesy the Ampleforth Journal
Fig. 149 Courtesy the Ampleforth Journal
Fig. 150 Courtesy the Ampleforth Journal
Fig. 151 Courtesy Frank Tribbe
Fig. 152 Courtesy Frank Tribbe
Fig. 154 Lennox Manton
Fig. 157 Courtesy Shroud Spectrum International
Fig. 158 John Gitchell

Fig. 160 Courtesy Dr. Alan Whanger and the Holy Shroud Guild
Fig. 161 Courtesy Dr. Alan Whanger and the Holy Shroud Guild
Fig. 162 Courtesy Dr. Alan Whanger
Fig. 163 Courtesy Dr. Alan Whanger
Fig. 164 Courtesy Dr. Alan Whanger
Fig. 165 Courtesy Dr. Alan Whanger
Fig. 166 Courtesy Giulio Fanti
Fig. 168 Courtesy Ian Wilson
Fig. 169 Copyright 1978, Ernest Brooks, II
Fig. 170 Grabar, la Sainte Face de Laon
Fig. 171 Grabar, la Sainte Face de Laon
Fig. 172 Courtesy Holy Shroud Guild
Fig. 173 Vernon Miller
Fig. 175 Copyright 1978 Barrie M. Schwortz Collection, STERA, Inc. All Rights Reserved
Fig. 177 Ian Wilson
Fig. 178 Courtesy Ian Wilson
Fig. 179 Courtesy Dan Scavone
Fig. 180 Courtesy Ian Wilson
Fig. 181 Courtesy British Museum
Fig. 183 Bibliotheque Nationale
Fig. 184 Courtesy Br. Bruno Bonnet-Eymard
Fig. 185 Copyright 1978-81, Ernest Brooks II, Vernon Miller, and Barrie Schwortz
Fig. 186 Copyright 1978, Vernon Miller
Fig. 187 Copyright 1978, Vernon Miller
Fig. 188 Courtesy Brooks Institute
Fig. 189 Courtesy Giulio Fanti
Fig. 190 Courtesy Giulio Fanti
Fig. 191 Copyright 1931 Giuseppe Enrie, All Rights Reserved
Fig. 192 Copyright 1978, Vernon Miller
Fig. 193 Courtesy Brooks Institute
Fig. 194 Copyright 2003 ODPF
Fig. 195 Fr. Francis Filas
Fig. 196 Courtesy Dr. Avinoam Danin and Dr. Alan Whanger
Fig. 197 Copyright 1978, Vernon Miller
Fig. 198 Copyright 1978, Vernon Miller
Fig. 200 Copyright 1978, Vernon Miller
Fig. 201 Copyright 1978, Vernon Miller
Fig. 202 Courtesy Dr. Zahi Hawass
Fig. 203 Copyright 2003 ODPF
Fig. 204 Copyright 2003 ODPF
Fig. 205 Copyright 2003 ODPF
Fig. 206 Copyright 2003 ODPF

Fig. 207	Copyright 1978, Vernon Miller
Fig. 208	Copyright 1978, Vernon Miller
Fig. 209	Courtesy Dr. Sebastiano Rodante
Fig. 210	Reconstruction drawing by Dr. Leen Ritmeyer
Fig. 211	Copyright 2003 ODPF
Fig. 212	Copyright 1978, Vernon Miller
Fig. 213	Copyright 2003 ODPF
Fig. 214	Copyright 1978, Vernon Miller
Fig. 215	Copyright 1978 Barrie M. Schwortz Collection, STERA, Inc. All Rights Reserved
Fig. 216	Copyright 1920 Ernest Brooks
Fig. 217	University of Toronto
Fig. 218	Copyright 1978 Barrie M. Schwortz Collection, STERA, Inc. All Rights Reserved
Fig. 219	Copyright 1978 Barrie M. Schwortz Collection, STERA, Inc. All Rights Reserved
Fig. 220	Courtesy Robert Rucker
Fig. 221	Copyright 2003 ODPF
Fig. 222	Courtesy Ian Wilson
Fig. 223	Courtesy Institute of Physics Publishing

Every effort has been made to obtain the necessary permissions with reference to copyright material, both illustrative and quoted. We apologize for any omissions in this respect and will be pleased to make the appropriate acknowledgements in any future edition.